Lecture Notes in Computer Science   15747

Founding Editors

Gerhard Goos
Juris Hartmanis

The series Lecture Notes in Computer Science (LNCS), including its subseries Lecture Notes in Artificial Intelligence (LNAI) and Lecture Notes in Bioinformatics (LNBI), has established itself as a medium for the publication of new developments in computer science and information technology research, teaching, and education.

LNCS enjoys close cooperation with the computer science R & D community, the series counts many renowned academics among its volume editors and paper authors, and collaborates with prestigious societies. Its mission is to serve this international community by providing an invaluable service, mainly focused on the publication of conference and workshop proceedings and postproceedings. LNCS commenced publication in 1973.

Manuel Egele · Veelasha Moonsamy ·
Daniel Gruss · Michele Carminati
Editors

Detection of Intrusions and Malware, and Vulnerability Assessment

22nd International Conference, DIMVA 2025
Graz, Austria, July 9–11, 2025
Proceedings, Part I

 Springer

Editors
Manuel Egele 🆔
Boston University College of Engineering
Boston, MA, USA

Veelasha Moonsamy 🆔
Ruhr University Bochum
Bochum, Germany

Daniel Gruss 🆔
Graz University of Technology
Graz, Austria

Michele Carminati 🆔
Politecnico di Milano
Milan, Italy

ISSN 0302-9743   ISSN 1611-3349 (electronic)
Lecture Notes in Computer Science
ISBN 978-3-031-97619-3   ISBN 978-3-031-97620-9 (eBook)
https://doi.org/10.1007/978-3-031-97620-9

Preface

On behalf of the Program Committee, we are pleased to present the proceedings of the 22nd Conference on Detection of Intrusions and Malware & Vulnerability Assessment (DIMVA 2025).

Over the past two decades, DIMVA has become a recognized venue for cutting-edge security research, attracting high-quality submissions and promoting collaboration among academia, industry, and government. DIMVA is organized by the Special Interest Group on Security, Intrusion Detection, and Response (SIDAR) of the German Informatics Society (GI).

This year, we received 103 valid submissions (11 desk-rejected), and accepted 25 full papers, resulting in a competitive acceptance rate of 24%. Each paper underwent a rigorous double-blind peer review process, with every submission reviewed by at least three experts. Each Program Committee member was assigned to review up to 7 papers. For the first time, the DIMVA proceedings include 11 poster papers selected through a single-blind review. For the third consecutive year, DIMVA followed a dual-deadline submission model. The acceptance rate was balanced across the two cycles: in the first cycle, 7 out of 40 submissions were accepted (2 directly and 5 with shepherding), while in the second cycle, 18 out of 63 submissions were accepted (8 directly and 10 with shepherding).

We extend our sincere gratitude to the Program Committee members for their tireless efforts in reviewing papers in both cycles, engaging in in-depth online discussions, and shepherding papers to completion. Across the reviewing phases, PC members exchanged over 900 comments.

We would also like to thank the Organizing Committee for their dedication and hard work in preparing this edition of DIMVA. We are grateful to our sponsors, including Genua and Springer, and our host institutions for their support.

Finally, we thank all authors for submitting and presenting their work, and all attendees for their participation. Your engagement and commitment continue to make DIMVA an impactful event. We look forward to your future contributions.

DIMVA 2025 was held in the Aula of the historic main building of Graz University of Technology. The venue offered a unique blend of historical atmosphere and academic prestige, creating an inspiring setting for insightful discussions and new collaborations. In addition to paper presentations, the program featured an engaging keynote session and a newly introduced poster session in the picturesque landscape of Southern Styria, providing attendees with the opportunity to discuss early-stage work.

Together, the program offered an excellent platform for discussion, collaboration, and social interaction in a relaxed setting in South-East Styria.

May 2025

Manuel Egele
Veelasha Moonsamy
Daniel Gruss
Michele Carminati

Organization

General Chair

Daniel Gruss   Graz University of Technology, Austria

Program Committee Chairs

Veelasha Moonsamy   Ruhr University Bochum, Germany
Manuel Egele   Boston University, USA

Publication Chair

Michele Carminati   Politecnico di Milano, Italy

Poster Chairs

Tarini Saka   Ruhr University Bochum, Germany
Flavio Toffalini   Ruhr University Bochum, Germany

Steering Committee Chairs

Ulrich Flegel   Infineon Technologies, Germany
Michael Meier   University of Bonn and Fraunhofer FKIE,
                                  Germany

Steering Committee

Magnus Almgren   Chalmers University of Technology, Sweden
Sébastien Bardin   CEA, France
Leyla Bilge   Gen Digital, France
Gregory Blanc   Télécom SudParis, France
Herbert Bos   Vrije Universiteit Amsterdam, Netherlands
Danilo M. Bruschi   Università degli Studi di Milano, Italy

Program Committee

Daniele Cono D'Elia	Sapienza University of Rome, Italy
David Klein	Technische Universität Braunschweig, Germany
Eleonora Losiouk	University of Padua, Italy
Emilio Coppa	LUISS University, Italy
Fabio Pagani	Binarly, Italy
Fabio Pierazzi	King's College London, UK
Flavio Toffalini	Ruhr-Universität Bochum, Germany
Hervé Debar	Télécom SudParis, France
Ilya Grishchenko	University of California, Santa Barbara, USA
Jan Wichelmann	Universität zu Lübeck, Germany
Johanna Ullrich	University of Vienna, Austria
Johannes Kinder	LMU Munich, Germany
Juan Caballero	IMDEA Software Institute, Spain
Juan Tapiador	Universidad Carlos III de Madrid, Spain
Kaan Onarlioglu	Akamai, USA
Kimberly Tam	University of Plymouth/Alan Turing Institute, UK
Konrad Rieck	TU Berlin, Germany
Mannat Kaur	Max Planck Institute for Informatics, Germany
Manuel Egele	Boston University, USA
Marco Cova	VMware, UK
Martina Lindorfer	TU Wien, Austria
Mathias Fischer	University of Hamburg, Germany
Michael Meier	University of Bonn and Fraunhofer FKIE, Germany
Michael Schwarz	CISPA Helmholtz Center for Information Security, Germany
Michalis Polychronakis	Stony Brook University, USA
Michele Carminati	Politecnico di Milano, Italy
Moritz Schloegel	Arizona State University, USA
Prashast Srivastava	Columbia University, USA
Ricardo J. Rodríguez	Universidad de Zaragoza, Spain
Roland Yap	National University of Singapore, Singapore
Seungwon Shin	KAIST, South Korea
Silvia Sebastián	CISPA Helmholtz Center for Information Security, Germany
Simon Koch	TU Braunschweig, Germany
Stefano Zanero	Politecnico di Milano, Italy
Stijn Volckaert	KU Leuven, Belgium
Sven Dietrich	City University of New York, USA
Tapti Palit	UC Davis, USA
Tiago Heinrich	Max-Planck-Institut für Informatik, Germany
Urko Zurutuza	Mondragon Unibertsitatea, Spain

Contents – Part I

Side Channels

Obfuscation

Contents – Part II

OS and Network

Resilient Systems

Web Security

ScamFerret: Detecting Scam Websites Autonomously with Large Language Models

Hiroki Nakano$^{(\boxtimes)}$, Takashi Koide, and Daiki Chiba

NTT Security Holdings Corporation & NTT Corporation, Tokyo, Japan
hi.nakano.sec@gmail.com

Abstract. With the rise of sophisticated scam websites that exploit human psychological vulnerabilities, distinguishing between legitimate and scam websites has become increasingly challenging. This paper presents ScamFerret, an innovative agent system employing a large language model (LLM) to autonomously collect and analyze data from a given URL to determine whether it is a scam. Unlike traditional machine learning models that require large datasets and feature engineering, ScamFerret leverages LLMs' natural language understanding to accurately identify scam websites of various types and languages without requiring additional training or fine-tuning. Our evaluation demonstrated that ScamFerret achieves 0.972 accuracy in classifying four scam types in English and 0.993 accuracy in classifying online shopping websites across three different languages, particularly when using GPT-4. Furthermore, we confirmed that ScamFerret collects and analyzes external information such as web content, DNS records, and user reviews as necessary, providing a basis for identifying scam websites from multiple perspectives. These results suggest that LLMs have significant potential in enhancing cybersecurity measures against sophisticated scam websites.

Keywords: Large Language Model · Human Psychological Vulnerability · Scam Website · Agent

1 Introduction

Scam websites have become an increasingly prevalent threat, causing significant financial losses and personal information compromise. In 2023, reported losses in the United States reached $12.5 billion, a 22% increase from the previous year [10]. While traditional phishing websites often mimic legitimate websites and can be detected through specific visual cues [8,20], modern scam websites have evolved to become highly sophisticated, making them challenging to identify even for security experts. These sophisticated scam websites exploit human psychological vulnerabilities, perpetuating deception and escalating financial losses, evolving into a significant societal issue that demands urgent attention.

© The Author(s), under exclusive license to Springer Nature Switzerland AG 2025
M. Egele et al. (Eds.): DIMVA 2025, LNCS 15747, pp. 3–25, 2025.
https://doi.org/10.1007/978-3-031-97620-9_1

Previous research has focused on developing machine learning models to detect scam websites using HTML content and domain name information [18, 37]. However, these approaches face several limitations:

- They require large labeled datasets for each scam type and language, which are time-consuming and costly to create.
- They demand complex feature engineering specific to each scam variant, limiting generalizability.
- They lack transparency in the detection process, making it difficult for users to understand the basis for decisions intuitively.

To address these challenges, we present ScamFerret, a novel agent-based system for analyzing diverse scam websites across multiple languages without requiring additional training on scam-specific data. ScamFerret leverages a large language model (LLM) that has already been trained on a broad corpus of Internet text, which likely includes some information about scams and fraudulent activities. This pre-existing knowledge allows the system to operate effectively without scam-specific fine-tuning.

ScamFerret uses the LLM to autonomously select appropriate information-gathering tools, collect useful information for website analysis, and perform contextual analysis to identify scam websites. By utilizing the natural language understanding capabilities and broad knowledge base of LLMs, ScamFerret can recognize subtle suspicious elements and provide explanations for its classifications, drawing on its general understanding of language, web content, and potential fraudulent patterns.

We evaluate ScamFerret on new datasets comprising four types of scam websites (fake online shopping, technical support scams, cryptocurrency scams, and investment scams) in English, as well as three languages (English, German, and Japanese) for fake online shopping websites. Our results demonstrate that ScamFerret achieves a mean classification accuracy of 0.972 across four scam types in English and 0.993 across three languages for online shopping websites when using GPT-4, outperforming both conventional machine learning-based detectors [3, 16] and simpler LLM-based approaches.

This paper makes the following contributions:

- We introduce ScamFerret, a system that autonomously collects and analyzes data to detect scam websites without requiring large, labeled datasets for each scam type, leveraging LLMs to recognize sophisticated scam websites.
- We demonstrate ScamFerret's effectiveness across multiple scam types and languages, achieving state-of-the-art accuracy of 0.972 for four scam types in English and 0.993 across three languages for online shopping websites using the GPT-4 model.
- We provide an analysis of ScamFerret's detection process, offering insights into how LLMs can be leveraged for explainable web security tasks.
- We share the code for the proposed system, the dataset used for evaluation, and the evaluation results at https://github.com/ScamFerret/artifact.

2   Scope and Goal

2.1   Scope of Scam Websites

This study focuses on four types of scam websites across three languages: English, German, and Japanese. We target fake online shopping, technical support scams, cryptocurrency scams, and investment scams, which have high victim rates and have been the subject of previous detection efforts [16,21,33,37]. These scams pose significant financial risks, with cryptocurrency and investment scams resulting in billions of dollars in losses annually, while online shopping scams exploit the rapidly expanding e-commerce market, leading to widespread consumer victimization [4,9]. Our language selection addresses the global nature of online scams [1]: English, the language with the highest number of reported fraud victims globally; German, the native language of Germany, which is frequently targeted for online fraud; and Japanese, the native language of Japan, which is the most affected language in information theft. This diverse set allows us to assess ScamFerret's effectiveness across different contexts. The characteristics of each scam type are as follows:

Fake Online Shopping. These websites mimic legitimate online shopping platforms, often advertising rare or discounted products. They use search engine optimization and create urgency to induce purchases, resulting in financial losses for victims.

Technical Support Scams. These websites falsely alert users to technical issues, prompting contact with attackers. They often use pop-ups or fake security warnings to direct users to fraudulent support pages, where attackers may request remote access or payment for non-existent services.

Cryptocurrency Scams. These websites typically employ phishing tactics through fraudulent trading platforms and wallet services. Attackers use fake celebrity and company accounts on social media to lure potential victims. Once users enter their credentials, attackers can steal their cryptocurrency funds.

Investment Scams. These websites promise high profits or risk-free investments. They use professional-looking designs and create urgency with limited-time offers. Once users invest, attackers refuse refunds and eventually cease communication.

2.2   Research Goal

Our primary goal is to develop a system that can analyze diverse scam websites from input URLs and provide clear justification for its classifications. As scam websites become increasingly sophisticated, conventional detection systems based on predefined blocklists and feature learning face limitations [3,16,18,37]. We aim to address three key challenges:

Elimination of Labeled Dataset Requirements. Rapidly evolving scam websites make preparing optimal labeled datasets time-consuming, requiring

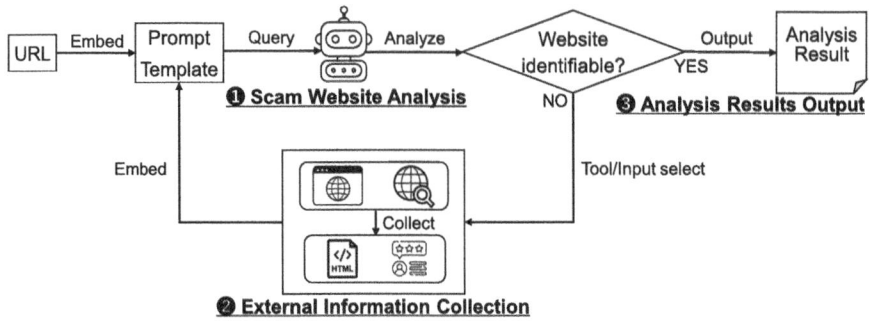

Fig. 1. Overview of ScamFerret

constant updates to keep pace with new scam tactics. We aim to introduce a system that can analyze various scam types and languages without pre-prepared labeled datasets, adapting in real-time to emerging threats.

Multi-type and Multi-lingual Detection with a Single Model. Conventional systems often use multiple models for specific scam types and languages, requiring frequent updates. We aim to develop a single, versatile model detecting various scam websites across types and languages, eliminating specialized models and updates.

Clear Verbalization of Detection Rationale. Conventional systems often lack transparency, classifying based on numerical values without clear justification. Our approach aims to verbalize the suspicious aspects of target URLs, enhancing reliability and understanding of new scam patterns.

3   Proposed System: ScamFerret

We propose ScamFerret, a novel system that addresses the three challenges outlined previously. Our approach leverages LLMs as autonomous agents to drive the information collection and analysis process, capitalizing on their text comprehension capabilities. Figure 1 provides an overview of the system workflow.

ScamFerret takes a URL as input and proceeds to autonomously collect and analyze relevant information. The system outputs a classification of whether the website is a scam, the specific type of scam (e.g., fake online shopping), and the rationale for this determination (e.g., non-existent operating company).

The core of ScamFerret's functionality relies on a carefully designed prompt template that guides the interactions between the system and the LLM. The template in Table 1 is crucial for eliciting appropriate responses from the LLM. The design of effective prompts has been an active area of research, with several studies exploring techniques to optimize LLM outputs for specific tasks [45,50]. In the following subsections, we describe each component of ScamFerret in detail.

Table 1. Prompt Template for Proposed System

Type	Prompt
Task Setting	I want you to act as a professional scam website detection expert. You are tasked with analyzing the content of URL given to you to determine if the URL is a scam website or not.
Characteristic Examples	Scam websites have the following features. 1. Unusually low prices and claims of free. 2. Claims to obtain an unusually large amount of money. 3. Websites contain texts targeting human psychological weaknesses. 4. Information on non-existent companies. 5. Handling different products from common e-commerce websites. 6. Inquiry phone number and email are unsuitable for business use. 7. Privacy of customer information notation is ambiguous. 8. Payment methods are not common and are unusual. 9. The information listed has not been updated.
Tool Definitions	You can access the following tools to help you answer the question: Tool Name 1: Tool Description (**Contents of Table 2**) ...
Analysis Method (ReAct)	Please follow the format below when answering the questions: Question: the question you must answer Thought: you should always think about what to do Action: the tool for information collection, should be one of [Tool Name 1, Tool Name 2, ...] (**Contents of Table 2**) Action Input: the input to the tool Observation: the result of information collection ... (You can repeat this Thought/Action/Action Input/Observation N times to derive your answer.) Thought: I now know the final answer Final Answer: the final answer to the original question You must derive your final answer based on no more than 10 actions.
Output Format	After the Final Answer is determined, output the analysis results in JSON format according to the following key: - result: True or False (result of URL scam determination) - scam_type: Fake online shopping website (specific type of scam) - reason: State your decision based on the scam website's features
Analysis Process	Begin! Question: Please analyze this URL https://example.com Thought: ... Action: ... Action Input: ... Observation: ... ... (**Repeat Thought/Action/Action/Input/Observation**) ...

3.1 ❶ Scam Website Analysis

ScamFerret analyzes input URLs for potential scams using a multi-step process. The system first evaluates the URL based on the information embedded in the template (i.e., the target URL for analysis) and the LLM's pre-trained knowledge, following the Task Setting and Analysis Process in Table 1. If the initial information is insufficient, ScamFerret performs External Information Collection (❷) and re-analyzes the website. This process iterates until a final determination is made.

Feature Analysis of Scam Websites. Research has shown that including specific features and cautions in LLM prompts improves performance [15]. We incorporate nine common scam website features into the prompts, leveraging the model's text comprehension abilities (Characteristics Examples, Table 1). These features include uncommon pricing, large monetary gifts, language targeting psychological weaknesses, and diverse product/service offerings. Additionally, we consider lack of company information, inappropriate business contact details, poorly written privacy policies, uncommon payment methods, and outdated information. This approach differs from traditional machine learning systems that rely on complex, scam-specific feature engineering. By describing these features in natural language, the LLM can analyze and identify suspicious elements in the collected information.

Tool Selection. While LLMs possess extensive knowledge, providing additional external information can enhance their performance. Studies have shown that external source information significantly improves response accuracy for challenging tasks [11]. We define a set of tools that provide useful information for scam website analysis (Tool Definitions, Table 1). Each tool includes a name, description, and required input information. The LLM uses this information to select appropriate tools and extract necessary inputs. Section 3.2 details the tools used in this study.

LLM Decision-Making. ScamFerret utilizes the REasoning and ACTing (ReAct) framework for its decision-making process [48]. ReAct is an innovative approach that combines reasoning and action, allowing AI systems to articulate their thought processes and adapt their actions based on new information, similar to how humans think and act. We chose ReAct because it enables LLMs to articulate their reasoning steps, which is crucial for analyzing potential scam websites. This verbalization of thought processes enhances the model's ability to explain its decision-making rationale. The decision-making process, implemented using LangChain (an open-source framework for building LLM-based applications that enables the creation of chains of actions for processing tasks) [17], follows these steps:

- Repeat the process until the given URL can be identified as a type of website (i.e., scam or legitimate).
- Scam Website Analysis (❶) is performed based on the information embedded in the prompt template.

– If the LLM determines that there is insufficient information for identification, it will perform External Information Collection (❷), embed this information into the template, and then conduct Scam Website Analysis (❶) again.
– If the LLM determines that the information is sufficient for identification, it will perform Analysis Results Output (❸) based on the results of all previous analyses.

To prevent infinite loops, ScamFerret imposes a limit of 10 tool selections per URL (Analysis Method, Table 1). This constraint ensures efficient processing while allowing for thorough investigation. This iterative approach enables ScamFerret to autonomously gather and analyze information, leading to accurate scam detection. By combining LLM-based reasoning with strategic tool usage, our system can adapt to various scam scenarios and provide detailed justifications for its conclusions. Even as the types of prevalent scam websites evolve, ScamFerret can flexibly respond without requiring major updates, as its analytical processes remain universal across various scam scenarios.

3.2 ❷ External Information Collection

ScamFerret collects external information using tools selected during the Scam Website Analysis (❶) phase. We designed these tools to capture the inherent characteristics of scams that attempt to deceive users, rather than focusing on specific scam types. Our approach is informed by previous studies on scam websites. The tools collect information from external sources that are likely to yield traces indicative of scam websites, enabling LLMs to analyze sophisticated scams effectively.

ScamFerret allows the LLM to determine which tools to use for information collection, meaning that not all tools are used in every analysis, and some may be used only once. The system may also select the same tool multiple times when needed, such as when analyzing multiple domain names found in the collected information. We have defined six categories encompassing a total of nine tools, as shown in Table 2, to collect information from various perspectives. These tools can obtain information commonly considered by human analysts when analyzing scam websites and that has been reported to be effective for detection in previous studies [3,16,18].

Web Content. Attackers often use text, images, and other web content to deceive users. We analyze these elements using three tools built with Playwright [29], which can render JavaScript and interact dynamically with websites. The Access URL tool takes a URL as input and retrieves the HTTP response status code. The Extract Text tool extracts strings (i.e., the innerText of HTMLElements) from the HTML content of the page accessed by the Access URL tool. To avoid including irrelevant strings, we target a maximum of three HTML tags in the same hierarchy. The Extract Hyperlink tool extracts combinations of the *href* attribute and the text within $< a >$ tags (e.g., (http://example.com/contact.html, Contact Page)) from the HTML content of the page accessed by

Table 2. List of Defined Tools

Tool Type	Tool Name	Description
Web Content	Access URL	A tool that accesses a URL to obtain a status code.
		This tool requires a URL as an argument.
	Extract Text	A tool that extracts text in the HTML.
		You must access a URL first before using this tool.
		This tool requires the URL as an argument.
	Extract Hyperlink	A tool that extracts a-tag hyperlinks and texts in the HTML.
		You must access a URL first before using this tool.
		This tool requires the URL as an argument.
Search Engine	Get Search Result	A tool to retrieve search results from a search engine.
		This tool requires a search query as an argument.
		You cannot use a URL as-is as a search query.
		Note that only the top 10 results will be retrieved.
Social Media	Search X/Twitter	A tool to retrieve posts containing a keyword from X/Twitter.
		This tool requires a search query as an argument.
		You cannot use a URL as-is as a search query.
		Note that only the latest top 10 results will be retrieved.
	Search Reddit	A tool to retrieve posts containing a keyword from Reddit.
		This tool requires a search query as an argument.
		You cannot use a URL as-is as a search query.
		Note that only the top five related posts and the top five
		associated comments will be retrieved.
WHOIS	Retrieve WHOIS	A tool to retrieve domain name information from WHOIS.
		This tool requires a domain name as an argument.
DNS Lookup	Retrieve DNS Record	A tool to retrieve DNS records using the dig command.
		This tool requires a domain name as an argument.
TLS Certificate	Retrieve Certificate	A tool to retrieve certificate information from crt.sh.
		This tool requires a domain name as an argument.
		Note that only the latest top 5 results will be retrieved.

the Access URL tool. To ensure relevance, we extract text content from the same level as the $< a >$ tag and one level below it in the HTML DOM tree structure.

These tools can be used recursively for detailed content analysis. For instance, after accessing the top page, if company information is not present in the extracted text, the hyperlinks can be analyzed to locate and access a dedicated company information page. While it's possible that the HTML of scam websites is obfuscated, these tools are not affected because they analyze the displayed character strings and $< a >$ tag elements to extract information.

Search Engine. Search engines provide valuable information about user-reported scam websites. We implemented the Get Search Result tool using Tavily [39], a commercial search engine API designed for LLMs that provides search results in an LLM-interpretable format. As illustrated in Fig. 1, this tool can collect information such as the reputation of a company operating the website under analysis. The tool takes a search query as input and returns relevant

URLs and web page content summaries, enabling comprehensive assessment of potential scam websites.

Social Media. Social media platforms can offer security-related information posted by various users [3,38]. The real-time nature of social media allows for quick gathering of scam reports and website reputations, often before they appear on dedicated review websites. We created two tools to retrieve posts by keyword search. The Search X/Twitter tool uses the X/Twitter API [46] to retrieve up to 10 latest posts containing the specified keyword. The Search Reddit tool uses the Reddit API [31] to retrieve up to 5 related posts and 5 associated comments with the specified keyword.

WHOIS Information. Attackers often launch websites shortly after acquiring domain names [28], resulting in recently registered domains. In contrast, legitimate websites typically have longer operational histories and well-defined management information. To analyze these characteristics, we implemented the Retrieve WHOIS tool. This tool uses the Linux Shell Command "whois" to retrieve information including the domain registrant, registration date, administrator, and managing organization.

DNS Record. DNS records can provide useful information for distinguishing between scam and legitimate websites. For example, NS and SOA record settings may differ significantly between scam and legitimate operations [13]. We implemented the Retrieve DNS Record tool, which uses the Linux Shell Command "dig [record type] [domain name] @8.8.8.8" to obtain DNS records such as A, AAAA, NS, SOA, TXT, and MX.

TLS Certificate. Modern scam websites often use TLS certificates, with potential biases toward specific certification authorities [14]. For instance, attackers may favor free certificates from Let's Encrypt or use the Subject Alternative Name to link multiple domain names to a single certificate. We implemented the Retrieve Certificate tool, which uses crt.sh [35] to search the Certificate Transparency log and retrieve a list of certificates associated with a given domain name.

Table 3. Ground-truth Dataset for Evaluation

Scam Type	Language	# of Scam Websites	# of Legitimate Websites
Online Shopping	English	200	200
	Japanese	200	200
	German	200	200
Technical Support	English	200	200
Cryptocurrency	English	200	200
Investment	English	200	200
Total	3 Languages	1,200	1,200

3.3 ❸ Analysis Results Output

After iteratively performing Scam Website Analysis (❶) and External Information Collection (❷), ScamFerret generates a final output based on the LLM's determination of whether the analyzed URL represents a scam or legitimate website. The analysis results comprise three key components:

Result. A binary classification indicating whether the URL is associated with a scam ("True") or a legitimate website ("False").

Scam Type. If the URL is classified as a scam, this field specifies the particular category or method of scam detected.

Reason. A detailed explanation of the LLM's decision-making process, outlining the key factors and evidence that led to the final determination.

These components are generated as part of Output Format in Table 1. The structured output allows for clear interpretation of the analysis results, providing both a concise classification and the underlying rationale. This approach enhances the transparency and interpretability of the system's decision-making process, which is crucial for both end-users and further research in the field of online scam detection.

4   Dataset

To assess ScamFerret's accuracy in detecting challenging scam websites, we created a new ground-truth dataset with verified labels. Existing public datasets [22, 47] were mostly inaccessible or unverifiable for our evaluation. Our dataset creation process involved collecting candidates, then establishing a reproducible ground-truth through four main steps.

4.1   Candidate Collection

Due to the challenges in detecting modern scam websites using traditional antivirus engines and services like VirusTotal [43], we created a custom dataset for evaluation. Our dataset comprises four types of English scam websites: Fake Online Shopping, Technical Support Scam, Cryptocurrency Scam, and Investment Scam. We collected these from five up-to-date public sources between April 1 and April 7, 2024 [26,27,34,40,47]. For Fake Online Shopping, we also included scam websites in German and Japanese, which were collected during the same period from two additional disclosure websites [25,44].

To assess ScamFerret's performance accurately, we compiled a corresponding dataset of legitimate websites. Unlike phishing websites, the scam websites in our study lack direct legitimate counterparts. We aimed to collect diverse legitimate websites to demonstrate that ScamFerret's classification is not based solely on website strings or domain names containing words like "shopping" or "support".

We utilized Curlie [7] and Trustpilot [41] to create our legitimate website dataset. Curlie, a manually compiled web directory, organizes multilingual websites into categories. Trustpilot is a user-driven review platform for business services and products. These sources have been effectively used in previous studies for domain name classification and creating legitimate website datasets [16,42]. From Curlie and Trustpilot, we collected information for four types of *legitimate* websites (Online Shopping, Technical Support, Cryptocurrency, and Investment) in English. For *legitimate* Online Shopping, we collected data in English, German, and Japanese.

4.2   Ground-Truth Dataset Creation

We create a ground-truth dataset of scam and legitimate websites through a four-step selection process from collected candidates.

Top List Filtering. We excluded websites listed in the top 100,000 domain names of the Tranco List [30], a widely used reference for legitimate websites in research. This step helped eliminate obviously legitimate websites that did not need to be analyzed in the LLM from the analysis target. Note that while we gathered data from Curlie and Trustpilot, most of the websites collected were minor sites, resulting in few that ranked within the top 100,000.

URL Accessibility Check. To detect active scam websites in real-time with high accuracy, we first excluded inaccessible URLs. We used Playwright to simulate common user access with a standard user agent (i.e., Mozilla/5.0 (Windows NT 10.0; Win64; x64) AppleWebKit/537.36 (KHTML, like Gecko) Chrome/122.0.0 Safari/537.36). URLs that did not return an HTTP status code of 200 were excluded from the dataset.

Manual Inspection. We manually verified each URL's appropriateness for our study and its proper categorization. Three security engineers examined each URL using search engines and analyzed the web content and screenshots. Due to individual language limitations, the evaluators collaborated to reach a consensus, excluding URLs that did not match the specified scam type.

Random Sampling. To create a balanced dataset, we randomly sampled an equal number of scam and legitimate websites for each type and language. This approach ensures an accurate evaluation of ScamFerret's classification performance. The final dataset comprises 1,200 scam websites and 1,200 legitimate websites, with 200 URLs for each language and type combination, as shown in Table 3.

5 Evaluation

We evaluate both the classification accuracy and explanation quality of Scam-Ferret using the ground-truth dataset described in Sect. 4.

5.1 Experimental Setup

Models. We compare the performance of three LLMs: GPT-3.5 (gpt-35-turbo-1106) and GPT-4 (gpt-4-1106-preview) from OpenAI, accessed via Azure OpenAI Service [23], and Gemini (Gemini 1.5 Pro) from Google DeepMind [12]. While LLMs have content filters to protect against harmful content and other issues, we disabled these filters for GPT-3.5 and GPT-4 in our experiments. It is important to note that these model performances are as of the experiments conducted in April 2024, and the models may have been updated since then.

Parameters. For each model, we configured two key parameters: the context size, which limits the input text length, and the temperature, which controls output diversity. Preliminary experiments showed that ScamFerret performed optimally with a context size of 128,000 (the maximum allowed) and a temperature of 0.7, which was found to be the most effective among tested values. We applied these settings in our main experiments.

Conventional Systems. We evaluate ScamFerret against three conventional systems:

Single-Turn Prompt: A system that uses a brief prompt to query LLMs for scam detection. This prompt includes the role of a scam detection expert, features of scam websites, web content, and specifies a JSON output format. Unlike ScamFerret, it analyzes the website only once.

Beyond Phish [3]: A binary classification system for online shopping scams, with publicly available implementation [22].

Scamdog Millionaire [16]: A supervised learning approach for binary classification of online shopping scam websites based on extracted web content features.

For the conventional systems, we created new training datasets to replicate their functionality, as the original training data was not available. We used 1,600 English websites (200 each for scam and legitimate across four types) that were not included in our ground-truth dataset. Features were generated based on information from the original papers, and models were trained to classify websites as scam or legitimate using URLs as input.

We evaluated the Single-turn Prompt on all scam websites in our ground-truth dataset, leveraging its multilingual capabilities. The two conventional systems, designed for English websites, were evaluated only on the English subset of our dataset.

Table 4. Summary of Binary Classification Results

		ScamFerret (Proposed System)			Single-turn Prompt		
		GPT-4	GPT-3.5	Gemini	GPT-4	GPT-3.5	Gemini
Overall Results for Four	Accuracy	**0.972**	0.938	0.887	0.833	0.803	0.781
Scam Types in English	TPR/Recall	**0.964**	0.913	0.848	0.790	0.786	0.676
	TNR	**0.980**	0.964	0.926	0.875	0.820	0.886
	Precision	**0.980**	0.962	0.920	0.863	0.814	0.856
	F1 score	**0.972**	0.936	0.882	0.825	0.800	0.755
Overall Results for Online	Accuracy	**0.993**	0.928	0.892	0.891	0.872	0.811
Shopping Websites in	TPR/Recall	**0.988**	0.872	0.840	0.847	0.858	0.688
Three Languages	TNR	**0.997**	0.985	0.943	0.935	0.885	0.933
	Precision	**0.997**	0.983	0.937	0.929	0.882	0.912
	F1 score	**0.992**	0.924	0.886	0.886	0.870	0.784

Table 5. Binary Classification Results for English Online Shopping Websites

	ScamFerret (Proposed System)			Single-turn Prompt			Beyond Phish [3]	Scamdog Millionaire [16]
	GPT-4	GPT-3.5	Gemini	GPT-4	GPT-3.5	Gemini		
Accuracy	**0.993**	0.923	0.870	0.873	0.863	0.790	0.883	0.915
TPR/Recall	**0.985**	0.880	0.820	0.805	0.825	0.625	0.815	0.890
TNR	**1.000**	0.965	0.920	0.940	0.900	0.955	0.950	0.940
Precision	**1.000**	0.962	0.911	0.931	0.892	0.933	0.942	0.937
F1 score	**0.992**	0.919	0.863	0.863	0.857	0.749	0.874	0.913

Table 6. Multi-class Classification Results

		ScamFerret (Proposed System)			Single-turn Prompt		
		GPT-4	GPT-3.5	Gemini	GPT-4	GPT-3.5	Gemini
Overall Results for	TPR/Recall	**0.860**	0.664	0.240	0.679	0.538	0.158
Four Scam Types	Precision	**0.977**	0.948	0.765	0.845	0.750	0.581
in English	F1 score	**0.915**	0.781	0.365	0.753	0.627	0.248
Overall Results for	TPR/Recall	**0.982**	0.767	0.830	0.827	0.858	0.633
Online Shopping Websites	Precision	**0.997**	0.981	0.927	0.929	0.882	0.905
in Three Languages	F1 score	**0.989**	0.861	0.880	0.874	0.870	0.745

5.2   Scam and Legitimate Website Classification Accuracy

We evaluate the binary classification performance (scam vs. legitimate) separately for each combination of scam type and language in our ground-truth dataset using standard metrics.

Evaluation Metrics. We use four main classification outcomes:

True Positive (TP): Correctly identified scam website.

True Negative (TN): Correctly identified legitimate website.

False Positive (FP): Legitimate website misclassified as scam.

False Negative (FN): Scam website misclassified as legitimate.

From these, we derive the following performance metrics:

Accuracy: The overall correct classification rate, calculated as $Accuracy = \frac{TP+TN}{TP+TN+FP+FN}$.

True Positive Rate (TPR) / Recall: The proportion of correctly identified scam websites, given by $TPR/Recall = \frac{TP}{TP+FN}$.

True Negative Rate (TNR): The proportion of correctly identified legitimate websites, expressed as $TNR = \frac{TN}{TN+FP}$.

Precision: The proportion of correct scam identifications among all identified scam websites, defined as $Precision = \frac{TP}{TP+FP}$.

F1 score: The harmonic mean of precision and recall, providing a balanced measure of the system's performance, calculated as $F1score = \frac{2 \times Precision \times Recall}{Precision + Recall}$.

Our evaluation employs two classification methods. For binary classification, the system's performance is evaluated based on the *Result* field output ("True" or "False"). For multi-class classification, the evaluation considers the system's *Scam Type* field, where semantically equivalent responses are considered correct (e.g., both "Fake investment site" and "Fake financial services site" are accepted for Investment scams). The evaluation uses TPR/Recall, Precision and F1 score as performance metrics, since our system only outputs the scam type when it classifies a URL as a scam website. With an equal distribution of URLs across scam types, we employ macro-averaging for evaluation. These metrics collectively provide a comprehensive assessment of the classifier's performance in distinguishing between scam and legitimate websites.

Summary of Binary Classification Results. Table 4 presents the classification accuracy comparison between ScamFerret and the Single-turn Prompt. ScamFerret (GPT-4) achieved mean scores of 0.972 (Accuracy), 0.964 (TPR), 0.980 (TNR), 0.980 (Precision), and 0.972 (F1 score) across four English scam types: Online Shopping, Technical Support, Cryptocurrency, and Investment. The results demonstrate that ScamFerret's multi-tool information gathering approach significantly outperforms the Single-turn Prompt method using only URLs and top page content, with TPR improving from 0.790 to 0.964 and TNR from 0.875 to 0.980.

For Online Shopping websites in English, German, and Japanese, ScamFerret (GPT-4) demonstrated robust performance with mean scores of 0.993 (Accuracy), 0.988 (TPR), 0.997 (TNR), 0.997 (Precision), and 0.992 (F1 score). While all models performed well in identifying legitimate Online Shopping websites (TNR > 0.940), ScamFerret's external information integration significantly improved TPR across the three languages (0.988 vs. 0.847 for Single-turn Prompt), confirming its effectiveness in classifying Online Shopping websites regardless of language. Compared to GPT-3.5 and Gemini, GPT-4 significantly improved TPR for German (0.990 vs. 0.810 and 0.790) and Japanese (0.990 vs. 0.925 and 0.910). These findings suggest that advanced LLMs can effectively perform expert-level analysis across multiple languages, potentially eliminating the need for language-specific expertise in scam detection.

Analysis of False Positives (FPs). We analyzed 18 false positive cases in ScamFerret using the GPT-4 model, resulting in an overall false positive rate of 1.5%. These cases were primarily in Cryptocurrency (16) and Online Shopping (German) (2). Three main characteristics were identified:

User-Inciting Phrases: 15 websites contained phrases like "free shipping" or "unusually large financial returns," which are common in both legitimate and scam websites, making it challenging for LLMs to differentiate.

Negative Reviews: A high number of negative posts on review websites, even for legitimate websites, led to false positives.

Privacy Protection Services: The use of WHOIS privacy protection was considered suspicious, despite being a common practice for both legitimate and scam websites.

To address these issues, including context about these features in the prompts could improve classification accuracy.

Analysis of False Negatives (FNs). We analyzed false negative cases in ScamFerret using the GPT-4 model, identifying 33 instances across various categories: Online Shopping (English) (3), Technical Support (10), Cryptocurrency (14), Investment (2), Online Shopping (German) (2), and Online Shopping (Japanese) (2). The overall false negative rate was 2.75%. Three main characteristics were observed:

Domain Status Changes: All 14 Cryptocurrency cases were related to subdomains of *stockfund.co*, which had a *pendingDelete* status during the experiment. This led to inconsistent LLM analysis results for related domain names, highlighting the need for caution in operational settings where LLMs may produce split judgments in ambiguous situations.

Inconclusive Evidence: 16 website judgments were based on suspicious aspects of website content and WHOIS information, but lacked definitive evidence. This suggests a need to enhance the tool to collect more relevant information for analysis.

Lack of Review Information: The absence of relevant URLs on review websites for reporting spam led to incorrect judgments. The LLM interpreted this lack of information as an indicator of a legitimate website. This highlights the need to improve prompts by considering that the absence of information should not be used as a sole basis for classification.

These findings indicate areas for improvement in both the information collection process and the LLM's decision-making capabilities. Future work should focus on refining the prompts to account for these scenarios and enhancing the tool's ability to gather more comprehensive and relevant data for analysis.

Comparison of Results with Conventional Systems. As shown in Table 5, when comparing the results for English Online Shopping websites, ScamFerret demonstrated superior performance compared to two conventional systems (Beyond Phish and Scamdog Millionaire). While the conventional systems achieved 0.883 and 0.915 accuracy rates through machine learning on structurally similar website datasets, ScamFerret outperformed them using LLM capabilities without requiring any additional training. These results suggest that the extensive knowledge base of LLMs provides a significant advantage over conventional machine learning models in classifying scam and legitimate Online Shopping websites.

Multi-class Classification Results. Table 6 presents the macro-averaged results of multi-class classification across Online Shopping (English, German, Japanese), Technical Support, Cryptocurrency, and Investment by scam type and language.

ScamFerret with GPT-4 achieved the highest performance with TPR/Recall of 0.860, Precision of 0.977, and F1 score of 0.915. Compared to binary classification (Table 4), TPR/Recall decreased from 0.964 to 0.860, mainly due to 9 Cryptocurrency and 24 Investment scams being misclassified as fake shopping sites. Gemini's TPR/Recall dropped significantly from 0.848 to 0.240, as it misclassified most Cryptocurrency and Investment scams as fake shopping sites. These results indicate that even GPT-4 struggles with accurate multi-class scam categorization.

For three-language Online Shopping classification, ScamFerret with GPT-4 maintained high performance (TPR/Recall: 0.982, Precision: 0.997, F1: 0.989). GPT-3.5 and Gemini also retained accuracy levels similar to their binary classification results, demonstrating effective fake shopping site detection across languages. The Single-Turn Prompt approach performed consistently lower than binary classification and failed to match ScamFerret's performance across all categories, confirming ScamFerret with GPT-4's superiority in multi-class classification.

Table 7. Number of Tools Selected and Usage per LLM

Tool	ScamFerret (GPT-4)		ScamFerret (GPT-3.5)		ScamFerret (Gemini)	
	# Selected	# Used	# Selected	# Used	# Selected	# Used
Access URL	2,724	100%	2,417	96.7%	2,471	77.8%
Extract Text	2,723	99.0%	2,398	95.3%	3,406	85.7%
Extract Hyperlink	1,018	40.5%	281	11.3%	1,088	34.6%
Get Search Result	1,798	67.6%	51	2.13%	552	17.5%
Search X/Twitter	1,060	43.5%	22	0.88%	270	9.83%
Search Reddit	1,545	63.5%	12	0.50%	196	7.12%
Retrieve WHOIS	2,479	99.5%	1,128	45.9%	419	16.3%
Retrieve DNS Record	617	25.5%	51	2.13%	129	5.04%
Retrieve Certificate	1,276	52.8%	85	3.54%	83	3.33%

Table 8. Selected Information Types and Keywords

Information Type	Keywords
Certificate Information	TLS, certificate, HTTPS, SSL
Company Information	company information, non-existent companies, non-existent company, physical address
Contact Information	email, phone number, contact information, toll-free number
Domain Name	WHOIS, registrant, privacy service, domain, DNS
Payment Method	payment, Bitcoin, cryptocurrency
Privacy Information	privacy policy, privacy notation, privacy policies
Social Engineering	psychological, lure, urgency, unrealistic, phishing tactic, scam tactic, short timeframe
Unusual Price	abnormal price, low price, discounts, free items, high return, guaranteed returns, free delivery, free shipping
User Review	social media, feedback, review, Twitter, Reddit, complaint, report, discussion, forum, low trust score, negative, indicators, social platforms
Website Status	update, copyright, outdated, up-to-date

5.3   Information Used for Website Analysis

We conducted a detailed analysis of the tools employed by the LLM and the key characteristics cited in its decision-making process.

Tools Used for Website Analysis. We analyzed the tools selected by the LLM to evaluate scam and legitimate websites. Table 7 shows the number of times each tool was selected (# Selected) and the percentage of the entire dataset in which the tool was used (# Used) for GPT-4, GPT-3.5, and Gemini. Tool selection and utilization varied significantly across models. Please note that the total number

Table 9. Information in Reasons for Website Decision

Information	ScamFerret (GPT-4) # Reasons	ScamFerret (GPT-3.5) # Reasons	ScamFerret (Gemini) # Reasons
Certificate Information	770 (32.1%)	43 (1.80%)	28 (1.17%)
Company Information	307 (12.8%)	405 (16.9%)	87 (3.62%)
Contact Information	621 (25.9%)	227 (9.46%)	233 (9.71%)
Domain Name	1,866 (77.8%)	952 (39.7%)	157 (6.54%)
Payment Method	339 (14.1%)	315 (13.1%)	81 (3.38%)
Privacy Information	379 (15.8%)	229 (9.54%)	45 (1.88%)
Social Engineering	796 (33.2%)	279 (11.6%)	108 (4.50%)
Unusual Price	1,104 (46.0%)	686 (28.6%)	501 (20.9%)
User Review	1,544 (64.3%)	52 (2.17%)	90 (3.75%)
Website Status	294 (12.3%)	165 (6.88%)	69 (2.88%)

of tool selections may exceed the dataset maximum of 2,400, as the same tool can be chosen multiple times within a single analysis.

For GPT-4, the most frequently used tools were Access URL (100%), Retrieve WHOIS (99.5%), and Extract Text (99.0%). GPT-4 demonstrated sophisticated tool combinations, such as accessing a given URL and extracting web content text upon confirming a 200 OK HTTP status. It also showed the ability to autonomously collect necessary information by recursively using Access URL and Extract Text when encountering relevant strings (e.g., "about", "privacy policy", "payment") on the top page. The model effectively used search engine and social media search tools to analyze websites based on external information, searching for "[domain name] review" and "[extracted company name]". GPT-4 demonstrated human-like analytical capabilities by selecting and combining various tools as needed, even when there was insufficient information for making a decision (e.g., information could not be obtained using a specific tool). In contrast, GPT-3.5 and Gemini were limited to using Access URL and Extract Text, unable to fully utilize other tools. This suggests that a certain level of text comprehension ability is necessary for effective tool selection in analyzing scam websites.

Future work could include tools for analyzing feature similarity to identify scam websites deployed by the same attacker. This would enhance the system's ability to detect coordinated scams.

Characteristics Included in Decision Basis. We analyzed the decisive factors used by ScamFerret in determining scam websites by examining the entire decision basis. We manually analyzed feature and keyword pairs that were decisive in the reasoning for 120 URLs (10 URLs per type and language from Table 3) correctly classified by ScamFerret using GPT-4. Table 8 presents the results, showing 47 keywords used across 10 information types. We then investigated the frequency of these 47 keywords in the overall basis for website judgments across the entire ground-truth dataset.

Table 9 shows the information types and their frequency in the website decision rationale for each LLM. The GPT-4 model, with its wide range of tool selection for acquiring external information, provided judgments from multiple perspectives. For Domain Name, which was the most common information type, the model often cited characteristics such as "suspicious due to recent domain registration" based on WHOIS information. User Review was the second most common, where the model assessed domain reputation using search engines and social media to incorporate public opinion. In the Unusual Price and Social Engineering categories, the model appropriately analyzed and identified statements targeting human psychological vulnerabilities or offering unusually inexpensive products or services.

This analysis demonstrates that referencing a wide range of external information enables multi-faceted judgment of website legitimacy, leading to improved detection accuracy and clearer explanations for the decision basis.

5.4   LLM Cost Analysis

API Usage Fees. We analyzed the cost per URL for ScamFerret using the 2,400 URLs in the Ground-truth Dataset. Using Azure OpenAI services, the total cost was \$497.39 for GPT-4 (\$0.207 per URL) and \$138.89 for GPT-3.5 (\$0.058 per URL). Gemini was available free of charge during the experiment (April 2024). However, if we calculate costs based on current prices (August 2024), Gemini would cost \$402.10 total, or \$0.168 per URL. It's noteworthy that OpenAI released GPT-4o on August 6, reducing the token cost to one-fourth of GPT-4's previous cost (\$0.01/1k tokens to \$0.0025/1k tokens). This trend suggests that as LLMs continue to develop, usage costs are likely to decrease further, potentially addressing current cost concerns in the near future.

Execution Time. We also analyzed the execution time of ScamFerret in the evaluation experiment in Sect. 5 using the 2,400 URLs in the ground-truth dataset. The total execution time when using GPT-4 was 48 h, 28 min, 4 s (1 min, 13 s per URL), GPT-3.5 was 7 h, 57 min, 29 s (12 s per URL), and Gemini was 29 h, 58 min, 13 s (45 s per URL). The execution time was divided between the interaction with the LLM and information retrieval by the tools. As a result, 79.2% of the total execution time for GPT-4 was spent interacting with LLM, 33.6% for GPT-3.5, and 80.7% for Gemini. In the current situation, the bottleneck in the analysis of scam websites is the time required for communication with LLM (excluding GPT-3.5, which has a high inference speed). In particular, GPT-4 achieved excellent results in terms of classification accuracy, but it takes longer to make inferences than other models, which is a major issue for practical application. However, the development of LLMs has been remarkable, and we believe that this problem will be solved by newly developed models (e.g., GPT-4o and Claude 3.5 Sonnet).

6   Discussion

6.1   LLM-Based Scam Website Detection

LLMs offer significant advantages over conventional machine learning approaches for scam website detection. They excel in understanding complex linguistic patterns and contextual nuances, identifying sophisticated scam tactics. Their adaptability allows effective detection of evolving scams across languages without extensive retraining. LLMs can autonomously utilize external information collection tools, iteratively gathering and analyzing data when lacking clear scam indicators. This self-directed process enables more accurate determinations about website legitimacy in challenging cases. LLMs provide human-interpretable explanations, improving system transparency and reliability. This approach increases detection accuracy, offers insights into emerging scam patterns, and potentially reduces false positives and costs associated with maintaining conventional machine learning models specialized for detection of each scam type. While LLMs offer many benefits, it is crucial to be mindful that their probabilistic nature may lead to inconsistent classifications when dealing with sophisticated scam websites that are difficult to identify at first glance.

6.2 Limitations

This study presents three primary limitations:

Cost Implications of Multiple LLM Uses. The repeated use of LLMs for thought processes increases token generation, significantly raising operational costs. While services like Azure OpenAI base their pricing on token usage, the extensive tool utilization in our proposed system may lead to higher-than-anticipated expenses in real-world applications. Mitigation strategies include designing tools for efficient token usage, carefully selecting URLs for analysis, and potentially employing locally executable LLMs like Llama3 for specific analysis targets and languages.

Detection Evasion by Attackers. Modern attackers employ sophisticated techniques to evade detection systems by manipulating external information sources that security tools rely on. They may attempt to influence the system's decision-making process by injecting false information or spreading misinformation across various platforms. For instance, attackers could artificially enhance a scam website's reputation through fake reviews or manipulated search engine results. However, ScamFerret's multi-perspective analysis approach, which evaluates websites through web content, DNS records, and search engine results, makes such evasion attempts impractical. The significant effort and resources required to consistently manipulate multiple information sources across different domains effectively prevent attackers from compromising the system's detection capabilities.

Image-Based Scam Attacks. Attackers increasingly employ image-based techniques to deceive humans while evading traditional detection systems. By embedding fraudulent content within images rather than text, attackers can bypass conventional security measures. While these image-based scams effectively deceive human users, automated detection systems struggle to identify malicious intent in images rather than machine-readable text. Recent advances in multimodal LLMs like GPT-4V show promise in analyzing visual content for fraud detection, but leveraging these capabilities for comprehensive image-based scam detection remains a future research challenge.

7 Related Work

Scam Website Analysis. Recent studies have focused on various types of scam websites [18,24,37]. Bitaab et al.'s "Beyond Phish" system achieved a 98.34% detection rate and 1.34% false positive rate for English scam e-commerce websites [3]. Kotzias et al.'s system for detecting fake online shopping websites achieved an F1 score of 0.973 [16]. *Our study extends beyond these by addressing multilingual and multi-type scams.*

Security Task-Specific LLMs. LLMs for security tasks have gained attention [2,15]. Li et al.'s "KnowPhish Detector" uses LLMs to extract brand information for phishing detection, achieving a 98.34% detection rate [19]. Roy et al.

demonstrated LLMs' potential to generate phishing content and proposed a BERT-based detection tool with 96% accuracy for phishing websites [32]. *Our approach differs by leveraging LLMs' text comprehension capabilities for scam website detection.*

LLM-as-a-Judge. Recent research has explored LLMs for evaluating LLM-generated content [36,49]. Chiang et al. used LLMs for text quality assessment, matching expert human evaluation [6]. Chan et al.'s "ChatEval" framework uses multiple LLMs for text generation quality assessment [5]. *Our study differs in that it analyzes detection rationale for classifying scam websites, rather than evaluating quality of LLM-generated text.*

8   Conclusion

This paper presents ScamFerret, an innovative agent system utilizing LLMs for scam detection without requiring additional training on scam website data. ScamFerret leverages LLMs' natural language interpretation to identify and analyze nuanced, context-dependent cues indicative of scam websites. Our evaluation demonstrates high classification accuracy: 0.972 for multiple scam types and 0.993 for multiple languages, providing clear decision rationales. Unlike traditional machine learning approaches, ScamFerret eliminates the need for additional training data, complex feature engineering, and frequent model updates. It autonomously collects information based on scam characteristics provided in natural language, enabling effective detection without conventional constraints. This work advances LLM applications in cybersecurity and opens new research directions.

References

1. Janssen, D.: How do the world's biggest countries deal with online fraud?. https://vpnoverview.com/internet-safety/cybercrime/how-countries-deal-with-online-fraud-and-cybercrime/ (2024)
2. Alfasi, D., Shapira, T., Barr, A.B.: Unveiling hidden links between unseen security entities (2024)
3. Bitaab, M., et al.: Beyond phish: toward detecting fraudulent e-commerce websites at scale. In: Proceedings of IEEE SP (2023)
4. CBS News: Cryptocurrency fraud is now the riskiest scam for consumers, according to bbb (2024). https://www.cbsnews.com/news/crypto-scam-risk-bbb-report/
5. Chan, C., et al.: Chateval: towards better llm-based evaluators through multi-agent debate (2023)
6. Chiang, D.C., et al.: Can large language models be an alternative to human evaluations? In: Proceedings of ACL (2023)
7. Curlie.org: Curlie - the collector of urls (2024). https://curlie.org/
8. van Dooremaal, B., et al.: Combining text and visual features to improve the identification of cloned webpages for early phishing detection. In: Proceedings of ARES (2021)

9. Fashion United: Online shopping fraud in the uk is a 2.3bn pounds crisis (2024). https://fashionunited.uk/news/retail/online-shopping-fraud-in-the-uk-is-a-2-3bn-pounds-crisis/2024111278529
10. FBI: Fbi releases internet crime report (2024). https://www.ic3.gov/Media/PDF/AnnualReport/2023_IC3Report.pdf
11. Gao, Y., et al.: Retrieval-augmented generation for large language models: A survey (2023)
12. Google DeepMind: Gemini (2024). https://deepmind.google/technologies/gemini/
13. Hao, S., et al.: Monitoring the initial DNS behavior of malicious domains. In: Proceedings of ACM IMC (2011)
14. Kim, D., et al.: Security analysis on practices of certificate authorities in the HTTPS phishing ecosystem. In: Proceedings of ACM ASIACCS (2021)
15. Koide, T., et al.: Chatphishdetector: detecting phishing sites using large language models. IEEE Access (2024)
16. Kotzias, P., et al.: Scamdog millionaire: detecting e-commerce scams in the wild. In: Proceedings of ACSAC (2023)
17. LangChain: Langchain (2024). https://www.langchain.com/
18. Li, X., et al.: Double and nothing: understanding and detecting cryptocurrency giveaway scams. In: Proceedings of NDSS (2023)
19. Li, Y., et al.: Knowphish: large language models meet multimodal knowledge graphs for enhancing reference-based phishing detection. In: Proceedings of USENIX Security (2024)
20. Lin, Y., et al.: Phishpedia: a hybrid deep learning based approach to visually identify phishing webpages. In: Proceedings of USENIX Security (2021)
21. Liu, J., et al.: Understanding, measuring, and detecting modern technical support scams. In: Proceedings of IEEE EuroSP (2023)
22. mbitaab: beyondphish (2024). https://github.com/mbitaab/beyondphish
23. Microsoft Azure: Azure openai service – advanced language models (2024). https://azure.microsoft.com/en-us/products/ai-services/openai-service/
24. Miramirkhani, N., et al.: Dial one for scam: a large-scale analysis of technical support scams. In: Proceedings of NDSS (2017)
25. Neoblood Corporation: Disclosure of information on fake website (2024). https://www.neo-blood.co.jp/
26. NISLabUGA: Tss esp23 (2024). https://github.com/NISLabUGA/TSS_ESP23
27. NOLA Defense: Nola defense (2024). https://www.noladefense.net/
28. Oest, A., et al.: Sunrise to sunset: analyzing the end-to-end life cycle and effectiveness of phishing attacks at scale. In: Proceedings of USENIX Security (2020)
29. Playwright: Fast and reliable end-to-end testing for modern web app (2024). https://playwright.dev/
30. Pochat, V.L., et al.: Tranco: a research-oriented top sites ranking hardened against manipulation. In: Proceedings of NDSS (2019)
31. reddit inc.: reddit.com: api documentation (2024). https://www.reddit.com/dev/api/
32. Roy, S.S., et al.: From chatbots to phishbots? - preventing phishing scams created using chatgpt, google bard and claude (2023)
33. Saad, B.A.M., et al.: Conning the crypto conman: end-to-end analysis of cryptocurrency-based technical support scams (2024)
34. ScamGuard^{TM}: Listings (2024). https://scamguard.com/reviews/
35. Sectigo: crt.sh — certificate search (2024). https://crt.sh/

36. Sottana, A., et al.: Evaluation metrics in the era of GPT-4: reliably evaluating large language models on sequence to sequence tasks. In: Proceedings of EMNLP (2023)
37. Srinivasan, B., et al.: Exposing search and advertisement abuse tactics and infrastructure of technical support scammers. In: Proceedings of WWW (2018)
38. Tang, S., et al.: Clues in tweets: twitter-guided discovery and analysis of SMS spam. In: Proceedings of ACM CCS (2022)
39. Tavily: Tavily (2024). https://tavily.com/
40. The Scam Directory: The scam directory (2024). https://scam.directory/
41. Trustpilot: Trustpilot reviews: Experience the power of customer reviews (2024). https://www.trustpilot.com/
42. Vallina, P., et al.: Mis-shapes, mistakes, misfits: an analysis of domain classification services. In: Proceedings of ACM IMC (2020)
43. VirusTotal: Virustotal (2024). https://www.virustotal.com/
44. Watchlist Internet: Fraudulent online stores (2024). https://www.watchlist-internet.at/liste-betruegerischer-shops/
45. White, J., et al.: A prompt pattern catalog to enhance prompt engineering with chatgpt (2023)
46. X Corp: Twitter api — products — twitter developer platform (2024). https://developer.twitter.com/en/products/twitter-api
47. Xigao Li: Double and nothing: Understanding and detecting cryptocurrency giveaway scams (2024). https://double-and-nothing.github.io/
48. Yao, S., et al.: React: synergizing reasoning and acting in language models. In: Proceedings of ICLR (2023)
49. Zheng, L., et al.: Judging llm-as-a-judge with mt-bench and chatbot arena. In: Proceedings of NeurIPS (2023)
50. Zhou, Y., et al.: Large language models are human-level prompt engineers. In: Proceedings of ICLR (2023)

Domain Name Encryption Does Not Ensure Privacy: Website Fingerprinting Attack With Only a Few Samples Using Siamese Network

Neriya Mazzuz and Asaf Shabtai[✉]

Department of Software and Information Systems Engineering,
Ben-Gurion University of the Negev, Negev, Israel
shabtiaa@bgu.ac.il

Abstract. In recent years, awareness of information security and the importance of protecting user privacy has grown significantly among Internet users. As a result, substantial effort is being invested to developing and deploying new protocols aimed at enhancing privacy and preventing the leakage of sensitive personal data. One of the most sensitive pieces of information at risk is the domain name, whose exposure can reveal a user's browsing history and habits. To address this privacy concern, various technologies have been introduced, including DNS over TLS, DNS over HTTPS, DNS over QUIC, Encrypted Client Hello, and Protected QUIC Initial Packets. However, despite these advancements, studies have demonstrated that these mechanisms do not provide a fully comprehensive solution, as attackers can still infer users' browsing activity under certain conditions. This is due to the fact that web pages are highly dynamic, with their content frequently changing. In this research, we propose an adaptive website fingerprinting attack based on a Siamese network model. We evaluate the effectiveness of the attack on both TLS and QUIC protocols and show that it can accurately infer domain names associated IP addresses using only a few traffic samples. Moreover, we demonstrate that the model maintains strong performance over time, enabling near real-time classification even several months after model training. The success of the attack and model's robustness over time highlight the ongoing privacy risks faced by users, as our attack provides adversaries with a novel tool to uncover users' browsing history and identify visited domain names.

Keywords: Encryption · Machine learning · Website fingerprinting

1 Introduction

With the rise in privacy violations by governments and large corporations in recent years [5, 8, 25], users have become increasingly concerned that their online activities are being monitored. As a result, there is a growing demand and need

© The Author(s), under exclusive license to Springer Nature Switzerland AG 2025
M. Egele et al. (Eds.): DIMVA 2025, LNCS 15747, pp. 26–45, 2025.
https://doi.org/10.1007/978-3-031-97620-9_2

for developing and adopting privacy-enhancing technologies (PETs) [17,19], and encryption-based protocols are at the forefront of novel measures aimed at protecting user's privacy. Browsers have adopted these protocols and implemented strict measures to enforce their use [20].

For example, the establishment of HTTPS protocol represent a pivotal milestone in reshaping Internet use, ensuring the encryption of traffic over the HTTP protocol. Its adoption has surged in recent years, with 95% of HTTP Internet traffic now encrypted.[1] This surge was fueled by the ease with which certificates can be automatically issued at no cost, user demand, and browsers initiatives. Despite the widespread adoption of HTTPS, sensitive information is still transmitted on the web in plaintext, including domain names and the names of visited websites, thereby exposing users' browsing habits to potential eavesdroppers.

There are three primary avenues of domain name leakage: unencrypted DNS requests, the Server Name Indication (SNI) extension in the TLS (Transport Layer Security) protocol, and certificates from servers to clients as part of the TLS protocol. TLS version 1.3 addresses the leakage issue on TLS certificates by encrypting certificates,[2] and its adoption has increased, with 60% of HTTPS traffic leveraging this version.[3] However, the domain name leakage also exists in unencrypted DNS requests and in SNI extension in the TLS protocol. While various solutions have been proposed, they are not widely used [6]. The proposed solutions include DoH (DNS over HTTPS), DoT (DNS over TLS), and DoQ (DNS over QUIC), which are respectively aimed at mitigating information leakage in DNS, and ECH (Encrypted Client Hello), which is designed to address SNI leakage in the TLS protocol.

Prior research has demonstrated that despite the use of the aforementioned solutions, attackers can still uncover users' browsing activities by constructing fingerprints based on IP addresses and statistical features. However, these studies overlooked the temporal dimension of fingerprints, and the fact that the evolving nature of web content can significantly affect fingerprints' accuracy over time.

In this study, we propose an adaptive website fingerprinting attack, which is a variation of the state-of-the-art triplet fingerprinting attack [29]. Our attack reveals the domain name of an encrypted session, and our evaluation shows that it is resilient to changes in website content, and its accuracy is maintained over time, even with limited sample sizes.

Our attack consists of two phases: the training phase and the classification phase. In the training phase, a Siamese model serves as an embedding model, which is used to generate proximate representations for similar websites and distant representations for dissimilar ones. The model undergoes extensive hyperparameters tuning to achieve optimal performance. In this phase, a database correlating domains with their respective IP addresses is also established; this is accomplished by sampling from top 100K websites in Alexa and their embedded websites. In the classification phase, which can be performed in real time with

---

[1] https://transparencyreport.google.com/https/overview.

[2] https://www.rfc-editor.org/info/rfc8446.

[3] https://almanac.httparchive.org/en/2021/security.

minimal processing overhead, a target's session sample is used for domain name extraction. After extracting the server's IP address from the target's session IP Header, it is cross-referenced against our database to derive a list of potential domains, which constitutes our anonymity set. If the set's size is singular, we obtain a straightforward one-to-one mapping, identifying the domain name; larger set sizes pose a more difficult challenge which our methodology aim to solve.

To execute the attack, we collect a limited number of samples for each domain using a crawler. The embedding model described above is used to compute the embedded representations of sessions, and used those representations to train a k-NN classifier to predict the domain for the target session. Our evaluation of the attack highlights its success in leaking domain names in networks protected by DoH, DoT, and ECH. However, it is important to note that as the size of the anonymity group increased, the performance decreased.

We conducted an analysis to understand the complexity of domain name leakage using IP-Domain correlation. We collected a dataset by crawling the top 100K websites on Alexa, performing name resolution for all domains associated with the 7M URLs obtained. Our analysis revealed that 72.3% of IP addresses were associated with a single domain, while 25.48% had anonymity set sizes between 2 and 10. Only 2.14% of the IPs were linked to anonymity sets larger than 10

To optimize our attack's performance, we tuned the Siamese model's hyperparameters. This involved using a dataset collected from Alexa's top 1,000 websites on the local network. The optimal hyperparameters yielded accuracy of 91.6% on the test set. We also analyzed feature importance to assess the impact of various feature groups on the results. Our findings showed that the packet direction-based feature group had the greatest impact and might be enough on its own.

We assessed our proposed attack using a dataset obtained by crawling the top 1,000 websites in Alexa. Our evaluation was performed on two protocols: TLS and QUIC. For the TLS protocol evaluation, we tested multiple anonymity set sizes—2, 3, 5, and 9—using different hyperparameter configurations. The corresponding test set accuracies were 90.0%, 92.4%, 86.9%, and 82.5%, respectively. For the QUIC protocol, an anonymity set size of 9 resulted in a significantly lower average accuracy of 53%.

We also examined the stability of our attack over time, in order to assess its ability to cope with the dynamic nature of website content. To this end, we used datasets collected four months apart to train the embedding model. We compared the models' performance for multiple anonymity set sizes (2, 3, 5, and 9) and found that there were negligible differences in the average accuracy on the test sets, with both models demonstrating consistently high accuracy.

To summarize, the main contributions of our research are:

1. We analyzed the complexity of domain name leakage using IP-Domain correlation. Our analysis showed that classification is necessary to leak the domain names of 28% of websites.

2. We introduced an adaptive website fingerprinting attack capable of near real-time execution which require only minimal number of samples. A comprehensive evaluation of the attack was performed on different anonymity set sizes and protocols (TLS and QUIC), consistently demonstrating its high accuracy.

2 Background

Website fingerprinting (WF) is a traffic analysis technique used to infer which websites a user visits by analyzing encrypted traffic patterns.

The first website fingerprinting attack, introduced by Cheng and Avnur [4], focused on identifying specific files accessed over SSL by examining file sizes. Hintz [12] expanded this by aiming to identify websites when the server was unknown, such as in anonymized browsing. Sun et al. [30] introduced Jaccard's coefficient to match websites by comparing observed traffic patterns with pre-collected data, even if the website sizes varied slightly. These early approaches demonstrated the viability of WF, primarily relying on web object sizes. However, they were based on the assumption of separate TCP connections for each request, which was applicable only to older HTTP versions. With the evolution of HTTP to versions that use persistent connections, pipelining, and multiplexing, distinguishing between individual requests became more challenging.

Bissias et al. [3] was the first to perform WF based on IP packet sizes and inter-packet arrival times instead of web object sizes. To further improve the attack, Liberatore and Levine [21] compared the effectiveness of Jaccard's coefficient and the Naïve Bayes classifier on SSH-protected channels. Lu et al. [22] showed that WF could be enhanced by considering information about the packet's order. Several other related studies did not focus on WF in particular but rather on the detection of other distinct characteristics of network traffic, e.g., the language of a Voice-over-IP (VoIP) call [38] or spoken phrases in encrypted VoIP calls [37]. Gong et al. [7] even showed the feasibility of remote traffic analysis (where the adversary does not directly observe the traffic pattern) by exploiting queuing side channels in routers.

Most WF research has focused on anonymous networks like Tor and I2P, since non-anonymous networks inherently leak website-related information through DNS requests and TLS handshakes, making site identification more straightforward [14,33]. The first WF attack on the Tor network, introduced by Herrmann [11], achieved low accuracy using normalized IP packet size distributions. Later, Wang and Ian [36] proposed using Tor cell direction data for richer traffic representation. Wang et al. [35] used a k-NN classifier and a comprehensive feature set including packet lengths, order, concentration, and bursts. Panchenko et al. [23] further enhanced accuracy by accumulating Tor cells as features, while Hayes et al. [9] combined random forest and k-NN classifiers for effective classification using a more compact feature set. Although it achieves quite good results, hand-crafted feature design is limited due to its high dependence on expert knowledge and lack of interaction with the classifier model.

Inspired by deep learning's rapid advancement in various fields, automated feature extraction has emerged as a powerful tool in WF attacks [1,2,26,28]. Various architectures, such as stacked denoising autoencoder (SDAE) [1], CNNs [26], VGG network [28] and ResNet [10] [2], have demonstrated superior performance over traditional WF methods, even when trained on limited data. In response to the growing adoption of domain name encryption protocols (such as, encrypted DNS and ECH in TLS), recent research has shifted focus toward non-anonymous networks where the IP layer remains unencrypted [15,24]. In this research, we explore the feasibility of an adaptive WF attack under the assumption of full deployment of domain name encryption. Unlike prior studies on anonymous networks, our approach focuses on the IP-based classification task. In addition, while most existing WF attacks rely on extensive training datasets, our proposed attack requires only a few samples to achieve effective performance.

A **Siamese neural network** (SNN) is a network architecture for learning similarities. It has been successfully used for metric learning in different domains such as person identification [32]. A SNN uses two or more identical sub-networks that share the same weights. SNNs have been extensively used in situations requiring a similarity criterion to be established between two or more training samples of the same type. Though computationally expensive, they perform better than other learning similarity techniques due to their ability to generalize to similar datasets [16]. To manage the trade-off between computational time and accuracy pretrained sub-networks can be used. Generally, the outputs of the two sub-networks are fed into another module, which produces the final output using a distance metric (e.g., L1 norm, L2 norm, and cosine similarity).

In this paper, we use an SNN for representation learning for learning the best representation of data to suit a specific space. In this approach, neural networks are trained to embed the input features in a lower and more discriminating space. In our case, we designed an SNN to learn the best representation of features before performing our WF attack.

3   Related Work

In recent years, several studies have been performed aiming at developing a means of bypassing domain name encryption. Patil et al. [24] conducted a measurement study to examine IP-based fingerprints' uniqueness. The authors construct the fingerprint based on union of the IP addresses that each website loads. They found that 95.7% of websites have a unique fingerprint. Trevisan et al. [34] classified encrypted TLS traffic using a simple random forest classifier. The authors extracted traffic features such as flow duration, byte counters, packet sizes, and packet inter-arrival times. For 80% of the domains, their model achieved an F1 score of over 0.8. However, these studies do not consider the temporal aspect of fingerprints. Over time, the dynamics of web content and IPdomain mappings can impact the fingerprints. We address this issue by using a few up-to-date samples to train the classifier used in our adaptive WF attack.

The privacy benefits domain name encryption offers was assessed by Hoang et al. [13] who considered the relationship between hostnames and IP addresses.

The authors found that 20% of the domains have a one-to-one mapping and that the size of the anonymity set of 30% of the domains is greater than 100. In addition, they found that only 7.7% of the domains change their hosting IP addresses daily.

In a later study, Hoang et al. [15] presented IP-based WF using IP address sequences. They constructed the fingerprint by concatenating the IP(s) of the primary domain name with a set of IPs obtained by resolving the secondary domains and successfully fingerprinted 84% of the selected domains. To further enhance the discriminatory capacity of the fingerprints, they considered the critical rendering path[4] to capture the approximate ordering structure of the domains contacted at different stages, which resulted in a 91% success rate. However, due to the high variability of website content and IP-domain mappings across time, their method's effectiveness decreased over time; therefore, to maintain their accuracy, their fingerprints need to be updated from time to time. Furthermore, different locations could resolve different IP addresses for the same website, and thus, the fingerprints are limited to a specific location. Our adaptive WF attack creates updated fingerprints during the classification phase with only a few samples required for each website.

In addition to research that explored IP and TLS traffic-based fingerprints and their ability to bypass domain name encryption, studies have been performed on encrypted DNS traffic-based fingerprints, examining whether they can bypass domain name encryption. Houser et al. [18] analyzed DoT traffic for WF and proposed a DoT fingerprinting method to better understand how much information can be deduced by analyzing DoT packet traffic. The authors used ML techniques to identify and classify individual websites into three main groups: dating, gambling, and health insurance websites. For unpadded DNS messages, they showed that DoT traffic could be accurately categorized. They also found that individual websites could be accurately identified, obtaining an FNR of 0.17 and a FPR of 0.005. Siby et al. [27] studied the effectiveness of traffic analysis attacks on DoH traffic and showed that features traditionally used for WF are unsuitable for DoH traffic analysis. Therefore, the authors engineered a new set of features and demonstrated that the proposed DoH traffic analysis effectively identifies web pages.

4   Proposed Method

In this section, we present our method for performing a WF attack, an attack aimed at discovering the domain name for an encrypted session. The proposed method is based on statistical features extracted from the encrypted session. We perform the attack using a Siamese model and a k-NN classifier.

4.1   Threat Model

We make the following practical assumptions regarding the adversary: (1) The adversary is a passive on-path attacker, which means they have access to the

---

[4] https://web.dev/articles/critical-rendering-path.

network traffic of the end user but cannot manipulate the data or inject packets; (2) All of the protocols mentioned (e.g., TLS, QUIC) encrypt the recorded
communication between the targeted online platform and the end user; and (3)
The headers of the IP layer of the traffic trace are not encrypted and thus may
be analyzed by the adversary. Background traffic (noise) from other websites or
protocols is therefore trivial to filter out, since the background traffic sessions
do not have the same IP address as the targeted server IP address.

Such adversaries come in many forms. Social Wi-Fi providers and government
agencies can essentially maintain a passive man-in-the-middle (MitM) position,
and can apply on-path attacks. The adversary is interested in identifying the
visitors of several websites, which we call monitored websites. We refer to all
other websites as unmonitored websites.

The adversary visits the websites on their own and creates a database of
traces, sequences of packets, and the timestamps generated when visiting the
websites. Once the database is created, the adversary monitors and collects users'
traces, which are classified as belonging to either the monitored or unmonitored
set of websites.

4.2   Method Overview

Our methodology comprises two distinct phases: training and classification. During the training phase, the objective is to train a Siamese model to serve as an
embedding model during the classification phase. In the classification phase, we
aim to predict the domain name for unknown sessions, using the trained embedding model and classifier.

4.3   Training the Siamese-Based Embedding Model

The process of training the embedding model is illustrated in Fig. 1. Each stage
is described below.

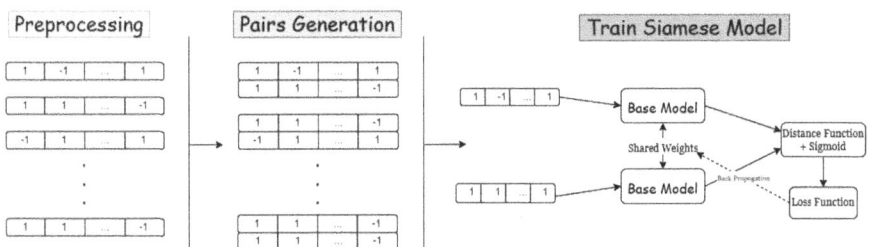

Fig. 1. Stages in the embedding model's training phase.

Preprocessing. The preprocessing steps vary depending on the base model
selected for the Siamese model. For all models except Var-CNN [2], network

sessions are transformed into sequences that represent the direction of each packet, where +1 and -1 indicate outgoing and incoming packets, respectively. These sequences are then adjusted to a fixed length of 1,000 packets by trimming or padding with zeros. The Var-CNN base model uses a data representation based on three components: packet direction, timing, and metadata. The direction component consists of sequences that represents traffic direction, as described above. The timing component is derived by calculating the time interval between consecutive packets, with sequences also adjusted to a fixed length of 1,000 packets. The metadata component comprises seven features: the total packet count, the number of incoming and outgoing packets, the ratios of incoming and outgoing packets to the total count, the total transmission time, and average transmission time per packet. Sessions with a length of less than 100 packets are filtered out to prevent model bias towards sessions with insufficient data.

Pair Generation. Pairs of samples are generated from the preprocessed data, with a distinction made between matching and non-matching pairs. Matching pairs will be assigned a label of 1, while mismatched pairs will be labeled as 0. Maintaining an equal number of matching and non-matching pairs is essential to ensure class balance and reduce the risk of overfitting.

Train Siamese Model. The pairs generated are used to train the Siamese model. The model receives a pair of samples as input and produces a similarity score as output. This score indicates the likelihood that the input pairs match (with values closer to one indicating greater similarity). The Siamese model includes a range of hyperparameters, which are described later in Sect. 6.2.

In addition to these steps, in our method's training phase, a database that maps domains to their corresponding IP addresses is generated, which is essential for training the model (described in Sect. 5).

4.4   Classification

The classification phase flow is illustrated in Fig. 2. The classification phase is performed based on the threat model described in Sect. 4.1. Each stage in this phase is described below:

Unknown Encrypted Traffic. The attacker captures the encrypted traffic of the victim in order to perform a WF attack, specifically targeting and extracting the session that they want to classify.

Extract IP Address. The attacker extracts the server's IP address (this can be done, since the IP headers are not encrypted).

Get Domains Set. The attacker uses the extracted IP address to fetch a set of potential (candidate) domains from the database established in the training phase. This set serves as our anonymity set.

Crawl Domains. We denote the domain's anonymity set as $\{D_1, D_2, ..., D_N\}$, where N represents the size of the set. To collect M samples per domain, we crawl each domain M times. The resulting dataset is represented as $\{SP_{i,j}|1 \leq$

$i \leq N, 1 \leq j \leq M\}$ where $SP_{i,k}$ denotes the j-th sample from the i-th domain. The $N \times M$ collected samples will be our anonymity domain dataset. Each sample in the anonymity domain dataset is then preprocessed as described in Sect. 4.3.

Train Classifier. First, all preprocessed samples in the anonymity domain dataset are converted into embedded vectors using the pretrained Siamese embedding model. Then, the given $N \times M$-generated embedding vectors are used as the dataset of the k-NN classifier.

Prediction. The unknown captured traffic is preprocessed as described in Sect. 4.3. Then the embedded vector is calculated using the pretrained Siamese embedding model. The embedding vector we built serves as input to the classifier, and the output is the domain name prediction.

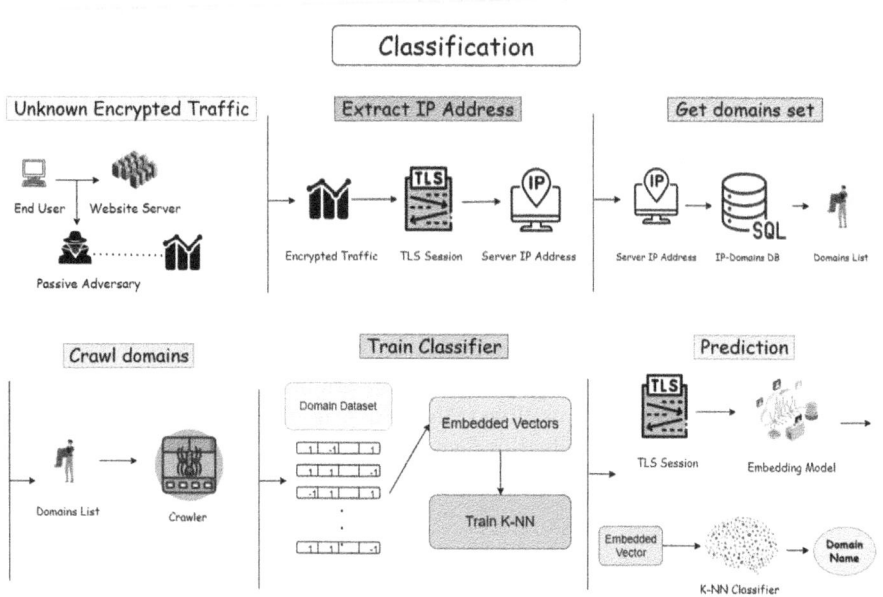

Fig. 2. Stages in the classification phase.

5 IP-Domain Collection and Correlation Analysis

In the classification phase of our method, the domain name is predicted from the anonymity set, which consists of all domains associated with a specific IP address. Classification is straightforward when a one-to-one mapping exists between the IP address and the domain. However, when the group size exceeds one, a classifier must be employed to distinguish among the domains.

In this section, we perform an IP-domain correlation analysis to examine the distribution of anonymity set sizes and assess the complexity of the classification task. As the size of the anonymity set increases, the classification task becomes more challenging.

To perform this assessment, we employed MIDA,[5] an adaptable web crawler based on Chromium and the Chrome DevTools Protocol. Using this tool, we visited the top 100K websites in Alexa,[6] collecting comprehensive information associated with those sessions, including detailed browser metadata, resource data, sub-query URL specifics, and hosted resources. On average, each site loaded approximately 81 different URLs from 17 distinct domains. The web crawl was executed in February 2022.

Then, we performed name resolution for all domains associated with the 7,093,035 URLs obtained, utilizing ZDNS,[7] a bulk DNS resolution tool. These lookups were executed from a single vantage point in approximately one hour. The resulting $(domain, IP)$ pairs were then stored in a local database for use in the classification phase. The attack preparation process can be expedited by populating the database with a reduced set of domains tailored to the attacker's interests. For example, if the attacker aims to identify domains from specific categories (e.g., financial websites), the database can be populated accordingly.

In addition, given that IP addresses can change regularly in real life and may vary by region, the attacker must account for these dynamics to maintain accurate correlations. To overcome this challenge, the attacker can periodically crawl the domain sets, refreshing the associated IP address data. This ensures that the database remains up-to-date and reflects any changes in IP address mappings, helping to maintain the effectiveness of the attack strategy over time.

In our dataset, which contains seven million sessions, we identified 198,899 unique domain names. Of these, we successfully resolved 191,762 (96.4%) domains, resulting in 146,456 distinct IP addresses. The complete dataset details are provided in Table 1. By examining the distribution of anonymity set sizes, we can assess the difficulty of the classification task. We found that 106,001 out of the 146,456 IPs (72.3%) are associated with a single domain, allowing for immediate inference of the domain name without requiring classification. Additionally, 37,321 IPs (25.48%) have an anonymity set size between 2 and 10, presenting a relatively manageable classification task. The remaining 3,134 IPs (2.14%) have an anonymity set size greater than 10. Figure 3 presents a histogram of these sizes.

6 Evaluation

6.1 Evaluation Setup

Dataset. To evaluate our proposed method, we created two datasets consisting of browsing traffic from popular websites, selected from Alexa's top 100K websites list. We used the MIDA tool for web crawling and captured the network

---

[5] https://github.com/teamnsrg/mida.
[6] https://www.alexa.com/.
[7] https://github.com/zmap/zdns.

traffic generated during each browsing session. Data collection was conducted on our local machine using an Internet connection provided by a residential ISP. The datasets used in our evaluation are as follows:

1. TOP1000ALEXA: This dataset contains crawling data from Alexa's top 1,000 websites. For each site, we conducted three crawls for the training set and two crawls for the test set. In total, we collected 40,001 TLS sessions across 3,376 distinct domains. Data collection for this dataset was completed in February 2022.
2. BIGTOP1000ALEXA: This dataset also targets Alexa's Top 1,000 websites but features a larger sample size to improve model performance. For each website, we performed five crawls for the training set and three for the test set. In total, we collected 76,486 TLS sessions from 3,753 unique domains. This dataset was generated in June 2022.

Evaluation Environment. Our base and Siamese models were implemented using Python libraries, with Keras serving as the front end and TensorFlow as the back end. The k-NN model was implemented using the scikit-learn library. All training and experiments were conducted within a Google Colab environment equipped with GPU support.

6.2   Evaluation Results

Siamese-Based Embedding Model Hyperparameter Tuning. We tuned the hyperparameters by exhaustively exploring all parameter combinations. The values tested and the top-10 best-performing combination are presented in Table 2. Given the large number of potential configurations (512), we utilized the smaller dataset TOP1000ALEXA to expedite the hyperparameter tuning process. All measurements were taken during the training phase.

During the pair generation stage, we created 9,352 matched pairs and 9,320 mismatched pairs for the training set, as well as 1,190 matched pairs and 1,179 mismatched pairs for the test set. The final step of the training phase involved

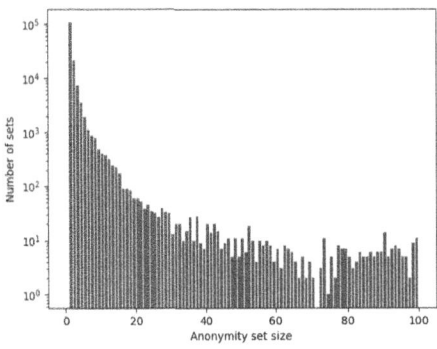

Fig. 3. IP anonymity set sizes.

Table 1. Details about our dataset.

Statistic	Value
Number of visited websites	100,000
Successfully visited websites	87,411 (87.4%)
Number of resources (URLs)	7,093,035
Average resources per request	81.1
Average distinct domains per crawl	17.8
Number of distinct domains	198,899
Successfully resolved domains	191,762 (96.4%)
Average IPs per domain	2
Number of distinct IPs	146,456

training the Siamese model with these pairs, with each iteration employing a different set of hyperparameters. Performance was evaluated based on accuracy achieved on the test set.

The base model parameter pertains to the sub-network of the Siamese network. In our evaluation, we examined two models commonly used in the WF domain (Deep Fingerprinting (DF) [28] and Var-CNN [2] models), along with two models from the image recognition domain (the GoogLeNet [31] and ResNet [10] models). Our evaluation result showed that the Var-CNN model outperformed the other candidates, and thus we selected it as our base model.

We examined two traditional metrics for the distance function: L1 and L2. Our analysis revealed that L2 yielded superior results compared to L1. Additionally, we examined various optimizers: SGD, Adam, Adagrad, and RMSprop. Across most scenarios, SGD demonstrated the highest performance, with 0.21.5% better accuracy than other optimizers. Consequently, we opted to use SGD as the optimizer during the model's training phase.

Table 2. Top-10 results in the hyperparameter tuning process.

base model	distance	batch size	embedding size	optimizer	learning rate	test accuracy	duration
var_cnn	L2	32	128	SGD	0.1	0.92	959.6
var_cnn	L2	32	256	Adagrad	0.1	0.92	1514.2
var_cnn	L2	32	512	SGD	0.1	0.91	999.1
DF	L2	64	256	SGD	0.1	0.91	191.1
var_cnn	L2	32	256	SGD	0.1	0.91	1019.3
DF	L2	64	256	SGD	0.01	0.91	383.4
var_cnn	L1	32	512	SGD	0.1	0.91	725.0
var_cnn	L1	32	128	Adagrad	0.1	0.91	798.9
var_cnn	L1	32	512	SGD	0.01	0.91	455.3
var_cnn	L2	32	256	SGD	0.01	0.90	1855.7

Feature Importance Analysis. As we described in Sect. 4.3, the preprocessing phase of the Var-CNN model differs from the other models in that the data consists of three different feature sets: direction, timestamp, and metadata (Illustrated in Fig. 4).

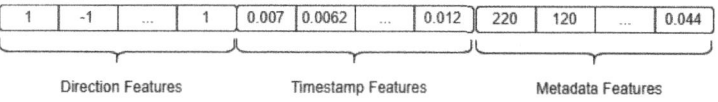

Fig. 4. Feature groups illustration for Var-CNN preprocessing.

Table 3. Feature group combinations.

Feature group combinations	Num. of features	Test accuracy	Duration
All	2007	0.92	1502.9
Direction + Timestamp	2000	0.91	959.3
Direction + Metadata	1007	0.89	649.5
Direction	1000	0.88	1156.3
Timestamp + Metadata	1007	0.90	567.9
Timestamp	1000	0.90	757.8
Metadata	7	0.61	15.0

We now describe the feature importance analysis we performed in order to examine each feature set's contribution to the model's accuracy. This analysis enabled us to determine whether an ensemble of feature sets and the use of all sets are essential, as recommended by Bhat et al. [2].

The feature importance analysis proceeded as follows: We considered all possible combinations of the feature groups (see Table 3). For each combination, we extracted the relevant features corresponding to these groups and trained the model accordingly. The model training utilizes the hyperparameters that yielded optimal results for the Var-CNN base model, employing the same dataset mentioned in Sect. 6.2.

The results, presented in Table 3, show that feature groups containing direction or timestamp information consistently yield good performance. In contrast, metadata-based groups demonstrate only satisfactory performance, indicating a relatively lower contribution to the model's accuracy. This limited contribution can be attributed to the significantly smaller size of the metadata feature set, which comprises only seven features, compared to the 1,000 features in the direction and timestamp groups.

Nevertheless, models that include metadata alongside other features exhibit improved performance compared to those excluding it. This indicates that, despite their limitations, metadata features provide complementary information that aids in distinguishing between websites under certain conditions. Based on these findings, we include all three feature groups in our experiments.

Website Classification. This experiment was aimed at comprehensively implementing the methodology outlined in Sect. 4. We performed both the training and classification phases in order to classify encrypted traffic sessions, thereby determining the domain accessed by the target. Two versions of this experiment were conducted: one for TLS sessions (over the TCP) and another for QUIC sessions (over the UDP).

We started with training the Siamese model, which was then employed as an embedding model for the subsequent classification phase. We used the best hyperparameters, as described in Sect. 6.2 and the BIGTOP1000ALEXA dataset. We opted for the larger dataset to ensure a more precise and less biased model, as a broader spectrum of sessions enables the model to glean insights into various aspects of website fingerprinting. Consistent with the approach taken in the experiment detailed in Sect. 4.3, we filtered TLS sessions with a length of 100 packets or more. Then, we performed the preprocessing step described in Sect. 4.3, using the Var-CNN base model. This was followed by the pair generation stage, in which 114,996 matched pairs and 114,843 mismatched pairs were generated for the training set.

For the test set, we generated 44,754 matched pairs and 44,588 mismatched pairs. Then, we trained the Siamese model using these pairs, achieving an impressive 95.16% accuracy on the test set. This notable improvement compared to the results presented in Sect. 6.2 (91.6% accuracy), can be attributed to the use of a substantially larger dataset in this stage.

Figures 5 and 6 present the accuracy and loss of the Siamese model over the epochs for both the training and validation sets. It is noteworthy that the validation accuracy slightly exceeds that of the test set, indicating minor overfitting.

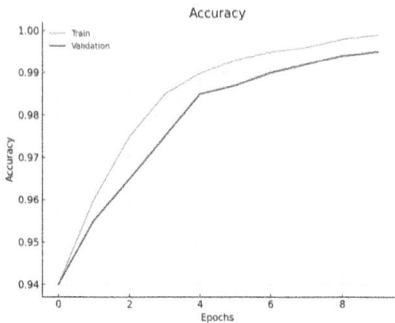

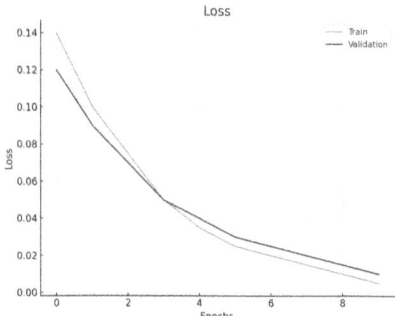

Fig. 5. Siamese model's accuracy as a function of the number of epochs, after model training.

Fig. 6. The Siamese model's loss as a function of the number epochs, after the model has been trained.

End-to-End Evaluation of the Proposed Method. The trained Siamese model was then used an embedding model in the classification phase (described in Sect. 4.4). To expedite the experiment, we randomly pre-selected four IP addresses from our IP-Domain dataset, encompassing various anonymity set sizes (2, 3, 5 and 9), enabling the preliminary execution of the crawl domains stage.

Recall that the subsequent train classifier stage (as described in Sect. 4.4) involves several hyperparameters: the anonymity set size, the size of the training set (i.e., the number of samples per website), and the k-value for the k-NN classifier. A larger anonymity set size typically correlates with a reduction in accuracy, since there are more classes to predict. Conversely, a larger training set size tends to augment the classifier's accuracy by providing more samples, facilitating the creation of more accurate clusters. Given the influence of these diverse hyperparameters, it is worth examining the combination which yields the best performance.

For every IP address, we conducted 20 crawls per site for the training set and five crawls per website for the test set. To mitigate location-based biases, we ensured that the samples for classification and the training phase were collected from a different geographical locations. Specifically, the data for the training phase originated from a local machine in Israel, while the data for the classification phase was gathered from an EC2 instance situated in US. For each IP address, a classifier was trained, with each training iteration involving diverse hyperparameters. The classification phase concludes with the prediction stage in which the output is the domain name prediction on the test set.

The results obtained are presented in Table 4. As can be seen in the table, high accuracy was achieved, particularly when the anonymity set size was 9, which represents the most challenging classification task examined. It can also be seen that for different k-values, the accuracy increases as the size of the training set grows, as expected. It's interesting to note that increasing the anonymity set size doesn't significantly affect accuracy, which suggests it's possible to achieve high success rates in fingerprinting even with large anonymity sets.

The experiments described above were replicated for the QUIC protocol. The QUIC protocol differs fundamentally from the TLS. Despite these differences, fingerprinting on QUIC traffic is feasible due to the statistical nature of our features, which are independent of the protocol's content. In those experiments, we employed the same Siamese model trained in TLS evaluation. It is important to note that this model was trained on the BIGTOP1000ALEXA dataset, which predominantly comprises TLS traffic; as a result, the performance of this experiment may not be optimal.

In this case, we began by pre-selecting an IP address associated with an anonymity set of size 9. For each website in this anonymity set, we performed ten crawls for the training set and five crawls for the test set. The embedding model was then used in the train classifier stage, and during each iteration of training, we varied the hyperparameters. Then, classification was performed on the test set in the prediction stage. As done in the experiment on TLS traffic, we randomly sampled subsets of varying sizes from the training set to evaluate the impact of the training set size on classification accuracy.

The results obtained are presented in Table 5. Although the accuracy values are lower than those observed in the TLS experiment, they significantly surpass the accuracy expected for random guessing (which, for a group size of 9, is 11.11%). Furthermore, it is evident that as the size of the training set increases, the average accuracy also rises, consistent with our expectations. These findings provide evidence that fingerprinting is feasible for QUIC traffic. Several strategies can be employed to improve the accuracy, including training a Siamese model exclusively on QUIC traffic and increasing the size of the training set.

Model Stability Over Time. As previously mentioned, the training phase involves the development of a model which is used as an embedding model in the subsequent classification phase. In the context of deploying the model in a production environment, it is presumed that the first phase occurs infrequently relative to the second phase, which is executed for each sample earmarked for classification. This experiment is aimed at assessing the stability of our embedding model over time. Potential fluctuations in model accuracy may arise due to

Table 4. Prediction stage results of TLS traffic for the pre-selected IPs with the different hyperparameters.

Train Size	K-Value	Accuracy			
		Anonymity Set Size			
		2	3	5	9
2	2	0.90	0.87	0.92	0.67
3	2	0.90	0.87	0.88	0.69
3	3	0.90	1.00	0.76	0.67
5	2	0.90	0.93	0.80	0.84
5	3	0.90	0.93	0.88	0.78
5	5	0.90	0.93	0.88	0.82
10	2	0.90	1.00	0.88	0.87
10	3	0.90	0.93	0.88	0.82
10	5	0.90	0.87	0.84	0.89
10	10	0.90	0.73	0.88	0.91
20	2	0.90	1.00	0.92	0.87
20	3	0.90	1.00	0.88	0.87
20	5	0.90	0.93	0.88	0.87
20	10	0.90	0.93	0.88	0.91
20	20	0.90	0.93	0.88	0.91

Table 5. Prediction stage results of QUIC traffic for the pre-selected IP 104.18.201.29 (anonymity set size 9) with the different hyperparameters.

Training Set Size	K-Value	Accuracy
10	3	0.60
10	10	0.58
10	5	0.56
3	2	0.53
5	2	0.53
2	2	0.51
10	2	0.51
5	5	0.51
3	3	0.49
5	3	0.47

the dynamic nature of websites, including updates to content and variations in server performance. The expectation is that the model should exhibit stability over time, as it was designed to discern dissimilarities across diverse websites. Rather than memorizing specific website behaviors, the model's objective is to differentiate between websites and identify similarities among comparable ones.

To assess the model's stability, we performed an experiment similar to those described above. The training phase was executed on the TOP1000ALEXA dataset, which was compiled in February 2022. Leveraging the optimal hyperparameters identified in Sect. 6.2, we attained a model with 92% accuracy, similar to the outcomes delineated in Sect. 6.2. Then, in the second stage of the classification phase the dataset compiled in June 2022 was used, as done in the previous experiments. This dataset comprises 20 samples per site for the training set and five for each for the test set. It's worth noting that there's a four-month gap between the data collection periods for training the embedding model and for classification. This experiment is set to provide valuable insights into the model's stability over time.

The results are illustrated in Table 6. Notably, the disparities between the results for the outdated data and the contemporary data are marginal in the majority of instances. Despite conducting classification on an embedding model trained using data procured four months earlier, commendable outcomes persist. These findings align with our premise that the embedding model adeptly discerns between websites, irrespective of specific content. Consequently, this experiment bolsters our methodology's suitability for deployment in a production environment. It underscores that the training phase can be infrequent while still enabling the model's efficacy in the classification phase, with negligible accuracy degradation over time.

Table 6. Comparison of the classification results for the fresh data (from June 2022) versus stale data (from February 2022) for different anonymity set sizes (2, 3, 5, 9). It can be seen that there is a minor difference in most of the results between the different times.

Train Set Size	K-Value	Accuracy							
		Anonymity Set Size							
		2		3		5		9	
		Fresh	Stale	Fresh	Stale	Fresh	Stale	Fresh	Stale
2	2	0.90	0.80	0.87	0.87	0.92	0.72	0.67	0.58
3	2	0.90	0.60	0.87	1.00	0.88	0.84	0.69	0.80
3	3	0.90	0.60	1.00	0.80	0.76	0.76	0.67	0.64
5	2	0.90	0.90	0.93	1.00	0.80	0.92	0.84	0.78
5	3	0.90	0.90	0.93	0.80	0.88	0.84	0.78	0.82
5	5	0.90	1.00	0.93	0.87	0.88	0.88	0.82	0.69
10	2	0.9	0.9	1.00	0.87	0.88	0.88	0.87	0.82
10	3	0.90	0.90	0.93	1.00	0.88	0.92	0.82	0.78
10	5	0.90	0.90	0.87	0.93	0.84	0.88	0.89	0.76
10	10	0.90	0.90	0.73	0.93	0.88	0.88	0.91	0.76
20	2	0.90	0.90	1.00	1.00	0.92	0.92	0.87	0.76
20	3	0.90	0.90	1	0.93	0.88	0.92	0.87	0.82
20	5	0.90	0.90	0.93	1.00	0.88	0.92	0.87	0.82
20	10	0.90	0.90	0.93	0.93	0.88	0.88	0.91	0.84
20	20	0.90	0.90	0.93	0.87	0.88	0.84	0.91	0.84

7   Conclusions

In recent years, significant efforts have been invested in the development of encryption solutions for domain names. Even with their deployment, there remains a high probability of domain name exposure. Notably, in both the TLS and QUIC protocols, the IP address is transmitted in plaintext, thereby narrowing down the potential domain name options. Consequently, the anonymity set size in such cases tends to be relatively small, enabling classification based on the methodology used in our study.

Our study underscores the potential inadequacy of these solutions, as our experiments show they do not fully safeguard domain name privacy. This research introduces a two-phase methodology in which a Siamese model is trained to serve as an embedding model, which is then used to classify the anonymity set inferred from the IP address of the remote server. Our approach demonstrates the feasibility of fingerprinting with high accuracy across different sizes of anonymity sets for both the TLS and QUIC protocols. Furthermore, our method exhibits scalability, allowing for extensive application across numerous traces, supported by our embedding model's demonstrated stability over time. These findings highlight the ongoing need for enhancements to existing solutions aimed at concealing domain names.

Training the Siamese embedding model can be performed by the attacker in advance using publicly available data, without requiring any interaction with the victim. For the classification phase, the attacker can construct an anonymity set focused on specific target websites, allowing for efficient and directed attacks. Importantly, our method requires only a minimal number of crawls to each targeted website, which greatly reduces the risk of detection and operational overhead. As a result, the attack can be carried out quickly and with minimal exposure, making it significantly more practical than previous approaches that depend on extensive and repeated crawling.

Two potential strategies emerge to address this anonymity risk. The first involves consolidating numerous websites under a single IP address, thereby enlarging the anonymity set and complicating classification efforts; however, this approach may prove impractical for website owners managing their hosting or utilizing smaller hosting services. Frequent IP address rotation presents another avenue, impeding the collection of IP-domain couplings and complicating an attacker's ability to determine the current anonymity set.

References

1. Abe, K., Goto, S.: Fingerprinting attack on tor anonymity using deep learning. Proc. Asia-Pacific Advanced Netw. **42**, 15–20 (2016)
2. Bhat, S., Lu, D., Kwon, A.H., Devadas, S.: Var-cnn: a data-efficient website fingerprinting attack based on deep learning (2019)
3. Bissias, G.D., Liberatore, M., Jensen, D., Levine, B.N.: Privacy vulnerabilities in encrypted HTTP streams. In: Danezis, G., Martin, D. (eds.) PET 2005. LNCS, vol. 3856, pp. 1–11. Springer, Heidelberg (2006). https://doi.org/10.1007/11767831_1
4. Cheng, H., Avnur, R.: Traffic analysis of ssl encrypted web browsing. University of Berkeley, Project paper (1998)
5. Fuchs, C., Boersma, K., Albrechtslund, A., Sandoval, M.: Internet and surveillance: the challenges of Web 2.0 and social media, vol. 16. Routledge (2013)
6. García, S., Hynek, K., Vekshin, D., Čejka, T., Wasicek, A.: Large scale measurement on the adoption of encrypted dns. arXiv preprint arXiv:2107.04436 (2021)

7. Gong, X., Borisov, N., Kiyavash, N., Schear, N.: Website detection using remote traffic analysis. In: Fischer-Hübner, S., Wright, M. (eds.) PETS 2012. LNCS, vol. 7384, pp. 58–78. Springer, Heidelberg (2012). https://doi.org/10.1007/978-3-642-31680-7_4

8. Grothoff, C., Wachs, M., Ermert, M., Appelbaum, J.: Toward secure name resolution on the internet. Comput. Sec. **77**, 694–708 (2018)

9. Hayes, J., Danezis, G.: k-fingerprinting: a robust scalable website fingerprinting technique. In: 25th USENIX Security Symposium (USENIX Security 16), pp. 1187–1203 (2016)

10. He, K., Zhang, X., Ren, S., Sun, J.: Deep residual learning for image recognition. In: Proceedings of the IEEE Conference on Computer Vision and Pattern Recognition, pp. 770–778 (2016)

11. Herrmann, D., Wendolsky, R., Federrath, H.: Website fingerprinting: attacking popular privacy enhancing technologies with the multinomial naïve-bayes classifier. In: Proceedings of the 2009 ACM Workshop on Cloud computing Security, pp. 31–42 (2009)

12. Hintz, A.: Fingerprinting websites using traffic analysis. In: Dingledine, R., Syverson, P. (eds.) PET 2002. LNCS, vol. 2482, pp. 171–178. Springer, Heidelberg (2003). https://doi.org/10.1007/3-540-36467-6_13

13. Hoang, N.P., Akhavan Niaki, A., Borisov, N., Gill, P., Polychronakis, M.: Assessing the privacy benefits of domain name encryption. In: Proceedings of the 15th ACM Asia Conference on Computer and Communications Security, pp. 290–304 (2020)

14. Hoang, N.P., Asano, Y., Yoshikawa, M.: Your neighbors are my spies: location and other privacy concerns in glbt-focused location-based dating applications. In: 2017 19th International Conference on Advanced Communication Technology (ICACT), pp. 851–860. IEEE (2017)

15. Hoang, N.P., Niaki, A.A., Gill, P., Polychronakis, M.: Domain name encryption is not enough: privacy leakage via ip-based website fingerprinting. Proc. Priv. Enhancing Technol. **2021**(4), 420–440 (2021)

16. Hoffer, E., Ailon, N.: Deep metric learning using triplet network. In: Feragen, A., Pelillo, M., Loog, M. (eds.) Similarity-Based Pattern Recognition, pp. 84–92. Springer International Publishing, Cham (2015). https://doi.org/10.1007/978-3-319-24261-3_7

17. Holz, R., et al.: Tracking the deployment of tls 1.3 on the web: a story of experimentation and centralization. ACM SIGCOMM Comput. Commun. Rev. **50**(3), 3–15 (2020)

18. Houser, R., Li, Z., Cotton, C., Wang, H.: An investigation on information leakage of dns over tls. In: Proceedings of the 15th International Conference on Emerging Networking Experiments And Technologies, pp. 123–137 (2019)

19. Kotzias, P., Razaghpanah, A., Amann, J., Paterson, K.G., Vallina-Rodriguez, N., Caballero, J.: Coming of age: a longitudinal study of tls deployment. In: Proceedings of the Internet Measurement Conference 2018, pp. 415–428 (2018)

20. Kraus, L., Ukrop, M., Matyas, V., Fiebig, T.: Evolution of SSL/TLS indicators and warnings in web browsers. In: Anderson, J., Stajano, F., Christianson, B., Matyáš, V. (eds.) Security Protocols 2019. LNCS, vol. 12287, pp. 267–280. Springer, Cham (2020). https://doi.org/10.1007/978-3-030-57043-9_25

21. Liberatore, M., Levine, B.N.: Inferring the source of encrypted http connections. In: Proceedings of the 13th ACM Conference on Computer and Communications Security, pp. 255–263 (2006)

22. Lu, L., Chang, E.-C., Chan, M.C.: Website fingerprinting and identification using ordered feature sequences. In: Gritzalis, D., Preneel, B., Theoharidou, M. (eds.) ESORICS 2010. LNCS, vol. 6345, pp. 199–214. Springer, Heidelberg (2010). https://doi.org/10.1007/978-3-642-15497-3_13
23. Panchenko, A., et al.: Website fingerprinting at internet scale. In: NDSS (2016)
24. Patil, S., Borisov, N.: What can you learn from an ip? In: Proceedings of the Applied Networking Research Workshop, pp. 45–51 (2019)
25. Penney, J.: Internet surveillance, regulation, and chilling effects online: a comparative case study. Regulat. Chilling Effects Online: Comparative Case Study **6**(2) (2017)
26. Rimmer, V., Preuveneers, D., Juarez, M., Van Goethem, T., Joosen, W.: Automated website fingerprinting through deep learning. arXiv preprint arXiv:1708.06376 (2017)
27. Siby, S., Juarez, M., Diaz, C., Vallina-Rodriguez, N., Troncoso, C.: Encrypted dns–¿ privacy? a traffic analysis perspective. arXiv preprint arXiv:1906.09682 (2019)
28. Sirinam, P., Imani, M., Juarez, M., Wright, M.: Deep fingerprinting: undermining website fingerprinting defenses with deep learning. In: Proceedings of the 2018 ACM SIGSAC Conference on Computer and Communications Security, pp. 1928–1943 (2018)
29. Sirinam, P., Mathews, N., Rahman, M.S., Wright, M.: Triplet fingerprinting: more practical and portable website fingerprinting with n-shot learning. In: Proceedings of the 2019 ACM SIGSAC Conference on Computer and Communications Security, pp. 1131–1148 (2019)
30. Sun, Q., Simon, D.R., Wang, Y.M., Russell, W., Padmanabhan, V.N., Qiu, L.: Statistical identification of encrypted web browsing traffic. In: Proceedings 2002 IEEE Symposium on Security and Privacy, pp. 19–30. IEEE (2002)
31. Szegedy, C., et al.: Going deeper with convolutions. In: Proceedings of the IEEE Conference on Computer Vision and Pattern Recognition, pp. 1–9 (2015)
32. Taigman, Y., Yang, M., Ranzato, M., Wolf, L.: Deepface: closing the gap to human-level performance in face verification. In: Proceedings of the IEEE Conference on Computer Vision and Pattern Recognition (CVPR) (2014)
33. Trevisan, M., Drago, I., Mellia, M., Munafo, M.M.: Towards web service classification using addresses and dns. In: 2016 International Wireless Communications and Mobile Computing Conference (IWCMC), pp. 38–43. IEEE (2016)
34. Trevisan, M., Soro, F., Mellia, M., Drago, I., Morla, R.: Does domain name encryption increase users' privacy? ACM SIGCOMM Comput. Commun. Rev. **50**(3), 16–22 (2020)
35. Wang, T., Cai, X., Nithyanand, R., Johnson, R., Goldberg, I.: Effective attacks and provable defenses for website fingerprinting. In: 23rd USENIX Security Symposium (USENIX Security 2014), pp. 143–157 (2014)
36. Wang, T., Goldberg, I.: Improved website fingerprinting on tor. In: Proceedings of the 12th ACM Workshop on Workshop on Privacy in the Electronic Society, pp. 201–212 (2013)
37. Wright, C.V., Ballard, L., Coull, S.E., Monrose, F., Masson, G.M.: Spot me if you can: uncovering spoken phrases in encrypted voip conversations. In: 2008 IEEE Symposium on Security and Privacy (sp 2008), pp. 35–49. IEEE (2008)
38. Wright, C.V., Ballard, L., Monrose, F., Masson, G.M.: Language identification of encrypted voip traffic: Alejandra y roberto or alice and bob? In: USENIX Security Symposium, vol. 3, pp. 43–54 (2007)

Making (Only) the Right Calls: Preventing Remote Code Execution Attacks in PHP Applications with Contextual, State-Sensitive System Call Filtering

Yunsen Lei[1,2] and Craig A. Shue[1(✉)]

[1] Worcester Polytechnic Institute, Worcester, MA, USA
{ylei3,cshue}@wpi.edu
[2] George Washington University, Washington, D.C., USA
yunsen.lei@gwu.edu

Abstract. PHP powers over 76% of websites worldwide, making security vulnerabilities in its applications particularly damaging. Unfortunately, such defects remain common: in 2021, nine of the top 15 most-exploited vulnerabilities identified by CISA involved remote code execution (RCE). Prior research has attempted to contain RCE through system call filtering (e.g., via `seccomp`), but these efforts are typically coarse-grained. They allow all system calls that could potentially be invoked anywhere in the application, providing attackers substantial opportunities for exploit.

We introduce a fine-grained, state-sensitive approach that builds an automaton for each PHP script, mapping different execution stages to carefully curated system call subsets. At runtime, our kernel module combines information from system call traces and PHP script-level events to apply these context-driven allow-lists. We demonstrate our method's effectiveness against real-world CVEs and against attackers crafting RCE payloads designed to mimic legitimate calls. Our model successfully detects these "stealth" attacks and maintains a low performance overhead of only 1%—a substantial improvement over the 5% overhead observed in prior work.

1 Introduction

PHP is a popular server-side language powering 75.4% of measured websites in 2024 [27]. PHP is the driving force behind WordPress, which itself powers 43.7% of measured websites worldwide [28]. Unfortunately, PHP faces the same security issues as other back-end server languages.

A critical and challenging to combat threat is remote code execution (RCE). An attacker typically exploits an application defect to deliver and execute malicious code that exceeds the application developer's intent, potentially spawning processes and accessing system resources without restriction. To address the risks posed by RCE, the research community has explored various mitigation strategies. A common approach involves a two-phase process of modeling and enforcement. During the modeling phase, the target application is analyzed to determine its intended behavior. By requiring the program's runtime execution to comply with the defined model, defenders can constrain the scope in which arbitrary code can be executed.

Defenders can use a system call allow-list to enforce a program's behavior. The allow-list is first obtained by analyzing the target program for a set of reachable system calls and is then enforced through the `seccomp` module in Linux. The `seccomp` module can be viewed as a deterministic finite automaton with a single approve and reject state. System calls defined in the allow-list consist of transition rules that keep the program in the "approve" state. Conversely, system calls not in the allow-list form transition rules that move the program to the "reject" state. The `seccomp` module can choose to terminate a program if a disallowed system call is captured.

The effectiveness of the system call allow-list approach is directly related to how tight the constraints are. Prior efforts have built allow-lists that work at the entire PHP application granularity [6,12] and at the HTTP request granularity (i.e., the script specified in the HTTP URL) [4]. Unfortunately, there are significant drawbacks to each: the first is too permissive, enabling attackers to exploit any allowed call; the second, which is built upon `seccomp`, requires restarting the PHP process or dedicating separate PHP engines per target script. Our investigation shows that even request-level profiles can permit stealthy "mimicry" RCE. We therefore re-examine the whole system call modeling and enforcement pipeline for PHP applications. We propose a protection system that constructs a detailed, stateful system call profile and enforces that profile using run-time sensors and kernel-mode system call filtering.

With this foundation, we explore important research questions: *How can we construct fine-grain profiles of PHP applications? To what extent can such fine-grained modeling and enforcement constrain an RCE attack on the PHP platform? What performance cost does this approach introduce to the existing PHP platform? Does such an approach negatively impact programs' legitimate execution?* In exploring these questions, we contribute the following:

A Design for Detailed PHP Application Profiling and Sensing: We design an automata-based approach for modeling the state of a PHP application. We design tools to construct the automata from PHP scripts and from the executables and libraries associated with the PHP engine (Sect. 3).

A System for Context-Aware, State-Sensitive System Call Filtering: We implement sensors and an enforcement module that recognizes application-level events that precede the underlying system calls. We use these to apply appropriate system call profiles to enforce (Sect. 4).

An Evaluation of the Security Effectiveness and Performance: Through real-world RCE vulnerabilities and legitimate workloads, we compare against `seccomp`-based defenses. Our approach completely stops the attacks across three tested classes (Sect. 5), adding under 3% overhead for script-level enforcement (Sect. 6).

2   Background for PHP Application Risks

In PHP applications, the risk of remote code execution is heightened by two intertwined factors. First, attackers have various techniques at their disposal to inject malicious code. Second, system call usage is prevalent in application scripts. Those prevalent system calls result in a larger list of permitted system calls, which can provide attackers with ample opportunities to mimic legitimate system operations and thus bypass existing security measures. In this Section, we first examine the common techniques for RCE on PHP applications and then discuss the impact of system call frequency.

2.1   Attack Techniques

Remote Code Execution (RCE) in PHP typically requires explicit injection of malicious code or data that the PHP engine then integrates into its execution, unlike intra-procedural attacks like return-oriented programming, which reuse existing instruction gadgets. We classify RCE based on how malicious code affects the system call profile across execution scopes. One category involves file inclusion or uploads, introducing new scripts and altering request-handling structures. Another relies on command injection or deserialization, executing malicious code within an existing script's scope.

```
1 <?PHP
2 $file_path = $_GET['file'];
3 include($file_path);
4 include('app_dir/' . 'dependency.php');
5 /* rest of the code ... */ ?>
```

Fig. 1. A file inclusion vulnerability through user-supplied path

File Upload and Inclusion. Attackers often exploit arbitrary file upload vulnerabilities to place malicious scripts within the PHP application's source directory. Then, the attacker can execute their own uploaded script by accessing it directly via a URL or with a local file inclusion vulnerability (line 3 in Fig. 1). For applications that need write access to its code directory (e.g., a WordPress updates its plugin), the attack may be more covert by overwriting existing scripts during the upload process (such as altering dependency.php on line 4 in Fig. 1).

Injection and Deserialization. This method exploits functions that process unvalidated user inputs, allowing attackers to embed and execute malicious code. For instance, an attacker might craft a payload that is dynamically evaluated via functions like eval or injected into a template engine. In addition, unsafe deserialization can instantiate objects with carefully crafted fields, triggering unwanted function calls.

Figure 2 illustrates a Server-Side Template Injection (SSTI) attack on the Twig PHP template engine [24]. In a misconfigured or older version of Twig, attackers can construct the controllable parameter to inject a template that registers a callback function, then invoke it through another parameter. For example, setting greeting using registerUndefinedFilterCallback and name using getFilter. Such attacks are typically the result of improper configuration or unsafe template features rather than an inherent flaw in Twig itself.

```
1 <?PHP
2 $twig->render("Hello " .$_GET['greeting'] .$_GET['name']);
3 /* rest of the code ... */ ?>
```

Fig. 2. A Minimal Server Side Template Injection Example

Figure 3 illustrates a deserialization attack involving a PHP class called LogWriter, which includes a destruct magic method that writes content to a file when the object is unset, or the script ends. The separate script (lines 15) retrieves data from a request parameter and calls process_data, appearing harmless at first. However, an attacker can submit a serialized LogWriter object, manipulating its filename and log_content properties to write arbitrary files. In large applications using multiple libraries, attackers can construct complex object chains to trigger a series of malicious actions. This is akin to Return-Oriented Programming but is known as Property-Oriented Programming [10]. Tools like those in [1,20] automate finding such "gadgets" in common PHP libraries.

```
1 <?PHP
2 require("dep.php");
3 $data = unserialize($_GET['data']);
4 process_data($data);
5 /* rest of the code ... */ ?>
-------------------------
6 <?PHP # dep.php
7 class LogWriter {
8     public $filename;
9     public $log_content;
10    public function __destruct() { file_put_contents(
11        $this->filename, $this->log_content);}
12 function process_data($data) {
13    $data->value += 1; } ?>
```

Fig. 3. Code vulnerable to Unsafe Deserialization attacks

2.2   System Call Pervasiveness and Mimicry

All of these attacks arise from software defects. Ideally, flawless code would prevent them, but in reality, security vulnerabilities keep surfacing in production software. Consequently, we need a cross-application mitigation strategy that does not rely on a "perfect" application. System call filtering can help, but to be effective, it must be stringent enough to limit attackers' options. Although web applications legitimately require file and network operations during request processing, these same capabilities can be exploited for malicious purposes.

A basic `seccomp`-style filter, for instance, lacks awareness of the specific PHP application it protects; it must permit every system call reachable from the PHP engine or its extensions. APIs like `shell_exec` pose a particular risk, since they rely on powerful system calls (`fork`, `exec`), leaving wide avenues for arbitrary code execution if an RCE vulnerability is discovered.

Recent work moves toward dynamic system call specifications, adapting the filter to different execution stages (e.g., initialization vs. serving [13], or updated each `execve` call [11]). In the PHP domain, Saphire [4] filters system calls on a per-request basis. However, even a request-level scope can still be too broad: a single request may involve multiple scripts, all of which share the same system call set. While the filter permits system calls that are essential for legitimate request components, it cannot distinguish these from identical system calls initiated by malicious code.

The attack examples we have described above illustrate the need for more state-aware filters to prevent RCE attacks. In file inclusion or upload attacks, the malicious code executes as a new, separate script within a request's scope. For injection and deserialization, the malicious code is executed within an existing script. Effective defense mechanisms must, therefore, perceive the scope in which a system call is issued and enforce a multi-state model, each aligning with a specific execution context or phase of a request.

3   Design: Automata-Based Enforcement

We build an automaton for each PHP script to precisely capture valid system call behavior. Figure 4 depicts how, in a trusted environment, we use two sensors: one examines the compiled PHP binary and its libraries, and the other inspects the application's PHP source code. Based on their outputs, we generate an automaton per script.

Once these automata are ready, we instrument the PHP executable so that specific points in its execution trigger software interrupts, which then get handled by the kernel. In production, when clients connect, these interrupts provide the kernel with execution context, which is tracked via a stack that represents the application's state. Each system call is then checked against the relevant script automaton and the current execution context. If the automaton permits the call in that state, it is allowed; otherwise, it is rejected.

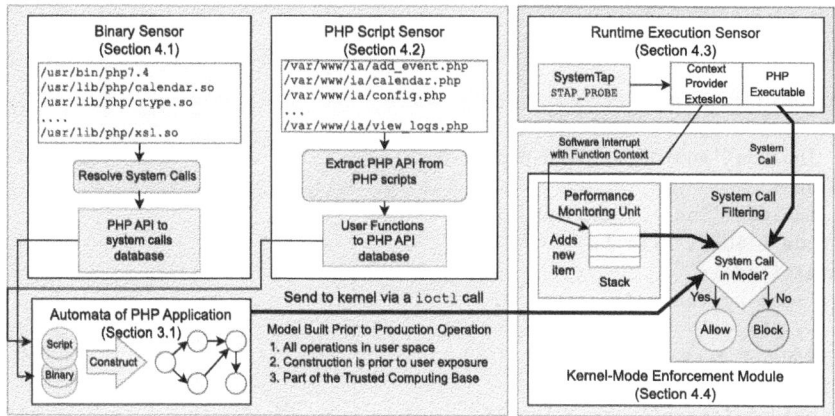

Fig. 4. The left side of the diagram shows the instrumentation and model construction that occurs prior to production operation. Through a set of sensors, we build an automaton for each PHP script. During operation (depicted on the right), we fuse that prior model with probe and system call data from the PHP executable to filter system calls that deviate from the model.

3.1 Threat Model: Trusted Models and Kernel

Like other approaches that aim to constrain untrusted PHP server applications, we rely on a trusted kernel for security decisions and assume we can examine the PHP binary and scripts in a trusted environment to build our models.

We trust that the constructed model can be read by the kernel without malicious alteration (e.g., from a trusted account on the server). We trust that the kernel can safely load our constructed model and that we can instrument the PHP executable before any client connections. Once a client connects, an attacker may attempt corruption through system calls.

The defender's goal is to block at least one necessary system call the attacker needs to compromise the environment. The attacker's goal is to corrupt the environment through a set of system calls without any of the required system calls being blocked by the defender. The attacker may utilize mimicry techniques to operate using system calls that are likely to be allowed by the module as desired, but they must eventually attempt to corrupt the environment to enact the attack.

3.2 From a PHP Script to an Automaton

When an HTTP server (e.g., Nginx or Apache) receives a request (e.g., GET, POST), it uses configuration data to determine which PHP script to run. We call this the *target script*. During execution, the target script may include other scripts—*included scripts*—which can likewise include additional scripts. Scripts with global-scope code run immediately upon inclusion, while others only define classes or functions.

```
1 <?PHP # req1.php
2 if (isset($_GET['admin'])) {
3   include('admin/'. 'dep2.php'); }
4 else {
5   include('inc/'. 'dep3.php'); }
6 $code = $GET['file'];
7 eval($code);
8 /* rest of the code ... */ ?>
---------------------------------
9  <?PHP # dep2.php
10 include(dep4.php);
11 logAdminAccess('access.log');
12 loadAdminTemplate();
13 /* rest of the code ... */ ?>
---------------------------------
14 <?PHP # dep3.php
15 include(dep4.php);
16 loadVisitorTemplate();
17 /* rest of the code ... */ ?>
```

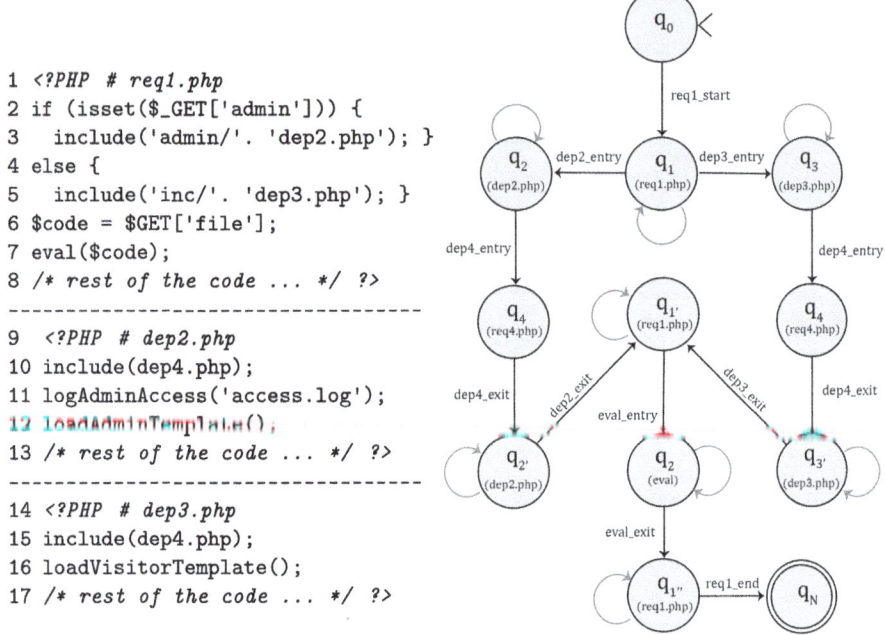

Fig. 5. The code on the left is an example of a vulnerable PHP application consisting of two request handling scripts: `Req1.php` and `Req2.php`. The diagram on the right shows how the program can be transformed into a DFA-based system called filtering that leverages the application context in its decisions.

We can represent the target script's inclusion and execution of a dependent script as a unique and identifiable execution stage. These execution stages can be conceptualized as a state in a Deterministic Finite Automaton (DFA). Formally, our DFA is defined as $(Q, \Sigma, \delta, q_0, q_N)$:

States Q. Each state q_i represents an execution phase in the script. We create a state for the global scope code of each script (including included scripts) and additional states for specific function scopes. Each state q_i is associated with a system call allow-list A_i. This variable state granularity definition enables flexible system call policies.

Alphabet Σ. This consists of both system call symbols (identified by their kernel call numbers) and application-level events (e.g., entering a script, exiting a script, or calling a function).

Transition Function δ. Given a state and an input event, it determines the next state. System calls allowed in the current state lead to self-transitions (e.g., $\delta(q_i, s_k) \rightarrow q_i$ for $s_k \in A_i$) while events like `script_entry` or `script_script_exit` move execution to different states.

Initial State q_0. The PHP engine is awaiting a request to process.

Final State q_N. Reached when the script completes processing the request.

Figure 5 shows a sample automaton for a target script `req1.php`. Depending on the `admin` parameter, `dep2.php` or `dep3.php` is included; each of these in turn includes `dep4.php`. Afterward, `req1.php` reads a `GET` parameter (`file`) and calls `eval`. The DFA starts in q_0, transitions to a script-specific state (e.g., q_1) upon request start, then moves to new states when it includes other scripts or enters function scopes. At each state, only the system calls listed in its allow-list are permitted. The final state q_N indicates the end of request handling.

The DFA approach naturally covers script-level system call specifications, but it can also apply at the function level. As shown in Fig. 5, we can enforce a dedicated filter for security-critical functions like `eval`, effectively blocking any system calls within those functions. Such granular policies help mitigate risks without unduly restricting legitimate functionality.

A more detailed DFA yields tighter state-level allow-lists, further limiting the system calls available to attackers. By confining powerful calls to fewer regions of trusted code, the defender reduces the chance an attacker can exploit vulnerabilities and successfully execute malicious operations.

3.3   Profiling: PHP APIs to System Calls

The automaton from the previous section can support fine-grained enforcement. To construct such a model, we need to understand when system calls will naturally occur in a PHP application. This is a two-step process, and this section describes the first step: discovering the system calls, if any, associated with each PHP API function.

The PHP API consists of internal PHP functions and methods, as well as additional functions and methods provided by dynamically linked shared libraries. These API implementations are written in lower-level languages and compiled into executable and shared object files. To profile the system calls invoked by each PHP API function or method, we analyze their implementations to determine the system calls they use. This analysis can be recursive since one API function or method may invoke others, forming a control flow graph (CFG) with inter-procedural calls. We annotate the CFG with any system calls discovered during the analysis and store this information in a mapping database for quick retrieval of system call details for each API function or method.

A significant challenge in this profiling process lies in resolving function references. At the PHP engine level, function calls may target other functions where the exact call target is initially unknown. This occurs when the call instruction uses registers or memory locations to specify the target function. We tackle these details in Sect. 4.1.

3.4   Profiling: User-Defined Functions to APIs

The second step in mapping a PHP application to its system calls involves identifying all PHP API calls made by the application's scripts. Unlike built-in APIs, functions and methods defined within scripts are written in PHP and referred to as *user-defined functions*. To capture all relevant PHP API calls, we recursively analyze these user-defined functions, as they may invoke other user-defined functions or PHP APIs.

To identify PHP API calls within scripts, we transform all application scripts into control flow graphs (CFGs). Each basic block in a CFG consists of a sequence of instructions represented as PHP Abstract Syntax Tree (AST) nodes. By analyzing these AST nodes, we perform recursive graph traversal to discover all referenced PHP internal APIs.

PHP can resolve function names and method functions dynamically, which our profiling must support. For the method names, we adopt a static analysis process that tries to backtrace a variable AST node's definition or initial assignment. This static analysis helps resolve a variable's class type, which gives us a full `class_name::method_name` association. We leverage dynamic profiling to resolve the function names that remain unknown. When constructing the DFA, we map these API invocations identified in each script with their underlying system calls. We add those calls to the relevant nodes associated with each script's corresponding state.

3.5 Enforcement: Context and Filters

The existing `seccomp` module lacks the visibility needed to match a system call to a specific scope in a process's execution. The module only knows the process ID that is associated with a system call. That module also statically associates a system call filter with a process, making the filter insensitive to the state within a process. In contrast, our design needs to keep the kernel enforcement module in lockstep with the state of the user space program. This requires synchronized context from the user space process to the kernel enforcement module.

Our approach uses a software interrupt to signal transitions. In addition, we pass an application-level event's specification to the kernel in advance. The specification tells the enforcement module how to collect an event's context data when a corresponding interrupt handler is invoked. To trigger the interrupt at the desired execution stage, we instrument a PHP extension to insert the interrupt's instruction. The PHP extension is developed to overwrite a set of PHP execution hooks. Each hook is essentially the entry and exit point of a PHP application-level event (e.g., a script entry or request start).

Before production usage, the filtering module registers callback functions that the interrupt handler will invoke to collect user context. To get a system call, we adopted the same process as `seccomp`, which tags the target process with a flag to indicate the need to examine a system call when invoked.

4 Implementation: Sensors and CFGs

Our automata-based enforcement design requires profiling work and detailed run-time sensors to operate. We now describe the details of implementing this design. We start with the instrumentation of the PHP executables and libraries and then describe the script-level instrumentation. We then describe our run-time sensors and the kernel enforcement.

4.1   PHP API to System Calls

Our goal is to map each built-in PHP API function or method to its associated system calls. The PHP engine can be built with various dynamically linked libraries; in the instance (PHP 8.4) we used for evaluation, there were 89 such libraries.

We implemented this mapping using the **angr** tool [23], specifically the **CFGFast** API, to build control flow graphs (CFGs) for the PHP binary and its linked libraries. These CFGs form the foundation of our binary analysis. First, we identify all function symbols defined in the PHP binary. We then locate basic blocks in the CFG that end with **call** instructions, which can be classified as either direct or indirect calls.

For direct calls, we resolve targets by directly identifying function definitions or through entries in the process linkage table (PLT). PLT entries reference dynamically linked functions and are typically resolved at runtime via lazy binding, where initial calls redirect to corresponding global offset table (GOT) entries. Using the GOT entry address, we access the relocation table (`.rela.plt`) to obtain the symbol type and an index into the dynamic symbol table (`.dynsym`). With this index, we identify the function's symbolic name, the owning library, and ultimately its offset within the library's symbol table. This method allows us to map PHP API functions to their dynamically-linked implementations.

Indirect calls typically reference memory locations or registers, often involving function pointers or dynamic resolution (e.g., via **dlsym**). During the static analysis phase, we label unresolved indirect calls and record their call site addresses. Then we use **DynamoRIO** [8] to instrument these calls to log the relative addresses of the call site and target, along with the module path. Running PHP's test suite lets us correlate this data to identify the target function's library and offset.

These methods allow us to establish an initial mapping between PHP internal APIs and their callee functions. We then traverse each callee function's CFG to identify system calls. On **x86_64** systems, system calls are invoked using the **syscall** instruction, which stores the system call number in the **eax** register. If the **eax** register is populated with an immediate value (e.g., **mov eax, 0x01**), determining the system call number is straightforward. In cases where the **eax** register is populated with a non-immediate value, we emulate the basic block containing the **syscall** instruction—or the preceding block—to identify the stored system call number. This information is then compiled into a database, mapping PHP internal APIs to their implementing system calls.

This mapping between PHP internal APIs and their associated system calls is independent of the API's arguments. Although the actual system calls invoked by a PHP API could vary depending on its parameters, our mapping captures the complete set of possible system calls for each API.

4.2   User-Defined Function CFGs to APIs

For PHP APIs, we examine each user-defined function to identify the internal PHP APIs it invokes. Our static analysis starts by using `php-cfg` [17] to build CFGs. We modify `php-cfg` to embed original source line numbers within the SSA nodes, making it easy to integrate dynamic profiling data.

Our analysis starts by traversing the abstract syntax tree (AST) to detect function and class definitions. We then inspect call-related AST nodes, extracting names for direct function calls and static methods. For dynamic method calls (e.g., `$a->method()`), we track each variable back to its last assignment to resolve its class. Specifically, we handle class resolutions arising from `new` instructions, resolvable function returns, or global variables. However, PHP's dynamic typing and callback registration mechanism can still obscure function names or class types. To resolve these, we employ a runtime profiler that logs user-defined and internal calls along with their script locations. This dynamic profiling complements our static analysis, resolving calls that cannot be determined statically.

In practice, we find that we can successfully resolve all calls in the PHP applications we evaluated, including popular and complex PHP applications like WordPress. In the event that the instrumentation fails to resolve a function, our implementation records the failure to enable manual resolution.

4.3   SysTap Probes for Runtime Context

We developed a minimal PHP extension called `context_collector` that uses SystemTap's `STAP_PROBE` macro to help track runtime events without modifying the core PHP engine.

This extension gathers runtime context by declaring pointers to memory locations for storing event-related data. Each variable is wrapped with a `STAP_PROBE` macro from the SystemTap tool [22]. This macro inserts an inline `NOP` instruction after the variable declaration and uses the `.pushsection` assembler directive to record probe specifications—such as the `NOP`'s address and register allocations—in the note section of the extension's ELF binary. The ELF binary is loaded as a shared library in the PHP engine.

The `context_collector` uses PHP's built-in hooks to intercept the original function handler. In our custom handler, we embed the `STAP_PROBE` macro to collect context such as the request URL, script name, and function name. This custom function handler still invokes the original handlers to maintain normal application behavior.

At runtime, probes are activated using the `perf_event_open` system call, which replaces the `NOP` instruction with a software interrupt (`INT3`). Probes can be disabled by closing their file descriptors, which restores the original `NOP` instruction. This design minimizes overhead when the PHP process does not enforce system call filtering.

4.4   Kernel Enforcement: Automata Checks

To collect context information from the software interrupts enabled in Sect. 4.3, we developed a kernel module that registers an interrupt handler for our custom STAP_PROBE probes with the Performance Monitoring Unit (PMU). The PMU manages the creation of the underlying probe structure, which invokes the handler function. The kernel module's header file is accessible to the context_collector extension, ensuring consistent encoding and parsing of context data between the two components.

The enforcement module processes two types of events: regular system calls and application events triggered by software interrupts. For system calls, the module queries the current system call profile to decide whether to allow or deny the operation. For application events, the module advances the automaton based on event attributes such as request URLs or script names.

To manage system call profiles, our implementation uses a fixed-size bitmap, mapping each system call to a specific bit. We utilize the kernel's DECLARE_BITMAP and bitmap_ APIs to create, set, and check the bitmap values. Each profile requires $\lceil NR\_syscall/8 \rceil$ bytes of memory, providing constant-time lookups and fixed memory usage. This approach is more efficient than traditional seccomp, where each seccomp_rule_add invocation increases the BPF bytecode size and results in linear growth for lookup times.

4.5   Artifacts Availability

The source code and scripts supporting this paper are publicly available at: https://github.com/yunsenlei/phpsys_filter.

5   Security Evaluation

We aim to answer two research questions: *To what extent can fine-grained modeling and enforcement constrain an RCE attack on the PHP platform? Does such an approach negatively impact programs' legitimate execution? To what extent can the approach protect itself from attacks?*

5.1   Prevention: Real-World Vulnerabilities

We tested our defense on two VMs running Nginx and PHP 7: one used our approach, and the other used Saphire. We focused on PHP application vulnerabilities using WordPress CVE-based exploits.

Attackers have significant flexibility after injecting malicious code. Attackers commonly use the system API to spawn new processes, enabling malicious actions beyond PHP's scope. A more advanced approach, system call mimicry, uses legitimate PHP APIs to replicate benign behavior at the system call level, avoiding obvious malicious calls like sys_exec.

Table 1 compares these strategies (labeled Process Launch and System Call Mimicry) across Saphire and our approach. Both block overt calls like sys_exec

Table 1. Comparison of Saphire and our approach across RCE attacks from reported WordPress CVEs.

	CVE	Saphire	Our Approach
Process Launch	CVE-2018-7602	Yes	Yes
	CVE-2018-7600	Yes	Yes
	CVE-2020-35729	Yes	Yes
	CVE-2023-39362	Yes	Yes
	CVE-2023-39147	Yes	Yes
System Call Mimicry	CVE-2018-12613	No	Yes
	CVE-2020-8644	No	Yes
	CVE-2021-26120	No	Yes
	CVE-2022-1329	No	Yes
	CVE-2023-28115	No	Yes

(already excluded from their allow-lists). However, our context-aware state machine better detects mimicry attacks, addressing subtle threats more effectively.

Saphire uses a request-level system call profile and blocks attacks in two scenarios: (1) the requested URL has an empty allow-list profile, automatically rejecting any system calls or (2) the attack uses system calls not included in the allow-list for that request. Scenario (1) applies to RCE attacks via file uploads where the malicious script, such as `uploaded_script.php`, is accessed directly through its URL. This script is unprofiled during the allow-listing phase, making such attacks easy to detect. Our evaluation primarily focuses on Scenario (2).

For instance, RCE via file inclusion allows attackers to embed malicious scripts into legitimate requests, exploiting permissive system call profiles. Using CVE-2022-1329 [19] as an example, a WordPress vulnerability permits non-admin users to modify plugin source code. This enables attackers to execute malicious scripts within any legitimate request. Our state machine model prevents such attacks by detecting script inclusions that violate defined transition rules and enforcing script-level system call profiles. This remains effective even if uploaded scripts overwrite legitimate ones.

For RCE via injection and deserialization, malicious code alters existing script behavior without introducing new scripts, often bypassing detection due to broad permissions. For example, CVE-2021-26120 targets the Smarty PHP template engine [26], exploiting a flaw in the `Smarty_Internal_Template` object that grants unauthorized access to its parent `Smarty` object. This allows attackers to overwrite template caches, executing malicious code when affected pages are reloaded. Our state machine detects the attack by monitoring function-level events, such as code evaluation and template rendering. This granularity distinguishes system calls triggered by these events, enabling our approach to detect and prevent malicious actions that bypass traditional defenses.

5.2   Profile Correctness: Legitimate Use

In a system call allow-list, the filter flags calls outside the list as "positive" and permitted calls as "negative." The previous section measured how each approach blocks unauthorized calls ("true positives") and avoids letting them through ("false negatives"). Here, we assess how they allow legitimate calls ("true negatives") and prevent mistaken blocks ("false positives"), indicating the profile's completeness. A complete profile correctly maps all valid system calls to the application or its states, preventing false positives.

We compare the system calls extracted by our approach with those from Confine [12], Sysfilter [6], and Saphire [4]. Confine automates extraction from containers, so we use a single PHP container for comparison. Sysfilter extracts calls from the program binary, while Saphire targets PHP's internal APIs. Like Saphire, we apply our filter only after PHP begins handling a request, using all mapped PHP internal API calls.

Table 2. A comparison of the system calls extracted across four approaches for the entire PHP engine. A lower number of allowed system calls may provide fewer attack opportunities.

Approach	Number of system calls extracted
Confine [12]	194
sysfilter [6]	159
Saphire [4]	123
Our work	119

In Table 2, we compare the number of system calls extracted by each approach. Confine and sysfilter generate more calls due to their broader analysis scope (a container or entire PHP process). By contrast, Saphire and our method attach filters when PHP handles a request, capturing only the calls triggered by scripts. Saphire's profile also includes additional epoll calls required between requests. Our 119 calls form a superset of all potential calls a PHP script can make before automaton-based modeling. In practice, scripts use fewer calls; for example, WordPress scripts average 32 calls in their allow-list.

In addition to the comparison, we tested our profiles for false positives by generating various request workloads targeting the application. A false positive would incorrectly block a legitimate request, resulting in an error response. To comprehensively assess this, we created multiple WordPress user accounts with varying privilege levels, from administrator to regular users. We employed automated scripts to execute actions permitted by each user role: administrative tasks included modifying site configurations, whereas normal user activities involved viewing, commenting, and posting content. Using Xdebug [7] with php-code-coverage [3], we measured code coverage. In our experiments with WordPress, benign requests achieved a code coverage of 54%, and no system calls were mistakenly rejected.

5.3   Potential Attacks on the Sensors and Enforcement Systems

Our approach relies on system call traces and user-level context events, which attackers may attempt to disrupt or compromise. With a trusted kernel threat model, attackers have limited means to prevent the kernel from detecting system calls. The threat model also assumes that effective attacks must use system calls, making it infeasible for adversaries to avoid them entirely.

Prior work has highlighted concerns for system call interposition frameworks, particularly regarding time-of-check/time-of-use (TOCTOU) issues. As with seccomp, we avoid pointers and evaluate the system call arguments (e.g., an immediate number) directly [25]. This avoids the conditions needed for a TOCTOU attack.

Attackers might also target mechanisms capturing user-level context events: two potential evasion strategies [21] include 1) **Memory Permission Alteration**: Changing Virtual Memory Area (VMA) flags to prevent the kernel's install_breakpoint function from operating. This can be done by modifying ELF binaries or remapping process memory. 2) **False Context Injection**: Using ptrace to manipulate the execution of a PHP application and inject false context data. Both strategies require root privileges, making them harder to execute in practice.

Our evaluation, assuming an attacker with root privileges, confirmed the feasibility of these evasion techniques. However, as these attacks require root access, they are considered beyond the scope of this study. An attacker with such privileges could execute most server-side operations without bypassing the detection system.

6   Performance Evaluation

This section explores the research question: *What performance cost does this approach add to the PHP platform?*

6.1   Experiment Setup

We consider four cases that use different filtering approaches:
1. No Filter: Base case: no system call filter used.
2. Request Aware: Our filtering approach is configured to sense the application's current request during runtime and filters system calls based on a per-request profile (as in Saphire).
3. Script Aware: Our filtering approach is configured to sense the interleaving of PHP script and filter system calls based on a per-script profile.
4. Saphire: This shows Saphire's system call filtering approach. It uses seccomp as the underlying enforcement module and filters system calls at the per-request level.

We conduct experiments on Ubuntu 22.04 with 4 CPU cores at 2.4 GHz and 8 GB of RAM, using PHP 7.4 for Saphire compatibility. We select WordPress [30] due to its popularity and realistic workloads, and use **Apachebench** to generate requests with varying concurrency and URL patterns. Our performance metric is the request response time.

6.2   Single URL Workload

In Saphire, when users concurrently access a single URL, the system call profile for a PHP worker is established during the first request. This profile remains unchanged for subsequent requests, leading to overhead primarily driven by the operations of the underlying `seccomp` module. Although the request is unchanged in this scenario, our approach still dynamically updates the system call profile for each incoming request or script event. We then apply this updated profile to enforce the appropriate system calls.

Table 3. Average request response time (in milliseconds)

Filtering Approach	Concurrent Users			
	10	25	50	100
No Filter	248	646	1302	2631
Request-Aware	252	657	1318	2667
Script-Aware	252	659	1339	2724
Saphire	260	677	1371	2765

Table 3 shows the request response time. Our request-aware filtering approach can generally keep up with the baseline across different concurrency levels, which only add 1% of overhead on average. Saphire adds around 5% overhead.

6.3   Complex Workload

When multiple users simultaneously request different URLs, each request must receive the correct system call profile. Our method automatically switches profiles at runtime using script entry/exit events. In contrast, Saphire must either restart the PHP process or dedicate separate worker pools to different request URLs. This can lead to resource imbalances. When configuring pools of PHP workers, a defender with Saphire needs to consider the size of the pool. Each script that can be requested must have a pool associated with it. It can be challenging for the defender to size the pool to optimally match the dynamics of user access patterns. In the worst-case scenarios, a small pool of workers is constantly active while large portions of workers in a different pool are idle and underutilized. For instance, if 50 concurrent users hit two URLs in a 1:4 ratio, one pool may be overused while another remains idle, slowing overall performance. Our approach instead loads all possible profiles into a single enforcement module, avoiding such inefficiencies.

7   Related Work

The risks of remote code execution (RCE) are well-recognized, and the research community has made significant efforts to understand PHP applications and mitigate these risks. We broadly classify related work into three areas: profiling and enforcing models for PHP applications, restricting system calls in applications, and uncovering vulnerabilities in PHP applications.

7.1   Modeling PHP Applications for Protection

Static and dynamic analyses, combined with runtime protection, are often used to detect or prevent RCE attacks. These methods require runtime behavior to adhere to a pre-constructed model. ZENIDS [14] exemplifies this approach by recording a PHP application's behavior with benign user inputs and building an execution profile using an inter-procedural control flow graph. While ZENIDS captures informative user-level context, such as request data, it lacks insight into system-level interactions, which are the primary focus of attackers in RCE scenarios. Saphire [4] takes a different approach by extracting a request-level system call profile for PHP applications to prevent RCE attacks. Saphire is effective when malicious actions include system calls not covered by a request-level profile. However, its coarse-grained approach struggles to handle more subtle attack strategies that require fine-grained detection.

Our work uses a similar approach to these modeling efforts, focusing on fine-grained restrictions for system calls during program execution. By integrating detailed runtime state sensing and a custom enforcement module, we leverage this context to constrain system calls more effectively.

7.2   Restricting System Calls in Applications

System calls mediate unprivileged user-space access to privileged system resources. Wagner and Dean [29] introduced four modeling approaches: (1) a basic allow-list, as in `seccomp`, which must include all calls a program might make; (2) a call graph model, incorporating control flow; (3) an abstract stack model; and (4) a digraph model capturing transitions between consecutive calls.

Most existing work targets general programs rather than PHP. For instance, SFIP [5] enforces digraph-based transitions and tracks system call origins (instruction addresses). However, this lacks sufficient context for PHP's interpreter-based calls, which often stem from a single internal API. Our approach tracks which specific PHP function or script initiates a system call, allowing more precise checks. Similarly, temporal system call specialization [13] refines profiles based on whether a server is initializing or serving requests. Other recent efforts use eBPF for programmable system call security [18], but still lack the fine-grained, script-level state awareness our method provides. In contrast, we automatically derive a detailed automaton for each PHP script and apply state-sensitive filters without manual labeling of program stages.

7.3 Discovering PHP Script Vulnerabilities

Prior work has identified RCE as a key threat in PHP applications. Huang et al. introduced WebSSAIR [16] to detect insecure information flows enabling script inclusion from user input. UChecker [15] focuses on file upload RCE by modeling and verifying exploit conditions via an SMT solver. Backes et al. [2] use a code property graph (CPG) that combines abstract syntax trees, control flows, and dependencies. By querying the CPG for patterns such as tainted inputs reaching sink functions, developers can detect vulnerabilities before deployment.

Our work is compatible with these prior efforts. The efforts in this section can be used to detect and correct the underlying errors in software. Until those vulnerabilities are identified and patched, our work can decrease the likelihood that an adversary can successfully exploit these vulnerabilities to implement an attack. Our approach can offer these benefits because, in contrast to the work in this section that aims to identify the software vulnerabilities, our script-level approach aims to detect and block system calls that do not match existing profiles. For more detailed sensitivity, developers can identify specific functions warranting heightened resolution (e.g., `eval`, template functions).

8 Discussion and Concluding Remarks

Applicability to Other Languages: The profiling approach is specifically designed for the PHP language for serving web applications. This allows us to fully explore any practical challenges of implementing the approach, such as unintended filtering or performance issues. However, our instrumentation and enforcement techniques generalize beyond PHP. In particular, we use the `perf_event_open` interface, which leverages Linux `uprobe` events. Many other frameworks (e.g., Node.js, Python) provide similar hooks at the HTTP request level, function entries, or script imports, making it straightforward to adopt our approach for other languages.

PHP Just-in-Time Compilation: PHP 8 introduced Just-in-Time (JIT) compilation as part of the Opcache extension. However, JIT-compiled code does not integrate seamlessly with runtime profiling tools that rely on intercepting PHP function calls. To address this limitation, third-party developers have created new tracing APIs [9]. In PHP 8.4, JIT compilation is not enabled by default, reducing its prevalence in most configurations. Our instrumentation extension remains compatible with the latest version of PHP when JIT is disabled.

Conclusion: Our work has explored a practical security problem in PHP applications. We have designed an automata-based approach that enables detailed, state-based filtering. In evaluation, we have found it is more sensitive and can prevent attacks that go unnoticed by current state-of-the-art techniques. We found the performance overheads for the approach are low, enabling practical deployment.

References

1. Ambionics Security: PHPGGC: PHP Generic Gadget Chains. https://github.com/ambionics/phpggc (2023), gitHub repository
2. Backes, M., Rieck, K., Skoruppa, M., Stock, B., Yamaguchi, F.: Efficient and flexible discovery of php application vulnerabilities. In: 2017 IEEE European Symposium on Security and Privacy (EuroS&P), pp. 334–349 (2017). https://doi.org/10.1109/EuroSP.2017.14
3. Bergmann, S.: php-code-coverage: Library for collecting test coverage statistics for php code. https://github.com/sebastianbergmann/php-code-coverage (2023), gitHub repository
4. Bulekov, A., Jahanshahi, R., Egele, M.: Saphire: sandboxing PHP applications with tailored system call allowlists. In: 30th USENIX Security Symposium (USENIX Security 21), pp. 2881–2898. USENIX Association, August 2021
5. Canella, C., Dorn, S., Gruss, D., Schwarz, M.: Sfip: coarse-grained syscall-flow-Integrity protection in modern systems (2022)
6. DeMarinis, N., Williams-King, K., Jin, D., Fonseca, R., Kemerlis, V.P.: sysfilter: automated system call filtering for commodity software. In: International Symposium on Research in Attacks, Intrusions and Defenses (RAID) (2020)
7. Rethans, D.: Xdebug: debugger and profiler tool for php. https://xdebug.org/ (2023), available: Xdebug Official Website
8. DynamoRIO Contributors: DynamoRIO: Dynamic Instrumentation Tool Platform. https://dynamorio.org/, Accessed 17 Dec 2023
9. Engineering, D.: PHP 8 observability (2021), https://www.datadoghq.com/blog/engineering/php-8-observability-baked-right-in/, Accessed 01 Dec 2024
10. Esser, S.: Utilizing code reuse or return oriented programming in php application exploits. In: Proceedings of the Black Hat Conference. Las Vegas, NV, USA (2010)
11. Gaidis, A.J., Atlidakis, V., Kemerlis, V.P.: Sysxchg: refining privilege with adaptive system call filters. In: Conference on Computer and Communications Security, p. 1964 1978. CCS 2023, Association for Computing Machinery (2023). https://doi.org/10.1145/3576915.3623137
12. Ghavamnia, S., Palit, T., Benameur, A., Polychronakis, M.: Confine: automated system call policy generation for container attack surface reduction. In: 23rd International Symposium on Research in Attacks, Intrusions and Defenses (RAID 2020), pp. 443–458. USENIX Association, San Sebastian, October 2020. https://www.usenix.org/conference/raid2020/presentation/ghavanmnia
13. Ghavamnia, S., Palit, T., Mishra, S., Polychronakis, M.: Temporal system call specialization for attack surface reduction. In: 29th USENIX Security Symposium (USENIX Security 20), pp. 1749–1766. USENIX Association, August 2020, https://www.usenix.org/conference/usenixsecurity20/presentation/ghavamnia
14. Hawkins, B., Demsky, B.: Zenids: introspective intrusion detection for php applications. In: 2017 IEEE/ACM 39th International Conference on Software Engineering (ICSE), pp. 232–243 (2017). https://doi.org/10.1109/ICSE.2017.29
15. Huang, J., Li, Y., Zhang, J., Dai, R.: Uchecker: automatically detecting php-based unrestricted file upload vulnerabilities. In: 2019 49th Annual IEEE/IFIP International Conference on Dependable Systems and Networks (DSN), pp. 581–592 (2019). https://doi.org/10.1109/DSN.2019.00064
16. Huang, Y.W., et al.: Securing web application code by static analysis and runtime protection. In: Proceedings of the International Conference on World Wide Web, p. 40 52. WWW 2004, ACM, New York, NY, USA (2004). https://doi.org/10.1145/988672.988679

17. ircmaxell: php-cfg: a library to build and work with a control flow graph in php. https://github.com/ircmaxell/php-cfg (2023), Accessed 19 Nov 2023
18. Jia, J., et al.: Programmable system call security with ebpf (2023). https://arxiv.org/abs/2302.10366
19. National Vulnerability Database (NVD): Cve-2022-1329 detail: Elementor website builder plugin for wordpress vulnerability. https://nvd.nist.gov/vuln/detail/CVE-2022-1329, April 2022, Accessed 19 Nov 2023
20. Park, S., Kim, D., Jana, S., Son, S.: FUGIO: automatic exploit generation for PHP object injection vulnerabilities. In: 31st USENIX Security Symposium (USENIX Security 22), pp. 197–214. USENIX Association, Boston, MA, August 2022. https://www.usenix.org/conference/usenixsecurity22/presentation/park-sunnyeo
21. Quarkslab: defeating ebpf uprobe monitoring. (2024). https://blog.quarkslab.com/defeating-ebpf-uprobe-monitoring.html, Accessed 20 Apr 2025
22. Red Hat, IBM, Intel: SystemTap Language Reference (2023), https://lrita.github.io/images/posts/systemtap/langref.pdf, Accessed 05 Feb 2024
23. Shoshitaishvili, Y., et al.: SoK: (State of) the art of war: offensive techniques in binary analysis. In: IEEE Symposium on Security and Privacy (2016)
24. Symfony: home - twig - the flexible, fast, and secure php template engine. https://twig.symfony.com/, Accessed 19 Nov 2023
25. The Linux Kernel Documentation: Seccomp BPF (SECure COMPuting with filters) (2024), https://docs.kernel.org/userspace-api/seccomp_filter.html, Accessed 07 Jul 2024
26. The smarty project contributors: smarty: a template engine for php. https://www.smarty.net/ (2023), Accessed 07 Feb 2024
27. W3Techs: usage statistics and market share of php for websites. https://w3techs.com/technologies/details/pl-php (2024), Accessed 2 Dec 2024
28. W3Techs: usage statistics and market share of wordpress. https://w3techs.com/technologies/details/cm-wordpress (2024), Accessed 2 Dec 2024
29. Wagner, D., Dean, R.: Intrusion detection via static analysis. In: Proceedings 2001 IEEE Symposium on Security and Privacy. S&P 2001, pp. 156–168 (2001). https://doi.org/10.1109/SECPRI.2001.924296
30. WordPress: Blog tool, publishing platform, and cms wordpress.org. https://wordpress.org (2023), Accessed 19 Nov 2023

Poster: Generating the WEB-IDS23 Dataset

Eric Lanfer[✉][iD], Dominik Brockmann[iD], and Nils Aschenbruck[iD]

Institute of Computer Science, Osnabrück Unviersity, Osnabrück, Germany
{lanfer,dobrockmman,aschenbruck}@uos.de

Abstract. Anomaly-based Network Intrusion Detection Systems (NIDS) require correctly labelled, representative and diverse datasets for an accurate evaluation and development. However, several widely used datasets do not include labels which are fine-grained enough and, together with small sample sizes, can lead to overfitting issues that also remain undetected when using test data. Additionally, the cybersecurity sector is evolving fast, and new attack mechanisms require the continuous creation of up-to-date datasets. To address these limitations, we developed a modular traffic generator that can simulate a wide variety of benign and malicious traffic. It incorporates multiple protocols, variability through randomization techniques and can produce attacks along corresponding benign traffic, as it occurs in real-world scenarios. Using the traffic generator, we create a dataset capturing over 12 million samples with 82 flow-level features and 22 fine-grained labels. Additionally, we include several web attack types which are often underrepresented in other datasets.

Keywords: Network Intrusion Detection · Machine Learning · Web Attacks

1 Introduction

The Development of anomaly-based network intrusion detection systems (NIDS) is still lacking properly labelled data with recent attacks. Therefore, it is important to create new datasets that can cover a diverse range of attacks and traffic characteristics, to enable researchers to develop and evaluate defense mechanisms, such as NIDS. However, it is very difficult to obtain or create such datasets. When collecting real-world traffic, for example in a company or university network, high privacy standards make it almost impossible to collect datasets. Besides this issue, the labelling problem of such datasets is even more impactful, resulting in small sample numbers of attacks. For training machine learning models, however, large sample numbers are needed to cover a wide variety and avoid overfitting. Uneven represented classes and too small sample numbers are a problem observed in popular NIDS datasets [3,5]. Large datasets ensure model robustness by covering diverse attack scenarios and reducing the likelihood of overfitting to specific traffic patterns.

© The Author(s), under exclusive license to Springer Nature Switzerland AG 2025
M. Egele et al. (Eds.): DIMVA 2025, LNCS 15747, pp. 66–72, 2025.
https://doi.org/10.1007/978-3-031-97620-9_4

In several popular datasets widely used in intrusion detection research, over-fitting is a big issue due to small sample sizes of certain attacks or the aggregation of such attacks into taxonomy super classes [4,7]. Without careful analysis and debugging of new models, such issues may not be visible, especially when the samples are only binary classified into benign and attack samples. Moreover, the particular attack-types with an accurate description of how the attacks were performed are needed to enable domain experts to perform an insightful model performance analysis [2].

Tackling the aforementioned issues, we developed a traffic generator, which is highly modular and configurable. We generate a dataset that includes a diverse range of attacks, fine-grained labels, and realistic traffic characteristics, address-ing the key limitations of existing datasets. Additionally, the dataset includes several web attack types that are typically underrepresented in other datasets [6], along with corresponding benign actions targeting the same network services.

2   Traffic Generation

We decided to develop our own generator that allows us to orchestrate multi-ple attacks and also according benign traffic. Considering a company network, users will have benign interaction with the services provided on a network, these benign traffic usually mixes up with the traffic of attackers trying to attack such services. This overlap creates a realistic challenge for intrusion detection systems to detect the attacks inside the normal, benign noise. We utilize the generator to execute benign actions that are closer to the attack patterns, e.g., utilizing the same web server endpoints. The generator supports multiple protocols, includ-ing HTTP(S), FTP, SMTP, SSH, ICMP, DNS, TCP, and UDP, ensuring broad coverage of typical network traffic scenarios. Additionally, the generator applies randomization techniques to introduce variability in request timings and traffic patterns. This helps mitigate model overfitting to features like precise timing patterns that follow the specific generation process. Currently, the traffic gener-ator is available on request to other researchers, it is planned to release it open source. Next we will describe the two primary modes, attack and benign, of the traffic generator.

Benign Mode: In benign mode, the traffic generator simulates normal user behavior across various protocols, including web browsing (HTTP(S)), file trans-fers via FTP, email exchanges through SMTP, and remote server interactions over SSH. The traffic generator simulates benign HTTP(S) interactions using automated bots and a web crawler to replicate realistic web browsing behavior. Bots perform randomized actions on an OWASP Juice Shop instance, such as browsing pages, submitting feedback, and logging in or out. The web crawler complements this by navigating internal links from a list of URLs, scraping page content, and updating a graph-based crawl frontier to emulate natural user nav-igation. Both components introduce variability through randomized actions and delays. Similarly, benign FTP interactions involve randomized directory naviga-tion, file uploads, and downloads, with variability in file names, sizes, permis-

sions, and login attempts. For SMTP, the traffic generator simulates randomized email exchanges by varying the subject, body length, and recipient count.
Finally, SSH interactions involve establishing remote sessions and executing random commands on servers, including delays and optionally simulated failures,
creating realistic server access behaviors.

Table 1. Implemented attacks in the traffic generator

Attack/Service	Alias	Tool(s)
FTP		
Fingerprinting	ftp_version	Metasploit (auxiliary/scanner/ftp/ftp_version)
Bruteforce	ftp_login	Metasploit (auxiliary/scanner/ftp/ftp_login)
HTTP/S		
Cross-site scripting	xss	Selenium WebDriver
SQL-Injection	sqli	Selenium WebDriver, python-requests, sqlmap
Denial of Service	dos	sqlmap
Bruteforce	bruteforce	Hydra
Server-side request forgery	ssrf	Selenium WebDriver
Reverse Shell	revshell	Selenium WebDriver, netcat
SMTP		
Fingerprinting	smtp_version	Metasploit (auxiliary/scanner/smtp/smtp_version)
User Enumeration	smtp_enum	Metasploit (auxiliary/scanner/smtp/smtp_enum)
SSH		
Bruteforce	ssh_login	Metasploit (auxiliary/scanner/ssh/ssh_login)
Misc		
Portscan	portscan	Nmap (-sS flag)
Hostsweep	hostsweep	Nmap (-sn and -Pn flags)

Attack Mode: In attack mode, the traffic generator simulates 13 different malicious activities across supported protocols. Table 1 shows the different attacks
grouped by the service. For HTTP(S), various web-based attacks are executed
on an OWASP Juice Shop instance, including SQL injection (targeting login
and search functionalities, with optional payload obfuscation), cross-site script-
ing (XSS) through feedback forms, denial of service (DoS) via sqlmap, brute
force login attempts using Hydra, server-side request forgery (SSRF) targeting
predefined URLs, and reverse shell exploits leveraging server-side template injection (SSTI) to execute commands on the victim server. The implemented FTP
attacks include fingerprinting via version detection and brute force attempts
with randomized or injected credentials to simulate successful and unsuccessful
login scenarios. For the attacks on the SMTP protocol, we focus on enumeration
of users and fingerprinting of the mail server configuration. Furthermore, SSH
attacks use brute force to compromise login credentials, incorporating successful breaches. Finally, miscellaneous attacks include host sweeps and port scans,

with parameters such as target ranges randomized for variability. Each attack dynamically adjusts parameters, timing, and target specifics to ensure realistic simulation.

2.1   Architecture

We utilize the traffic generator to generate a large dataset within a virtualized cloud environment based on OpenStack. Figure 1 depicts the architecture of the testbed. It consists of two virtual networks interconnected via a router running Zeek[1] to record all traffic. The external zone contains five clients, while the internal zone hosts five servers. Both networks are connected to the internet to include realistic web traffic in the dataset. To prevent client fingerprinting and ensure data authenticity, each client operates in both normal and benign modes across different iterations.

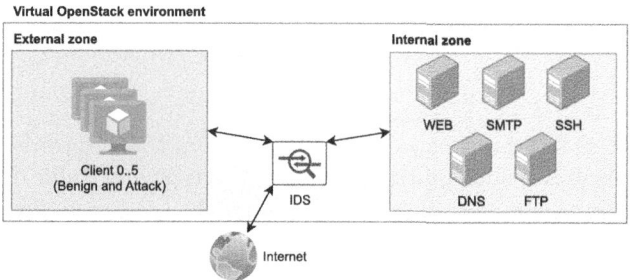

Fig. 1. Virtual data capturing testbed

3   Dataset

We process the traffic recordings to create flow-level features. This results in a total of 82 features capturing various aspects, such as packet counts, inter-arrival times, payload characteristics as well as metadata about service type and duration. Each sample is labeled based on the logs generated by the traffic generator and result in 22 classes, including 21 attack classes and one benign class. In total, the dataset includes 12,059,749 samples. The dataset can be downloaded via the osnaData Repository [1].

Table 2 shows the distribution of the samples over the different classes. Among the attack classes, *portscan* and *hostsweep_Pn* make up a large proportion of the dataset. In total, the dataset includes 825,187 benign samples that do not cover an attack. For each of the six web attacks (*xss, sql, dos, bruteforce, ssrf, revshell*), classes exist in both HTTP and HTTPS versions. For classification

---

[1] https://zeek.org.

purposes, these classes might be kept as they are or merged, to apply classification independent of the protocol. For *ssh_login*, classes exist for successful as well as unsuccessful attempts, in the classes *ssh_login* and *ssh_login_succesful*, respectively.

Table 2. Class Distribution

Class	Count	Class	Count	Class	Count
portscan	5,046,406	sql_injection_http	74,300	revshell_http	8,549
hostsweep_Pn	3,492,290	ssh_login	34,279	ssrf_https	6,656
bruteforce_http	912,503	ssh_login_successful	34,246	ssrf_http	5,509
bruteforce_https	865,126	dos_https	33,216	xss_http	4,558
benign	825,187	hostsweep_sn	22,637	xss_https	4,533
ftp_login	468,275	ftp_version	11,688	smtp_enum	7
sql_injection_https	102,584	smtp_version	11,353		
dos_http	86,443	revshell_https	9,404		

We extract 80 flow-level features from the network traffic recordings using the *Zeek FlowMeter* tool[2], and additionally the service type and traffic direction for a total of 82 features. For most metrics, statistics such as the minimum, maximum, total, average, and standard deviation within a flow are computed. Features like *flow_duration*, the inter-arrival time of packets (e.g., *fwd_iat.min* and *fwd_iat.avg*) and the idle time capture temporal characteristics. Packet counts and directional statistics are given as the total packets in forward/backward direction, also in relation to the flow time and the ratio between forward and backward packets. The packet payloads are extracted as statistics about the payload length (e.g., *fwd_pkts_payload.avg* and *payload_bytes_per_second*). Additionally, features about the header size, TCP control flags and bulk statistics are generated. A more detailed description of each feature can be seen in the repository of the *Zeek FlowMeter* tool.

4   Notes on Interpreting Classifier Results

Some attacks in this dataset are undetectable by inspecting a single flow, and correct detection by a flow-based classifier often indicates overfitting. Detecting attacks like `revshell` and `Server-side request forgery` requires at least two flows, as a single flow's encrypted payload resembles a benign one. Identifying these attacks requires either the inspection of unencrypted payloads or the inspection of resulting flows. However, during capturing the dataset we did not include a mechanism, that would be able to indicate, whether a stream resulted from a previous stream. With some uncertainty, it probably could be inferred,

---

[2] https://github.com/zeek-flowmeter/zeek-flowmeter.

by inspecting the flows and matching on the flow that goes from the victim's IP to the attacker's IP, after a `revshell` samples is observed. For `ssrf` the target was randomly chosen out of a list of 5 hosts on the public internet, typically our victim server is not performing web requests. Therefore, this could be used as an indicator.

5   Conclusion

In this technical report, we provide details on the creation of the WEB-IDS23 dataset. The dataset is publicly available and generated by a traffic generator developed by ourselves in a virtual environment. The aim of creating this new dataset is to encounter the lack of datasets with low numbers of attack samples to enable the community a balanced learning. Moreover, we include a set of web attacks that are used in real attack scenarios. We mixed the attack traffic with realistic benign traffic utilizing the same services as the attackers to form realistic scenarios.

The dataset is also pruned to some limitations, one big issue that it is synthetically generated and does not contain real-world traffic. Furthermore, only certain protocols are included, due to the high effort needed in implementing such actions into a traffic generator. Additionally, as described before in Sect. 4, some attacks are only detectable by inspecting two flows, we did not implement a mechanism to trace back which flow was the result of a previous flow.

In future work, when recording new datasets, it would be advisable to record the unencrypted payloads. They might be the only way to make some attacks differentiable from real traffic. This would also help to detect attacks that result in a second stream. By inspecting the payloads, e.g., a revshell could be detected.

Acknowledgment:. We would like to thank Marty Schüller for his work on the traffic generator and for helping in recording and curating the dataset.

References

1. Lanfer, E., Brockmann, D., Aschenbruck, N.: WEB-IDS23 Dataset (2025). https://doi.org/10.26249/FK2/MOCIY8
2. Lanfer, E., Sylvester, S., Aschenbruck, N., Atzmueller, M.: Leveraging explainable AI methods towards identifying classification issues on ids datasets. In: 2023 IEEE 48th Conference on Local Computer Networks (LCN), pp. 1–4 (2023)
3. Leung, K., Leckie, C.: Unsupervised anomaly detection in network intrusion detection using clusters. In: Proceedings of the Twenty-Eighth Australasian Conference on Computer Science, vol. 38, ser. ACSC 2005. Australian Computer Society, Inc., Australia, pp. 333–342 (2005)
4. McHugh, J.: Testing intrusion detection systems: a critique of the 1998 and 1999 DARPA intrusion detection system evaluations as performed by Lincoln Laboratory. ACM Trans. Inf. Syst. Secur. (TISSEC) **3**(4), 262–294 (2000)

5. Portnoy, L., Eskin, E., Stolfo, S.: Intrusion detection with unlabeled data using clustering. In: In Proceedings of ACM CSS Workshop on Data Mining Applied to Security (DMSA-2001, pp. 5–8 (2001)
6. Ring, M., Wunderlich, S., Scheuring, D., Landes, D., Hotho, A.: A survey of network-based intrusion detection data sets. Comput. Secur. **86**, 147–167 (2019)
7. Zoghi, Z., Serpen, G.: Unsw-nb15 computer security dataset: analysis through visualization. Secur. Priv. **7**(1), e331 (2024)

Vulnerability Detection

Sourcerer: Channeling the `void`

Nicolas Badoux[1], Flavio Toffalini[1,2], and Mathias Payer[1(✉)]

[1] EPFL, Lausanne, Switzerland
[2] Ruhr Universität Bochum, Bochum, Germany

Abstract. Type confusion vulnerabilities occur when a program misinterprets an object as an incompatible type. Such errors result in undefined behavior and can lead to illegal memory accesses undermining security. For compatibility reasons, the C++ programming language tolerates insecure type conversions, delegating the responsibility for assuring an object's type to the developer. Sanitizers help developers detect and patch vulnerabilities during dynamic testing, *i.e.*, before they reach production environments. However, current type confusion sanitizers either incur prohibitive runtime overheads, or fail to check all casts. In particular, casts from `void*` have historically been overlooked due to challenges in recognizing the underlying object's type, thus leading to incomplete type coverage.

We introduce Sourcerer, a new sanitizer that correctly and fully traces and recognizes *all* type confusions, in particular, casts from unrelated types and `void*`. Sourcerer enriches the classes involved in a cast with runtime type information to perform precise runtime checks. When compared with the state-of-the-art, Sourcerer expands type coverage to *all* cast operations, 8,507M additional casts on the SPEC CPU2006 and CPU2017 benchmarks—a 118% increase—with reasonable average performance overhead of 5.14%. Additionally, we conduct an ablation study to understand what causes this runtime overhead and showcase a fuzzing campaign finding six bugs, highlighting the improved bug-finding capabilities when Sourcerer is deployed.

Keywords: Sanitizer · type confusion · C++

1 Introduction

C++ offers speed and flexibility at the compromise of not enforcing strong type and memory safety guarantees. For example, the lack of type safety allows an object to be converted to any type at zero cost, trading versatility for the risk of *type confusion*, which occur when a program misinterprets an object as an incompatible type, leading to undefined behavior. Type confusion vulnerabilities have been consistently exploited in the wild (e.g., CVE-2024-30357 in Foxit, CVE-2025-24129 in iOS, CVE-2024-9859, or an \$11,000 vulnerability #329781390 in Chromium) and are classified under Common Weakness Enumeration CWE-704 [22]. Identifying and fixing such violations early in the development cycle is

M. Egele et al. (Eds.): DIMVA 2025, LNCS 15747, pp. 75–95, 2025.
https://doi.org/10.1007/978-3-031-97620-9_5

key to prevent their active exploitation. To this end, automatic software testing (*fuzzing*) is a strong ally in identifying unexpected behavior. However, type confusions are challenging to detect as they may not trigger a crash and the compiled code no longer has any notion of the *C++ types*. Even though recognizing type confusion is complicated, their detection is crucial for overall software security. To address this, *sanitizers* are used during automatic software testing to report unintended software behavior. For C++, sanitizers exist for many memory-related vulnerabilities, however, systematically detecting *all type confusions* remains a challenge for which no sanitizer has been widely deployed so far.

The underlying difficulty of designing a robust type confusion sanitizer for C++ stems from the language type conversion mechanisms. These are divided into three categories: (i) implicit conversion, *e.g.,* `bool` to `int`, (ii) user-defined copy constructors, *e.g.,* `To(From&)` for classes `From` and `To`, or (iii) explicit casting operators, *e.g.,* `reinterpret_cast`. While implicit and developer-defined conversions are safe, explicit casts allow for arbitrary conversions, potentially compromising type safety. In practice, aside from `dynamic_cast`, for which the standard mandates runtime checks, the other explicit cast operators, `static_cast` and `reinterpret_cast`, rely on developers to guarantee the conversion's validity. Due to a lack of security awareness, outdated fear of performance overhead, or C-style programming, current codebases still contain potential incorrect type conversions. To validate a type conversion, the relationship between the source and destination types must be checked at runtime. Static checks cannot be certain of the source type as C++ allows for type aliasing either through `void` pointers used as a generic pointer type or via polymorphism (which allows referring to derived objects through their parents). While these patterns allow for flexible code, they hinder the possibility of static validation of type conversions and increase the type system complexity, therefore putting the program at risk of type confusion. As C++ does not provide type information at runtime, most objects remain glorified C-style structs. Only a subset of all objects, those with virtual functions, are equipped with a runtime type identifier as mandated by the C++ standard. Indeed, only those objects can benefit from runtime checks (*e.g.,* `dynamic_cast`) to guarantee type safety while conversions of other object types remain prone to type confusion.

Different approaches exist to prevent type confusions. First, migrating to a type- and memory-safe language like Rust [21] removes many vulnerability classes, including type confusions. However, rewriting a project is costly and risks introducing new bugs such as logic ones. While safe languages is the preferred solution for new projects, porting old code bases is not always feasible due to constrained development resources. Switching to a C++ dialect with more safety guarantees, like the security profiles in the C++ Core Guidelines [26] is less demanding but still requires a deep understanding of the dialect and significant code changes. For example, to prevent type confusions, the C++ Core Guidelines mandate the removal of all unsafe cast operators [27]. If a rewrite is out-of-scope, the alternative is to detect and/or prevent type confusions at runtime. Throughout the years, many type confusion sanitizers [10,14] tried to trace and check object lifetimes. However, they incur high runtime costs and are prone to false positives [23]. LLVM Control Flow Integrity (CFI) [28] takes another approach by relying on Runtime Type Information (RTTI), which incurs a low runtime overhead but offers an incomplete detection as it is restricted to polymorphic objects. type++ [1], a C++ dialect, extends LLVM-CFI's detection to all derived casts with only small modifications of the codebase. After deep inspection, however, we observe that type++ only partially supports unions due to design limitations in the RTTI initialization. Thus, existing tools lack a viable method to detect all cast operations. Most importantly, current sanitizers are limited to derived casts, *e.g.*, between a parent and a child, while leaving the majority of unrelated casts, if not all, unverified (*e.g.*, from void*). For example, type++ leaves 7,151M casts unprotected—(45.67% of all unrelated casts across both SPEC CPU2006 and CPU2017). In contrast, EffectiveSan [7] addresses all cast operations by encoding type information as low-fat pointers but requires heavy-weight modifications of all allocators and checks the type at each pointer dereference (which, compared to casts, is a very frequent operation), causing an unnecessarily high runtime overhead of 49%. *Reliably detecting all type confusions with a low runtime cost remains, therefore, an open challenge.*

In this paper, we propose Sourcerer, a novel sanitizer capable of detecting *all* type confusions by enforcing a runtime type check at every explicit cast operation. The novelty of Sourcerer lies in embedding inline Runtime Type Information (RTTI) into all object types involved in unrelated casts, overcoming the restriction to derived casts that limits state-of-the-art solutions, *e.g.*, type++ and LLVM-CFI. To achieve this goal, Sourcerer introduces *RTTIInit*, a new C++ object initializer responsible solely of setting inline RTTI. With the introduction of RTTIInit, Sourcerer alleviates porting issues and overcomes the limitations of previous dialects, which forcibly injected default constructors and conflicted with the C++ standard. Furthermore, the introduction of RTTIInit solves the type initialization at union activation, another unsupported feature in existing dialects. As a result, Sourcerer performs runtime type validation for each cast from a non-generic types (*i.e.*, developer-defined types) and generic pointers (*i.e.*, void* or integral types), finally unlocking *full type testing* in C++ programs.

Sourcerer's pipeline consists of two stages. First, a static analysis identifies all classes involved in cast operations and helps the developer to adhere to the dialect. Then, our compiler generates an executable with the initializer responsible for setting the inline type information and the necessary type checks. We evaluate Sourcerer on the SPEC CPU2006 [11] and SPEC CPU2017 [3] benchmarks[1] and conduct a fuzzing case study on a leading C++ project, OpenCV [2], showcasing the ability to use Sourcerer to find illegal unrelated casts. Overall, Sourcerer detects all cast operations in the SPEC CPU benchmarks and incurs, on average, only a 5.14% performance overhead. Supporting objects involved in unrelated casts requires additional code changes. We deem the necessary adaptation of 453 out of 2,040K LoC reasonable with respect to the 4.95× more

---

[1] When using SPEC CPU, we refer to the SPEC CPU2006 and CPU2017 benchmarks.

classes instrumented compared to the instrumentation of only derived casts. In our evaluation, we conduct an ablation study to understand the root cause of the observed overhead. Our study concludes that the *Application Binary Interface* (ABI) change is the main cause of overhead highlighting the possibility of drastically improving performances by modernizing specific casts. In terms of security impact, Sourcerer identifies 152 type confusions in the SPEC CPU2006 and CPU2017 benchmarks. Finally, we conduct a fuzzing campaign in which we deploy our sanitizer on OpenCV [2], discovering six unrelated type confusion bugs, all missed by the state-of-the-art competitors.

Overall, the main contributions of Sourcerer can be summarized as follows:

– Quantifying *unrelated* casts so far hidden from state-of-the-art sanitizers.
– Proposing an extension to the type++ dialect to check *all* cast operations.
– A new initializer, RTTIInit, that sets object's type information, increasing the compatibility with the C++ standard and reducing the runtime cost.
– A thorough evaluation of Sourcerer against the state-of-the-art.
– A case study showcasing how Sourcerer can be used to detect vulnerabilities in real-world software through fuzzing campaigns on well-tested software.

We release the source code of Sourcerer and the documentation to replicate our experiments as open-source at `github.com/HexHive/Sourcerer`.

2   Background

In this section, we introduce the key concepts behind Sourcerer's approach, focusing on how C++ handles type conversion and distinguishes unrelated from derived casts. Additionally, we introduce the concept of type confusion vulnerabilities (CWE-704 [22]).

2.1   C++ Casting

Casting—adjusting the type of objects—allows for flexible and generic code. While implicit conversions (*e.g.,* conversions defined by the C++ standard like `char` to `int`) or the developer-defined ones are safe, explicit casts pose the risk of undefined behavior. Specifically, unsafe casts have two origins: *derived or down casts* occur when the resulting type inherits from the source class while *unrelated casts* encompass conversions between two types not related by inheritance (*e.g.,* `void` or other generic pointer type). The reverse of a derived cast, *i.e.,* from a base to a derived type, is known as an *upcast* and is implicit and therefore safe.

Both unrelated and derived casts can be safe. For example, casting from `void*` is well-defined if the pointed memory was previously cast from the desired type (*e.g.,* Line 27 in Listing 1). Similarly, a derived cast is legal if the object is of the derived type or one of its descendant (*e.g.,* Line 19 in Listing 1). Outside these cases, the result of the cast is undefined. For example, in Line 24 in Listing 1 a `Sibling` object referenced as a `Base` is illegally cast through `static_cast` to a `Drvd` pointer. After such a cast, the program might try to access the `y` attribute

```
 1    class Unrelated {char w;};
 2    class Base {
 3    public:
 4      int x;
 5      virtual ~Base()=default;
 6    };
 7    class Drvd:public Base{
 8      int y[3];
 9    };
10    class Sibling:public Base{
11        double z;
12    };
```

```
13    int main(){
14      Unrelated* u = new Unrelated();
15      Drvd* d = new Drvd();
16      Sibling* s = new Sibling();
17      // Safe casts: no need for verification //
18      Base* b = static_cast<Base*>(d);
19      Drvd* d2 = dynamic_cast<Drvd*>(b);
20      if(d2 == nullptr) return 1; //if cast fails
21      Base* b2 = static_cast<Base*>(s);
22      void* v = d; // Implicit generic cast
23      // ------------ Derived cast ---------- //
24      d2 = static_cast<Drvd*>(b2); // Illegal
25      d2 = reinterpret_cast<Drvd*>(b);
26      // ---------- Unrelated cast -------- //
27      d2 = static_cast<Drvd*>(v);
28      d2 = reinterpret_cast<Drvd*>(s);// Illegal
29      Unrelated* u2 = (Unrelated*) d; // Illegal
30      return 0;
31    }
```

Listing 1. Examples of derived (lines 24 & 25) and unrelated casts (lines 27–29). We omit statistics on safe casts (upcast and dynamic_cast, lines 18 & 19).

of the Drvd object which is out-of-bound as Sibling objects are smaller. Such illegal casts are referred to as *type confusions* and might lead to memory corruption. To avoid such errors, C++ developers need to maintain a mental model of the actual object types. To do a type conversion, the C++ standard mandates the use of one of the following explicit cast operators if the conversion is not inherently safe—*e.g.*, neither implicit nor defined by the developer:

- dynamic_cast handles only derived casts. It queries, at runtime, the type information of the object to verify the conversion's safety. Therefore, the source and destination type must be polymorphic to ensure the presence of RTTI. dynamic_cast is the *only* safe runtime cast operator in C++.
- static_cast provides only compile-time checks. For derived casts, the compiler checks that the source and destination types are part of the same type hierarchy. Notably, this fails to prevent all illegal derived casts. Finally, static_cast can convert void pointers to another type, without any checks. In contrast to dynamic_cast, static_cast lacks type safety guarantees.
- reinterpret_cast allows developers to interpret the bytes of an object as a new type. This is inherently unsafe and breaks type safety guarantees as no runtime checks are executed. Typically, this operator is used for unrelated casts, but it also supports derived cast.

Additionally, const_cast allows changing the qualifiers but not the object's type. This might result in undefined behavior but is not linked to type confusions. Lastly, C++ supports C-style casts which are translated to the above compatible cast operator with the highest safety guarantees. We will, therefore, not explicitly consider C-style casts in the rest of the paper.

2.2   type++ Limitations

Research for detecting type confusions has evolved with diverse mechanisms being explored. Badoux et al. [1] propose a new C++ dialect mitigating, by design, type confusions in derived casts. By extending all derived cast objects with runtime type information (RTTI), type++ can protect each derived cast, maintaining their type safety throughout the program execution. Adding RTTI into these objects changes the *Application Binary Interface* (ABI) resulting in some (limited) porting effort. To set the RTTI, their implementation artificially injects default constructors to all classes involved in down-to casts. This solution conflicts with some C++ idioms like if it is marked as `deleted` or when constructor calls are actively avoided (§5.5).

Crucially, type++ stops short of being *completely type safe* as it only partially protects unrelated casts—slightly less than half in the case of the SPEC CPU benchmarks. Moreover, instrumenting only a subset of cast classes breaks the implied compatibility between some classes resulting in unnecessary porting effort. We hypothesize that the reliance on the constructor to set RTTI is the root cause of these porting issues.

3   Threat Model

Sourcerer is a sanitizer for detecting *all*, derived and unrelated, type confusions in C++. In particular, when a type confusion bug is triggered during testing, we expect Sourcerer to detect the type safety violation. We assume a correct implementation of our compiler and the program to be correctly ported to the type++ dialect. Specifically, Sourcerer is not designed for an adversarial scenario due to possible false negatives if an attacker can leak and set type information. We refer to Sect. 9 for a thorough discussion. In summary, our threat model aligns with the ones from previous type confusion sanitizers [14].

4   Challenges

Upon careful evaluation of related works, we identified several unsolved challenges for type confusion sanitizers. First, the state-of-the-art, *e.g.,* type++ and HexType, offer incomplete type safety as they miss most unrelated casts. Only EffectiveSan offers theoretically full type safety but at an unnecessary high runtime cost, creating the second challenge we aim to address. Lastly, we identified only partial support for unions in type++ and EffectiveSan while Sourcerer provides complete support.

- **Unrelated casts**: To detect all type safety violation, a sanitizer needs to cover *all* cast operations, including unrelated casts.
- **Performance overhead**: For a sanitizer to be widely adopted, it should have a minimal performance overhead.
- **Union support**: C++ unions allow different types to refer to the same memory. This is a problem for type confusion sanitizers as they need to track the union type in memory.

```
                                      1   mov     $0x10,%edi
                                      2   call    47340 <malloc@plt>
                                      3   call    46f80 <_ZN9RTTIInitUnrelated>
1   mov     $0x1,%edi                 4   ...
2   call    47340 <malloc@plt>        5   <_ZN9RTTIInitUnrelated>:
                                      6   lea     0x1ad9(%rip),%rax #vtable address
                                      7   mov     %rax,(%rdi)
                                      8   ret
```

Listing 2. Assembly code of the `malloc`-ation of an `Unrelated` object from Listing 1 without and with RTTIInit.

5   Sourcerer's Design

In this section, we introduce the core concepts allowing Sourcerer to address these challenges. First, we lay out which classes need type information to truly check *all* casts. Then, we describe *RTTIInit*, an optimized inline type information initializer reducing type++ dialect divergence and lowering the performance overhead by 3× in comparison with the other complete type confusion sanitizer, EffectiveSan. Additionally, we explain the key properties of RTTIInit allowing the support of unions. Finally, we list the idioms unsupported by earlier work that Sourcerer handles, like templates for EffectiveSan.

5.1   Classes to Instrument

Sourcerer's core contribution is to check all casts. Specifically, Sourcerer instruments all the classes involved in any derived or unrelated casts, thus extending Property 2 laid out in the type++ paper [1]. Formally and in line with the properties defined in the type++ paper, we refer to this specialization as *Explicit Runtime Types For All Casts*:

Property 1 (Explicit Runtime Types For All Casts). Given all classes CS of a program P, Sourcerer associates a unique type T to each class $A \in CS$ if A is either the destination or the type of the source object of an explicit cast.

This new property allows for the verification of all type casts while keeping the number of classes to instrument at a minimum. Due to the ubiquity of casts from `void*`, the sanitizer needs to instrument an increased number of classes— 1,043 additional ones in the SPEC CPU benchmarks, a 5× increase compared to previous works [1] targeting only derived casts. The key insight to implement this property lays in RTTIInit, that allows transparent type information initialization in objects without interfering with the C++ constructors.

5.2   RTTIInit

The ability to type check an object at runtime depends on the presence of type information. Typically, approaches using external metadata to track object types

```
1   using namespace std;          12    int main() {
2   union Union {                 13        BiggerUnion bu; // No member active
3       int i;                    14        bu.x.u.i = 65; // bu.x and bu.x.u.i active
4       char c[2];                15        // u should only be accessed as integer.
5   };                            16        cout << bu.x.u.c[0] <<endl; // UB
6   struct X { Union u;};         17        bu.x.u.c[0] = 0x42; // u is active as char[]
7   union BiggerUnion {           18        bu.x.u.c[1] = 0x41;
8       X x;                      19        bu.c = OtherClass(); // Direct assignment
9       OtherClass c;             20        return 0;
10  };                            21    }
```

Listing 3. Abbreviated snippet showing valid and invalid union member accesses. The assignment at line 14 activates `bu.x` and `u.i`. Access to `u` through a non-active member leads to undefined behavior (*e.g.*, line 16) [5]. Sourcerer's RTTIInit is called at each access while type++ misses instrumenting line 17 as calling the constructor would overwrite `bu.x` and `u.c`.

struggle to follow all object lifetime events (*e.g.*, copy). Inline metadata, as pioneered by LLVM-CFI and type++, is more robust as the type information is stored in the object and not disjoint from the object such as for TypeSan [10] and HexType [14]. The type information will be carried in the different lifetime events. Therefore, only the object creation requires careful handling to ensure the type information is correctly initialized. Typically in C++, object creation happens through **new** or direct assignment, which both call the object constructor which also sets RTTI when required. The type++ implementation followed this approach, by forcing constructor calls for object creation not relying on any initialization but only allocation, *e.g.*, **malloc**. This approach breaks different idioms in the C++ standard like explicitly **deleted** constructors or **const** objects where initialization should occur only once. To avoid these issues, Sourcerer introduces a new initializer, *RTTIInit*, uniquely focused on setting RTTI as exemplified in Listing 2. By interacting only with the RTTI field, Sourcerer avoids incompatibilities with **const** qualifiers and minimizes the performance cost of setting the RTTI. Additionally, RTTIInit does not interfere with the remaining object content, allowing for complete support of unions as detailed in Sect. 5.3. Lastly, our initializer, as a new language feature, does not conflict with existing constructors or the lack thereof which was a limitation of type++ that complicated its deployment.

5.3   Support for Unions

Unions, in C++, use the same memory to store objects of different types, allowing, however, only a single type to be active at a time. The union switches type when a member is activated, either through a direct assignment (Line 19 in Listing 3) or by setting a field of a union member. As the core property of Sourcerer is to maintain inline type information throughout the program execution, Sourcerer needs to ensure that, upon activation, the type information is correctly updated in memory. Direct assignments do not require further handling as the incoming object's RTTI is already set. As a field assignment can

```
1    template <class _Tp>
2    struct __list_node : public __list_node_base<_Tp> {
3      // Starting the lifetime of nodes without initializing in order
       ⌙ to be allocator-aware.
4    private:
5      _ALIGNAS_TYPE(_Tp) char buf[sizeof(_Tp)];
6
7    public:
8      _Tp& __get_value() {
9        return *__launder(reinterpret_cast<_Tp*>(&buf));
10     }
```

Listing 4. Challenging idioms in the libc++ 19.0.0 `list` implementation. A node is allocated through a `char` array later cast to the desired type. The constructor is later called via `construct_at`. type++ calls the constructor at line 5 thereby unfortunately disabling the constructor homing optimization [17]. Sourcerer, on the other hand, can either allow-list the cast at line 9 or call the RTTI initializer at line 5 without breaking the optimization.

occur when the object is already activated, calling a constructor would overwrite the stored object content (*e.g.,* Line 17 in Listing 3). Sourcerer, however, calls RTTIInit at each field assignment, which sets the RTTI without modifying the remaining object content.

5.4   Dialect Simplification

Sourcerer relaxes two dialect requirements introduced by type++. First, the type++ compiler requires a default constructor to be defined for each instrumented class and has to relax the `deleted` attribute in case the default constructor is defined as such. Sourcerer's instrumentation, on the other hand, does not use constructors and retains the intention of the developer. Sourcerer, additionally, remove type++ changes for `const` variables initialization. Relying on a constructor to set RTTI imposes a second initialization step, breaking the single initialization requirement of `const` variables. Conversely, as RTTIInit is not counted as an actual initialization step, it averts any limitation for `const`.

5.5   Unsupported Idioms in Earlier Work

While deploying Sourcerer, we encountered a C++ idiom that type++ could not support. Specifically, libc++, the LLVM C++ standard library, explicitly avoids calling the object constructor when allocating a tree node to allow for constructor homing [17], an optimization reducing the amount of emitted debug information. Instead, they allocate a `char` array and then cast it to the desired type as shown in Listing 4. type++ either breaks the optimization or cannot set the RTTI, disregarding type safety. Sourcerer, on the other hand, can instrument this peculiar allocation pattern without breaking the optimization.

When evaluating EffectiveSan, we identified two unsupported idioms. First, Custom Memory Allocators (CMA) need to be replaced, incurring many LoC modifications. In comparison, Sourcerer only requires a list of CMAs and handles their instrumentation automatically. Secondly, similarly to the type++ authors, we encountered a false positive due to incomplete handling of C++ templates, an issue not faced by Sourcerer.

6   Implementation

Sourcerer needs to add inline metadata to the necessary classes and instrumentation to verify for all casts. We implement the Sourcerer prototype on top of the modular compiler toolchain, LLVM 19.0.0 [16]. Specifically, we port the class collection, custom allocator logic, as well as the warning analysis from type++ to LLVM 19.0.0. For RTTInit, we copy the logic of the default constructor, but trim it down to set only type information *i.e.*, the vtable and the RTTI. The resulting ABI change is similar to the one in type++—any function interacting directly with the object size might be problematic. For verification, we rely on LLVM optimized type checks. Moreover, compared to type++, we add support for multiple allocator edge cases such as zero-size allocation and frees through `realloc`. Overall, our implementation totals 9K LoC and consists of two compilation passes, first to gather the types and then to instrument them. We open-source Sourcerer and the evaluation at github.com/HexHive/Sourcerer.

7   Evaluation

Sourcerer's evaluation targets the following research questions:

- **RQ1:** What extra efforts are required to check all unrelated casts?
- **RQ2:** What is Sourcerer runtime overhead compared to the state-of-the-art?
- **RQ3:** Which source causes the performance overhead introduced by Sourcerer?
- **RQ4:** How effective is Sourcerer at detecting type confusion vulnerabilities?
- **RQ5:** How does Sourcerer perform in a real-world bug-hunting scenario?

Experimental setup. Our evaluation runs in Ubuntu 20.04 Docker containers on a server with two Xeon E5-2680v4 @ 2.4 GHz and 256 GB of RAM.

Evaluation Targets. Sourcerer's evaluation is twofold: first, we compare it to the state-of-the-art, then we demonstrate Sourcerer effectiveness on current large-scale projects. As such, we evaluate Sourcerer on the SPEC CPU2006 [11] and SPEC CPU2017 [3] benchmarks as they are the common benchmarks across the state-of-the-art. For both, we select all the C++ programs and compile them with -O2 optimization level. Overall, we evaluate Sourcerer on 2,040K LoC lines of code across the 16 programs of the SPEC CPU suites. Additionally, we conduct a fuzzing campaign against OpenCV [2] (commit `796adf`), the state-of-the-art computer vision library, showcasing the ability to use Sourcerer to find illegal unrelated casts. We choose OpenCV as it is a large, 1.3M LoC, and popular C++ project with a fuzzing setup readily available from OSS-Fuzz [25].

State-of-the-Art Competitors. We compare Sourcerer against a representative set of the state-of-the-art type confusion sanitizers. Specifically, we choose type++ [1] as it is the most recent type confusion protection and the cast checker of LLVM-CFI [28] due to its use in industry. Finally, we report numbers from EffectiveSan [7] as it is the only other tool claiming to protect unrelated and derived casts. Despite our efforts, we were unable to run EffectiveSan as the cast checking configuration is neither present in the source code nor in the documentation. Lastly, EffectiveSan checks types at pointer access, and, therefore, does not report the number of cast operations protected. This discrepancy makes a quantitative security comparison of Sourcerer and EffectiveSan meaningless.

For each tool, we report the performance overhead, averaged across five runs, and compare it to the corresponding LLVM vanilla version—13.0.1 for type++, and 19.0.0 for LLVM-CFI and Sourcerer. To assess the effectiveness of Sourcerer, we report the number of runtime cast operations checked, similarly to type++. Every configuration is run with Link-Time Optimization (LTO) enabled as required by LLVM-CFI cast checking.

7.1   Porting Effort

Sourcerer follows the type++ dialect specification, and, therefore, requires similar porting efforts to translate C++ code into its dialect. Starting from type++'s open-source patches, we address the additional warnings caused by the classes involved in unrelated casts. As a first metric, we report the number of extra classes that need to be instrumented to check unrelated casts. For these extra classes, we break down the kind of unrelated cast causing their instrumentation in the column *Unrelated* in Table 1. On average, Sourcerer instruments and monitors $4.95\times$ more classes than type++ as our checks stretch beyond derived casts. Some programs like POV-Ray have few classes, *e.g.*, 12, involved in derived casts but extensively use unrelated casts with 161 classes cast. SoPlex experiences a less dramatic increase, *e.g.*, $3\times$ more classes, with most classes cast as part of libc++ data structures headers (*e.g.*, vector). This reduces the overall effort as porting the library is amortized across the different programs. In Table 1, we report the classes requiring explicit instrumentation. The total number of types instrumented is a superset as some classes inherit the instrumentation and templates are counted once and not per specialization.

The actual porting effort caused by the new classes is small. For the SPEC-CPU benchmarks, we only modify 120 LoC on top of the type++ patches, for a total of 453 LoC patched. The new changes are relatively minor, for example, an initialization procedure (*e.g.*, `placement_new`) in NAMD 2017. Around 20%

Table 1. Breakdown of the number of classes instrumented by Sourcerer as well as the number of RTTI initialization. *Total* shows the number of classes that Sourcerer instruments which is broken down into the cause of the instrumentation. *Derived* indicates classes involved in derived cast and, therefore, already instrumented by type++. Then, we report the additional classes involved in unrelated casts either from a specific type (*i.e.*, *class*) or from generic pointers (*e.g.*, void*). The last two columns indicate the number of instrumented objects and the percentage which are initialized through RTTIInit.

	Program	LoC	Instrumented classes				# RTTI Init.	% Through init
			Total	Derived	Unrelated			
					class	void*		
SPEC CPU2006	NAMD	4K	27	9	4	14	460K	100.0
	deal.II	95K	63	12	4	47	15,875M	98.32
	GoPlex	20K	39	13	6	20	726M	5.01
	POV-Ray	79K	161	12	23	126	6,243M	99.97
	OMNeT++	27K	53	13	4	36	2M	65.82
	Astar	4K	31	9	4	18	6,024M	91.97
	Xalan-C++	264K	149	46	5	98	1,407M	99.86
SPEC CPU2017	cactuBSSN	63K	31	11	5	15	334K	81.19
	NAMD	6K	26	9	4	13	0	–
	Parest	359K	120	26	4	90	20,384M	98.53
	POV-Ray	80K	161	12	23	126	25,179M	100.0
	Blender	616K	57	12	20	25	70M	74.3
	OMNeT++	86K	125	14	5	106	3M	38.93
	Xalan-C++	291K	203	46	7	150	35,276M	99.61
	Deep Sjeng	7K	27	9	4	14	15M	0.0
	Leela	31K	34	11	4	19	4,491M	99.53
	Total	2,040K	1307	264	126	917	–	–

of the LoC changed are fixes for type confusions. They are necessary because the ABI changes do not always allow Sourcerer to recover from the subsequent memory corruption. For example, in Povray 2017, a parent object is created by malloc-ing enough memory but storing the returned pointer in a variable of a larger child type as exemplified in Listing 5. As Sourcerer identifies the returned memory as a child object, it calls RTTIInit which set information out of the allocated bounds, leading to memory corruption. The biggest changes were in Blender, which interacts heavily with C code resulting in two challenges. First, as some structs are defined in headers included in both C and C++ code, we had to mimic the presence of RTTI information in the C code to ensure a compatible ABI. The second issue arises when, in C code, an instrumented C++ object is cast to a pure C struct. As Sourcerer only instruments C++ code, the cast is not checked nor does the destination type expect the RTTI field. We modified the C struct to be aware of the presence of type information. Moreover, instrumenting

Table 2. Performance and security evaluation of Sourcerer compared to the state-of-the-art. Under "%"' we report the performance overhead compared to the tool's baseline. Then, in the *Casts* column, we report how many unrelated casts were checked at runtime. The two Δ columns show the extra operations verified by Sourcerer on top of the competitors. The last two columns report the average and peak memory overhead of Sourcerer in terms of the working set size.

	Program	LLVM-CFI		type++		Sourcerer		Δ Casts		Memory	
		%	Casts	%	Casts	%	Casts	LLVM-CFI	type++	Avg.	Max.
SPEC CPU2006	NAMD	0.41	1	0.27	0	0.53	1	0	1	0.45	0.45
	deal.II	−1.15	649K	1.95	122M	4.33	128M	127M	5M	−2.17	1.95
	SoPlex	−0.01	206K	0.22	27M	12.64	30M	30M	4M	66.79	66.83
	POV-Ray	0.92	1M	1.60	1,342M	7.93	1,345M	1,344M	4M	10.39	10.39
	OMNeT++	3.42	3	0.67	270K	3.04	284M	284M	283M	1.43	1.43
	Astar	0.85	0	0.90	0	16.99	4M	4M	4M	104.89	82.03
	Xalan-C++	0.63	5K	−0.54	5K	−1.29	161K	156K	156K	9.28	2.85
SPEC CPU2017	cactuBSSN	0.47	0	−0.71	100	−0.21	21K	21K	21K	0.30	0.08
	NAMD	0.52	5	−0.68	0	0.71	0	−5	0	0.23	0.22
	Parest	0.36	24K	0.07	85M	2.46	106M	106M	21M	2.11	1.99
	POV-Ray	0.08	39K	1.73	5,370M	9.59	5,381M	5,381M	11M	10.98	10.96
	Blender	0.69	0	2.92	7M	9.62	7,643M	7,643M	7,636M	2.78	3.01
	OMNeT++	1.31	6M	0.34	6M	4.11	501M	495M	495M	2.21	2.12
	Xalan-C++	−0.10	4K	−0.24	198M	15.38	243M	243M	45M	4.94	4.43
	Deep Sjeng	−0.14	0	−1.09	0	0.45	1	1	1	16.37	16.37
	Leela	−0.53	0	−0.27	1K	−0.19	416K	416K	414K	15.69	5.69
	Avg./Total	0.26	8M	0.43	7,158M	5.14	15,666M	15,658M	8,507M	–	–

more classes showed some unexpected benefits as we could remove a patch from type++ for SoPlex as layout similarity is restored between two types involved in an unrelated type confusion.

EffectiveSan preserves the C++ ABI but struggles with custom allocators. In their artifact, the changes necessary for SPEC CPU2006 totaled a non-trivial 297 LoC. In contrast, Sourcerer needs to modify 453 LoC while incurring, in the worst case, only a third of EffectiveSan's average runtime overhead.

Overall, we conclude that the porting efforts for Sourcerer are reasonable and in line with the efforts necessary for similar tools.

7.2   Performance Overhead

Since speed is key for automatic testing, sanitizers should incur a limited performance overhead (§4). Below, we quantify the performance cost of deploying Sourcerer on the SPEC CPU benchmarks and compare it to the state-of-the-art.

Table 2 details the performance overhead of the different tools in the columns marked "%". Each value is the average performance overhead compared to a binary compiled with vanilla Clang—version 13.0.1 for type++, 19.0.0 for both LLVM-CFI and Sourcerer. We observe a negligible standard deviation across the five runs. As expected, the additional classes instrumented increase the overhead of Sourcerer compared to LLVM-CFI and type++. While LLVM-CFI and type++ are mitigations, Sourcerer's higher overhead is outstanding for a sanitizer deployed in a testing environment. For example, UBSan [4] manifests slowdowns of up to ~1.7x. More precisely, Sourcerer incurs a 5.14% performance penalty but covers all casts in a program. When looking only at SPEC CPU2006, the target set of EffectiveSan, Sourcerer overhead is limited to 6.47% compared to the 49% overhead reported by EffectiveSan's authors. Consequently, Sourcerer reduces the runtime overhead for a complete sanitizer by a factor of seven. In fact, Sourcerer shows a similar overhead to HexType [14] but additionally checks unrelated casts and avoids false positives by design.

Looking at individual programs, we observe that the overhead is not uniform. As shown in Table 2, programs such as cactuBSSN, NAMD and Deep Sjeng experience virtually no overhead but also check the fewest casts. On the other hand, SoPlex and Xalan-C++ 2017 suffer from an overhead of around 15% due, in part, to caching being undermined by bigger objects on hot paths. To conclude, a complete type confusion sanitizer is practical if the checks occurs at cast time and not at the frequent dereference sites, as implemented in EffectiveSan.

7.3 Source of the Performance Overhead

In this section, we study the performance cost of Sourcerer instrumentation. In particular, we conduct an ablation study on the three following elements: the type checks, the RTTI initializer, and the ABI changes. First, we disable the type checks but leave the instrumentation intact. Then, we also remove the call to RTTIInit, leaving only the changes to the ABI in place. Each experiment is compared to the same vanilla Clang baseline. We present the results in Table 3.

Comparing the columns *Full* and *W/o type checks* shows that the verification cost can be important, *e.g.,* Blender. Upon closer inspection, Blender exhibits a high ratio of failing type checks which is a slow path. Indeed, in a testing environment, we expect the program to be terminated upon encountering a type confusion while in this evaluation we continue to assess the program performance. Fixing these type confusions would reduce the cost of the checks to a similar level as in Xalan-C++ 2006 which proceeds to more, but successful, type checks. Overall, the numerous optimizations implemented in LLVM type checks [20] allow for this limited performance cost.

Inspecting the column *ABI change only*, we observe that altering the object size negatively affects the caching behavior. Indeed, in Astar, we observe that the class `pointt` is instrumented by Sourcerer. As a single byte object, adding RTTI double its size, negatively impacting caching in the tight loops of the `flexarray::add` function. Reducing this overhead could be achieved by removing all casts of `pointt`, and, therefore, Sourcerer instrumentation. This would return Astar to the original caching performance and a minimal overhead. A similar issue is observed in SoPlex. The cost of the ABI change varies a lot across programs and use-cases. Compared to type++, this effect is magnified in Sourcerer by the additional classes instrumented.

Lastly, Table 3 allows us to estimate the overhead of Sourcerer's RTTIInit as it is the only change between the columns *W/o type checks* and *ABI change only*. Disabling RTTIInit does not lead to a significant change in performance, highlighting the effectiveness of Sourcerer's RTTI initialization design.

Overall, this study highlights the cost of the ABI change, which is strongly dependent on the program and its use-cases. Nonetheless, the average overhead of Sourcerer is lower than other sanitizers while detecting all type safety violations.

7.4   Security Effectiveness

Sourcerer is a sanitizer deigned to detect *all* type confusions. In this section, we compare the number of cast checks against similar tools and describe the type confusions that Sourcerer identified which were missed by previous works. From a theoretical point of view, both EffectiveSan and Sourcerer provide complete

```
1   class X {}; // Instrumented
2
3   class A {
4     int x; // A is
        instrumented
5   };
6
7   class B : public A {
8     int y;
9     X x; // Need to set x RTTI
10  };
11
12  int main() {
13    A* a;
14    a=(B*)(malloc(sizeof(A)));
15    free(obj);
16    return 0;
17  }
```

Listing 5. Simplified excerpt of a type confusion in POV-Ray 2017. Sourcerer calls RTTIInit at line 14 right after the call to `malloc`. Sourcerer assumes, from the cast, that the underlying object is of type B while the allocated size is only sufficient for an A object. The call to A's RTTIInit will try to set x's RTTI information (line 9), resulting in an out-of-bounds write.

Table 3. Ablation study of Sourcerer performance overhead. The column `Full` lists the overhead of Sourcerer compared to the baseline. The third column shows the overhead once the type checks are disabled. For the last column, we additionally disable calls to RTTIInit, highlighting the cost of the ABI change. Comparing `Full` and `W/o type checks` shows the overhead of the cast validation. Finally, comparing the last two columns allows for an estimation of the overhead induced by RTTIInit.

	Program	Full	W/o type check	ABI change only
SPEC CPU2006	NAMD	0.09	0.15	-0.15
	deal.II	3.89	5.36	3.53
	SoPlex	12.88	12.87	8.80
	POV-Ray	8.58	5.20	5.41
	OMNeT++	6.28	1.37	0.05
	Astar	16.51	16.86	15.53
	Xalan-C++	-1.83	1.42	0.48
SPEC CPU2017	cactuBSSN	0.18	-1.01	-0.30
	NAMD	-0.15	-0.06	0.75
	Parest	1.77	2.32	2.07
	POV-Ray	9.73	5.68	5.95
	Blender	9.16	1.00	1.26
	OMNeT++	5.96	2.29	2.90
	Xalan-C++	14.64	13.97	15.34
	Deep Sjeng	0.15	0.06	0.16
	Leela	-0.50	-0.82	-0.51

coverage of all cast operations. We do not provide statistics about EffectiveSan checks as they do not happen at cast time but at every object dereference. However, similarly to the type++ authors, we observed false positives in EffectiveSan due to incompatibilities with templates.

Table 2 reports the number of casts checked by LLVM-CFI, type++, and Sourcerer. For each program, we list the number of unrelated casts verified during the benchmark execution. The difference between Sourcerer and both HexType and LLVM-CFI is listed in the two columns headed by Δ *Casts*, respectively. LLVM-CFI checks a mere 8M unrelated casts, less than 1% of all unrelated cast operations, while type++ already verifies slightly less than 50% due to classes being involved in both derived and unrelated casts. Sourcerer covers the remaining 50% of casts, reaching all 15.7B unrelated casts, highlighting the wide attack surface left unchecked by previous research. The increase in verified casts is dominated by Blender, but almost all programs benefit from the additional type confusion detection capabilities offered by Sourcerer. NAMD 2017 is a special case where type++ and Sourcerer report fewer casts due the changes necessary for the dialect which removed the only cast triggered in the execution.

In regards with the security impact, Sourcerer discovered 152 type confusions in the SPEC CPU benchmarks—30 more than the state-of-the-art. Most of these type confusions expect classes to have identical layouts. Despite the lack of guarantee from the C++ standard, C++ compilers rarely optimize object layouts thus, reducing the risk associated with these type confusions. Nonetheless, relying on such undefined behaviors is unsafe despite its widespread use.

7.5   Sourcerer as a Sanitizer for Fuzzing Campaigns

In this section, we showcase the effectiveness of Sourcerer as a sanitizer by using our prototype in combination with AFL++ [9]. We conduct the first fuzzing campaigns targeting specifically type confusions and compare its performance with a state-of-the-art memory sanitizer, AddressSanitizer (ASan) [24].

As target, we select OpenCV [2], the leading computer vision library, due to its ubiquity and the availability of fuzzing drivers as part of the OSS-Fuzz project [25]. We unleash AFL++, a state-of-the-art fuzzer, on the seven drivers and conduct five 24-hour fuzzing campaigns for both ASan and Sourcerer. We replay the inputs on a binary instrumented with SanCov [19] to have collision-free coverage and avoid different numbers of edges due to the instrumentation.

Thanks to Sourcerer's increased compatibility with the C++ standard, few changes are necessary to support the 1.34M C++ LoC of OpenCV. A peculiar issue is how the characteristics of matrices elements (*e.g.*, depth, size, and number of channels) are stored and later used to compute the offset between elements, shunning `offsetof`. As Sourcerer modifies this offset, matrices traversal results in RTTI being interpreted as matrix elements. Indeed, the type++ dialect is incompatible with hard-coded object sizes. More importantly, this is also less portable and needs support through `#ifdef` and header files. We leave out this code modernization as it involves modifications of the core components of the library. Instead, we use Sourcerer flexibility to disable the instrumentation of the five matrix element types—out of 668 classes involved in casts—trading a slight reduction of the findable type confusions for easier deployment of Sourcerer.

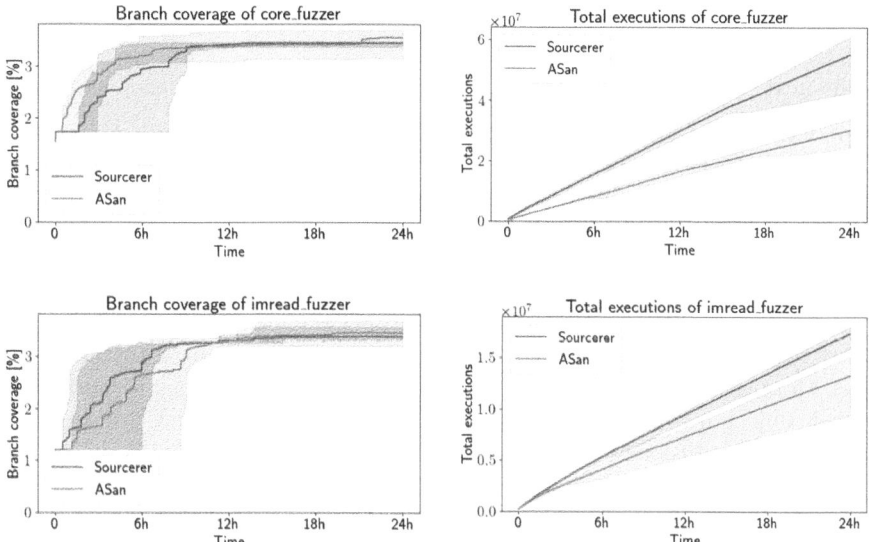

Fig. 1. Fuzzing campaign results. The left figures show the branch coverage throughout the 24h fuzzing campaign. On the right, Sourcerer allows for more executions due to reduced overhead compared to ASan

Figure 1 shows the performance of the three fuzz drivers. In shaded colors are the minimum and maximum values achieved across the five repetitions. The left graphs highlight that Sourcerer achieves a similar branch coverage to ASan, despite being hindered by type confusion crashes. On the right, the total number of executions of the fuzz drivers shows that Sourcerer instrumentation is more lightweight than ASan, allowing to test more inputs.

In terms of crashes, the two drivers, `imread_fuzzer` and `core_fuzzer`, highlighted in Fig. 1, triggered three type confusion bugs. The crashes are caused by similar unrelated casts to a `PaletteEntry` object at three different code locations. Our testing found three additional type confusions in code assuming indistinguishability between an array of `type`, *e.g.,* `float`, and a `Vec<type>` objects, representing a vector. However, as the array is oblivious to the `Vec` RTTI field, the `reinterpret_cast` results in shifted values and incorrect RTTI. Sourcerer's instrumentation makes this type confusion apparent but blocks the program execution as the object is corrupted, hindering fuzzing progress. To highlight the capabilities of Sourcerer, we investigate if type++ can identify the errors. Since type++ does not instrument `PaletteEntry` and `Vec`, type++ was unable to detect these errors, further showcasing the effectiveness of Sourcerer.

Previous type confusion sanitizers never conducted fuzzing campaigns likely due to false positives (*e.g.,* HexType) or the expensive porting effort (*e.g.,* CMAs in EffectiveSan). Sourcerer, therefore, is the first to demonstrate the feasibility and effectiveness of fuzzing campaigns with a complete type confusion sanitizer.

8   Related Works

In the next paragraphs, we discuss relevant works for type confusion sanitizers.

Type Confusion Defenses. TypeSan [10] and HexType [14] check derived cast by tracking object types throughout their lifetime in an external data structure. The complexity of C++ lifetimes leads to prohibitive cost and a high rate of false positives [23]. EffectiveSan [7] encodes, through fat pointers, the type and bounds of an object. At each pointer dereference, they perform a bound check and a type check causing a high runtime cost. Multiple dialects exist for C++ to prevent by design certain classes of vulnerabilities. Ironclad C++ [6] banned unions and added type information to every object requiring large changes to the source code. More recently, type++ [1] proposed limited code changes to add RTTI to each object involved in a derived cast. Finally, Uncontained [15] identifies derived type confusions in C containers, particularly in the Linux kernel.

Type Check Pruning. To avoid type confusions, developers implement their own type identifiers and checks. Recent works investigated disabling such checks when a sanitizer is deployed. In particular, Zhai & et al. [29] automatically remove type checks redundant with HexType to improve performance. Orthogonally, HTADE [8] removes derived casts that are never dereferenced.

9   Discussion

In the following, we detail Sourcerer's limitations and possible extensions.

Custom Allocator Identification. Similarly to previous tools, Sourcerer tracks object lifetime and, therefore, their allocation. While `new` and direct initialization sets RTTI automatically, C style allocations (*e.g.*, `malloc`, `calloc`, and `realloc`) require to explicitly initialize the RTTI through RTTIInit. Additionally, *Custom Memory Allocators* (CMA) wrap the standard allocation functions to provide extra features like memory pool or allocation metadata (*e.g.*, ASan). To initialize RTTI, Sourcerer as other sanitizers, must be aware of these allocators and is, therefore, configurable through an allow-list. Orthogonally, CMAsan [12] recently automated CMAs identification in C++ projects.

Identification of Allocation Type. To correctly set type information after an explicit allocation, Sourcerer needs to know the allocated type. Assuming the presence of a cast to the desired type right after returning from the allocation has proven sufficient in our evaluation. A sounder static analysis would be able to follow allocation until their first actual assignment.

Reliance on Link-Time Optimization. Sourcerer relies on the type checks implemented by LLVM-CFI which leverages the LLVM `type.test` function [20]. To allow optimized checks, it leverages Link-Time Optimization (LTO) to optimize inheritance hierarchies. During libc++ instrumentation, we discovered that some casts were unchecked by LLVM-CFI as their LTO-visibility attribute is set to expose symbols to external compilation units, preventing LLVM from emitting type checks. Clang provides an option, `-fsanitize-cfi-cross-dso`, to check externally visible objects across Dynamic Shared Objects (DSO), at the cost of lower performance and extra compilation requirements (*e.g.*, position-independent code) [18]. We leave evaluating this option as future work.

Future Work. Accessing a union through a non-active member is undefined in C++ [13]. Practically, the illegal access is identical to a `reinterpret_cast`. To solve this issue, Ironclad C++ [6] banned unions in their dialect. Adding type information to the types used in unions would allow Sourcerer to check the validity of union access at the cost of additional porting effort and performance overhead. We leave verifying union access correctness as future work.

Sourcerer is a sanitizer and, therefore, is not suited for an adversarial scenario. Multiple improvements are necessary to mitigate type confusions. First, all source objects need to be typed, as otherwise, an attacker controlling the object content could forge RTTI values, resulting in false negatives. As casts to integral types are widespread (*e.g.*, a pointer passed as a `void*` argument), it would result in more instrumentation. Static analysis might reduce the number of classes to instrument by inferring if a `void` pointer is ever cast in its scope. Adoption would also require Sourcerer's performance overhead to be reduced through, for example, type check pruning [29] or type check removal [8] (Sect. 8). Code modernization, *e.g.*, removing casts on hot paths or the source of instrumentation of classes heavily cached, has the highest improvement potential Sect. 7.3.

10 Conclusion

We introduce Sourcerer, a novel type confusion sanitizer that checks *all* casts at runtime. By combining inlined type information with our optimized RTTI initializer *RTTIInit*, Sourcerer checks all casts explicitly while reducing the divergence of the type++ dialect with the C++ standard. Additionally, RTTIInit supports unions and other idioms which were missing from existing tools.

We evaluate Sourcerer on the SPEC CPU benchmarks. Our tool checks twice as many unrelated casts compared to type++. On average, Sourcerer incurs 5.14% overhead, six times lower than the other complete type confusion sanitizer, EffectiveSan. Our ablation study identifies the cause of the overhead to be the required ABI changes. Lastly, during our fuzzing case study targeting specifically type confusions, we identify six new type confusion bugs and many code locations that would benefit from code modernization efforts.

Acknowledgments. This work was supported, in part, by the European Research Council (ERC) under the European Union's Horizon 2020 research and innovation program (grant agreement No. 850868), SNSF PCEGP2 186974, and the German Research Foundation (DFG) under Germany's Excellence Strategy – EXC 2092 CASA – 390781972.

References

1. Badoux, N., Toffalini, F., Jeon, Y., Payer, M.: type++: prohibiting type confusion with inline type information. NDSS **34**(4), 1–17 (2025)
2. Bradski, G., et al.: Opencv. Dr. Dobb's J. Softw. Tools **3**(2) (2000)
3. Bucek, J., Lange, K.D., v. Kistowski, J.: SPEC CPU2017: next-generation compute benchmark. In: Companion of the 2018 ACM/SPEC International Conference on Performance Engineering, pp. 41–42 (2018)
4. Clang: UBSan. http://clang.llvm.org/docs/UndefinedBehaviorSanitizer.html
5. cppreference.com: Union. https://en.cppreference.com/w/cpp/language/union
6. DeLozier, C., Eisenberg, R., Nagarakatte, S., Osera, P.M., Martin, M.M., Zdancewic, S.: Ironclad C++ a library-augmented type-safe subset of C++. ACM SIGPLAN Not. (2013)
7. Duck, G.J., Yap, R.H.: EffectiveSan: type and memory error detection using dynamically typed C/C++. In: Proceedings of the ACM SIGPLAN Conference on Programming Language Design and Implementation (PLDI) (2018)
8. Fan, X., Long, S., Huang, C., Yang, C., Li, F.: Accelerating type confusion detection by identifying harmless type castings. In: Proceedings of the 20th ACM International Conference on Computing Frontiers, pp. 91–100 (2023)
9. Fioraldi, A., Maier, D., Eißfeldt, H., Heuse, M.: AFL++ : Combining incremental steps of fuzzing research. In: 14th USENIX Workshop on Offensive Technologies (WOOT 20). USENIX Association (2020)
10. Haller, I., et al.: TypeSan: practical type confusion detection. In: Proceedings of the 2016 ACM SIGSAC Conference on Computer and Communications Security, pp. 517–528
11. Henning, J.L.: SPEC CPU2006 benchmark descriptions. ACM SIGARCH Computer Architecture News (2006)
12. Hong, J., Jang, W., Kim, M., Yu, L., Kwon, Y., Jeon, Y.: CMASan: custom memory allocator-aware address sanitizer. In: 2025 IEEE Symposium on Security and Privacy (SP), pp. 74–74. IEEE Computer Society (2024)
13. ISO C++ Standards Committee and others: Standard for Programming Language C++. Working Draft N4950. Technical report, ISO IEC JTC1/SC22 (2023)
14. Jeon, Y., Biswas, P., Carr, S., Lee, B., Payer, M.: HexType: efficient detection of type confusion errors for C++. In: CCS (2017)
15. Koschel, J., Borrello, P., D'Elia, D.C., Bos, H., Giuffrida, C.: Uncontained: uncovering container confusion in the linux kernel. In: 32nd USENIX Security Symposium. USENIX Association (2023)
16. Lattner, C., Adve, V.: LLVM: a compilation framework for lifelong program analysis & transformation. In: International Symposium on Code Generation and Optimization, CGO 2004, pp. 75–86. IEEE (2004)
17. LLVM: Constructor type homing. http://blog.llvm.org/posts/2021-04-05-constructor-homing-for-debug-info/

18. LLVM: LLVM Control Flow Integrity: Shared library support. https://clang.llvm.org/docs/ControlFlowIntegrity.html#cfi-cross-dso
19. LLVM: SanitizerCoverage. https://clang.llvm.org/docs/SanitizerCoverage.html
20. LLVM: Type Metadata. https://llvm.org/docs/TypeMetadata.html
21. Matsakis, N.D., Klock, F.S.: The Rust language. ACM SIG Ada Lett. (2014)
22. Mitre Corporation: Common Weakness Enumeration (CWE) 704: Incorrect Type Conversion or Cast. http://cwe.mitre.org/data/definitions/704.html
23. Payer, M.: Type Confusion: Discovery, Abuse, Protection (2017). http://hexhive.epfl.ch/publications/files/18SyScan360-presentation.pdf
24. Serebryany, K., Bruening, D., Potapenko, A., Vyukov, D.: AddressSanitizer: a fast address sanity checker. In: 2012 USENIX Annual Technical Conference (2012)
25. Serebryany, K.: OSS-Fuzz - google's continuous fuzzing service for open source software. USENIX Association, Vancouver (2017)
26. Stroustrup, B., Sutter, H., et al.: C++ Core Guidelines (2018)
27. Sutter, H., Stroustrup, B., et al.: C++ Core Guidelines (2015). https://github.com/isocpp/CppCoreGuidelines/
28. Tice, C., Roeder, T., Collingbourne, P., Checkoway, S., Erlingsson, Ú., Lozano, L., Pike, G.: Enforcing forward-edge control-flow integrity in GCC & LLVM. In: 23rd USENIX Security Symposium (2014)
29. Zhai, Y., et al.: Don't waste my efforts: pruning redundant sanitizer checks by developer-implemented type checks. In: 33rd USENIX Security Symposium (2024)

CodeGrafter: Unifying Source and Binary Graphs for Robust Vulnerability Detection

Saquib Irtiza[✉], Mahmoud Zamani, Shamila Wickramasuriya,
Kevin W. Hamlen, and Latifur Khan

The University of Texas at Dallas, Richardson, TX 75080, USA
{saquib.irtiza,mahmoud.zamani,scw130030,hamlen,lkhan}@utdallas.edu

Abstract. CodeGrafter is a novel framework for detecting security vulnerabilities in compiled C/C++ programs by integrating source- and binary-level code features into a unified Cross-Domain Code Property Graph (CDCPG). By combining the high-level semantic insights from source code with the detailed low-level information from compiled assembly, CodeGrafter uncovers vulnerabilities that are not detectable via source analysis or binary analysis alone. By combining both, it examines compiler decisions, such as dead code elimination, build-environment-dependent semantics (e.g., macros and pragmas), and compiler-generated interface code, to avoid false positives and false negatives in its analysis. For example, it can detect Points of Interests (POIs) where vulnerability severity is influenced by compilation-specific factors, such as stack layouts that place critical data near buffers. To streamline vulnerability detection, CodeGrafter represents these POIs as graphs and leverages Graph Neural Networks (GNNs) to significantly reduce manual auditing effort. Evaluations on six real-world applications demonstrate that CodeGrafter outperforms prior works that rely solely on source or binary-level representations alone, achieving an F1-score of 0.937 and a recall of 0.945 in identifying vulnerable functions.

Keywords: static analysis · code property graphs · vulnerability detection · function classification · source code · binary code · graph neural networks · contrastive learning · generative learning

1 Introduction

Timely identification of high-risk vulnerabilities in large software systems is critical, as nearly one-third of enterprise network attacks exploit software flaws, with over half classified as high-risk or critical [31]. The mean time to remediation (MTTR) of these flaws often stretches for months [14], exacerbated by the increasing complexity of software and inefficient detection methods. Traditional

---

S. Irtiza, M. Zamani and S. Wickramasuriya—Equal contributions.

M. Egele et al. (Eds.): DIMVA 2025, LNCS 15747, pp. 96–117, 2025.
https://doi.org/10.1007/978-3-031-97620-9_6

audits rely on manual or semi-manual source reviews, requiring significant expertise, while binary analysis is even rarer due to the difficulty of mapping binary
flaws to source-level patches.

Traditional source auditing faces practical challenges–among companies using
source auditing, only 36% of them find it effective, while 42% of the companies
without such processes cite personnel shortages as a barrier to adoption [10].
Many software products remain unaudited due to cost constraints and vast
attack surfaces. The growing reliance on open-source software worsens the issue,
with 96% of modern software incorporating open-source components, 84% of
which contain at least one vulnerability and nearly half of which are categorized
as high risk [37]. Additionally, rapid software evolution allows critical vulnerabilities to emerge post-audit. For example, flaws in vendor-specific Linux kernel
extensions, such as WEXT [33] and various Android device implementations [5]
have led to major security incidents, underscoring the need for scalable, automated vulnerability detection with fewer person-hours of effort.

Software abstraction layers are a particularly difficult source of challenges for
vulnerability detection. Low-level languages such as C and C++, which pervade
commercial off-the-shelf (COTS) software products, can expose programmers to
unsafe operations (e.g., unchecked type casting, pointer arithmetic), resulting in
a dense landscape of potential bugs. To mitigate these risks, modern compilation
environments have evolved to introduce abstraction layers that protect these
operations with automatically generated sanity checks. For instance, common C
programming practices replace many dynamic allocations and dereferences with
macros that expand during compilation into architecture-specific code sequences
designed to safeguard the underlying operations. However, this added complexity
exacerbates the auditing process, as effective source-level auditing tools must
navigate a labyrinth of macros, compiler directives, inline assembly, and security-
relevant optimizations to accurately analyze even simple programs.

To address these challenges, we propose a novel approach for semi-automated
code review by introducing *Cross-Domain Code Property Graphs (CDCPGs)*.
These integrate Code Property Graphs (CPGs) [39] with semantic data from
both source and binary code into a unified, searchable structure. By combining information from both domains, CDCPGs enable a more comprehensive
approach to discovering, analyzing, and diagnosing vulnerabilities in legacy software written in unsafe languages like C and C++. This cross-domain analysis
allows security-relevant queries that span both levels, such as identifying the
exploitability of a return-oriented programming (ROP) attack by tracing how a
buffer pointer (defined at the source level) is dereferenced and assigned to the
return address on the call stack (defined at the binary level). By integrating
syntactic, control-flow, and dataflow analysis, CDCPGs uncover critical vulnerabilities that are difficult to detect due to flow and context sensitivity.

CodeGrafter is a new scalable, automated analysis tool that implements
CDCPGs for efficient vulnerability detection. It extracts contextual information
from source-binary code pairs, enabling queries that span both domains. By
identifying POIs—code segments likely to be vulnerable—CodeGrafter assists

Table 1. CWE Classes of interest

CWE Class	Description
CWE-119	Improper Restriction of Ops w/i Bounds of Buffer
CWE-120	Buffer Copy without Checking Size of Input
CWE-200	Exposure of Sensitive Information
CWE-416	Use After Free
CWE-476	NULL Pointer Dereference
CWE-676	Use of Potentially Dangerous Function
CWE-690	Unchecked Return Value to NULL Pointer Dereference

auditors in prioritizing security-relevant sections for review. These POIs are further used to train learning models for automated vulnerability detection, improving accuracy and scalability. Our key contributions include:

– **Source-Binary Cross-Domain CPGs:** We develop CPGs that integrate source and binary-level data, enabling vulnerability detection beyond traditional source-only or binary-only analysis.
– **Enhanced Vulnerability Detection:** By merging source and binary semantics, CodeGrafter reduces false positives from undecidable code properties (e.g., reachability). It reveals compiler's answers to these questions, improving security analysis accuracy, independent of compiler correctness.
– **Improved Auditing Scalability:** Our new graph type, CDCPG, when used to train a learning model, allows auditors to focus on automatically flagged vulnerabilities, improving auditing efficiency and addressing scalability challenges due to the limited availability of skilled analysts.
– **Pipeline Implementation and Data collection:** We prototype Code-Grafter and test it on 180 vulnerable functions across six real-world applications, achieving an F1-score of 0.937 and recall of 0.945, outperforming traditional graph-based approaches. The code can be found here https://github.com/Saquibirtiza/CodeGrafter.

2   Overview

2.1   Motivation

The goal of our research is to provide human auditors a list of ranked code-points in C source code programs that might offer malicious exploitation opportunities to adversaries, to focus their efforts. Table 1 lists vulnerability classes of interest. This distinguishes our work from general debugging in that we intentionally omit or deprioritize code errors that are unlikely to present a risk within one or more of these CWEs. This reduces the search space and the false positive rate.

For example, Listing 1.1 shows C and assembly code that is *not vulnerable*, though many source-based analyzers falsely flag it as buffer overflow. The

Listing 1.1. Non-vulnerable C and assembly code example.

```
char buf[sizeof(pthread_t)];      sub rsp, 24
for (int i=0; i < 12; ++i)        ...
  buf[i] = 0x7F;
...



                                  mov rax, [rsp+4]
                                  xor rax, <canary>
                                  jnz __stk_chk_fail
return;                           ret
```

Listing 1.2. Compiler-introduced vulnerability

```
memset(secret, 0, sizeof(secret));     nop
...                                    ...
return;                                ret
```

size of `buf` is difficult to statically predict, since it depends on a system-defined structure's size containing system-specific members, `pthread_t`. However, the assembly code reveals that the compiler reserves 24 bytes of space for it (possibly including padding), which makes the loop bound of 12 safe. Moreover, it also reveals that the compiler has inserted a stack canary guard into the function epilogue, making it difficult to exploit this code even if a buffer overflow were present. This highlights how combined source-binary analysis can safely deprioritize such code, unlike source-only methods which may misclassify it.

Listing 1.2 exhibits the opposite problem. The source code zeros `secret` to prevent information leaks, but the assembly code reveals that the compiler has optimized the redaction away as a write-without-read. Hybrid source-binary analysis can therefore detect this compiler-introduced vulnerability, as neither source nor binary analysis alone has enough information to reliably succeed.

2.2 Challenges

The examples above highlight the value of integrating source and binary semantics for vulnerability detection. While CPGs have been leveraged to effectively capture source-level information for security analysis [13,39,42], prior works have not combined semantic data from multiple computational models of the same program, presenting significant challenges:

Relating Semantic Domains. To assess the vulnerability of potential exploits, CPG queries must access both source- and binary-level semantic information. For instance, queries that look for control-flow paths from source line ℓ_1 to line ℓ_2, where the (binary-level) `rsp` register is overwritten, cannot separate source and binary information into distinct graphs.

Simply unioning source and binary CPG edges is insufficient, as it creates unsound ambiguities. For example, a union of source and binary CFGs would contain two distinct paths from ℓ_1 to ℓ_2–one through source and one through binary code–rather than a single path representing both semantics. CDCPGs therefore introduce new *relational edges* to CPGs, which indicate that two nodes or edges represent the same code entity in different semantic domains. These relational edge sets are not one-to-one, since some binary entities have no source representation (e.g., compiler-introduced code), and some source-level entities have no binary-level representation (e.g., code that is optimized away).

Path Explosion. Linking CPGs across semantic domains with relational edges risks path explosion. We address this by lazy construction of cross-domain portions, generating subgraphs based on vulnerability scores. Binary-level symbolic analysis is restricted to regions where source analysis identifies high-risk POIs requiring exploitability and security impact assessment. This lazy approach is inspired by bottom-up function evaluation for duplicate analysis avoidance. We further optimize by analyzing only function subsets for each vulnerability class. For example, dynamic allocation analysis begins with functions calling *alloc* family functions, then traces forward to verify size propagation to callers and bounds-checking implementation.

Graph Analysis. Software graph analyses can be intra- or inter-procedural. Intra-procedural approaches [39] cannot detect flawed interactions between functions, so are unsuitable for many classes in Table 1. Traditional pattern-based analyses only identify universally path-quantified issues (like variables uninitialized on all paths) rather than issues on specific paths, and they detect missing sanitization but not incorrect implementations [39]. CodeGrafter addresses these limitations by combining inter-procedural binary CFGs with intra-procedural source CFGs, evaluating lazily for faster queries, and employing symbolic interpretation to verify sanitization correctness.

False Positives. Many CPG-based tools are ineffective for vulnerability discovery in large codebases because of high false alarm rates. This is in part because they search for likely code errors without assessing exploitability, which is left to the user. Exploitability is notoriously difficult for humans to assess since it often requires binary-level analysis. CodeGrafter reduces the false positive rate by leveraging binary semantics in CDCPGs to automatically assess exploitability, filtering or deprioritizing POIs with low security impact. Listing 1.3 shows code to which CodeGrafter assigns high priority in reports to auditors, because the compiler's placement of security-sensitive variable `authenticated` locates it adjacent to overflowable array `buffer` in the local stack frame. This binary information indicates a high exploitability potential for the flaw, elevating its priority above other questionable buffer operations in a large program.

False Negatives. To reduce false negatives, many prior works resort to overly permissive syntax-based queries. For example, syntax-based queries for missing bounds checks often classify as safe any dereference operation that is dominated by an inequality check that contains the correct bound [39].

Listing 1.4 shows a bounds check that this strategy incorrectly classifies as safe (false negative). The check on line 2 appears safe syntactically, but the value assigned to `sz` in line 1 is incorrect because it uses 32-bit instead of 64-bit multiplication. Such errors are typical of high severity security flaws; Listing 1.4 was exploited in 2015 as part of the Stagefright attacks against Android [6]. CodeGrafter supplements syntax-based queries with semantic queries evaluated by symbolic interpretation to reduce false negatives.

Parameter Analysis. Many C code vulnerabilities involve unsafe dereferences of pointers passed as function parameters. These are difficult for prior works to

```
1 char token[20];
2 char buffer[60];
3 int authenticated;
4 recv(socket_con, buffer, 80, 0);
5 if (!strncmp(buffer, token, strlen(token)))
6    authenticated = TRUE;
7 ...
8 if (authenticated != FALSE)
9    ...
```

Listing 1.3. Buffer Overflow

```
1 uint64_t sz = mSampleTime * 2 * sizeof(uint32_t);
2 if (sz > SIZE_MAX) return ERROR_OUT_OF_RANGE;
```

Listing 1.4. Incorrect bounds check

diagnose since safeguarding pointers to caller-allocated data structures requires knowledge of the binary-level layout of the entire call stack, not merely the local frame. CodeGrafter is the first CPG analysis tool that models full binary-level stack layouts of call chains, affording detection of pointer-parameter vulnerabilities that slip by other approaches.

3 Related Works

Prior research in software vulnerability detection can be broadly categorized into static, dynamic, and hybrid techniques.

Static methods encompass rule-based analysis [11] and code similarity detection [21], analyzing source code through representations such as Abstract Syntax Trees (AST), Control Flow Graphs (CFG), Program Dependency Graphs (PDG), and Code Property Graphs (CPG). While CPGs [39] offer scalable static analysis by combining ASTs, CFGs, and dataflow dependencies into a unified structure, static techniques often produce high false positives [19]. Source-level CPGs particularly struggle with precision, yielding high false negatives for precise queries and high false positives for broader ones. Tools like Leopard [13] have improved detection by ranking vulnerable functions using source-level metrics, while binary-focused approaches like Vyper [4] analyze compiled code for specific vulnerability classes such as stack/heap overflow and information leaks. Our approach integrates both source and binary features within the CPG framework to improve detection precision at file and function levels.

Dynamic techniques, such as fuzz testing [30] and taint analysis [34], detect vulnerabilities during runtime execution but often face high false negative rates due to limited code coverage. Machine learning (ML) approaches have also evolved to leverage various supervised and unsupervised techniques [28]. Binary analysis achieves higher precision ($\approx 75\%$) and recall ($\approx 20\%$) [28] compared to source-based models but presents limited developer insights. Our approach addresses this by leveraging binary semantics to complement source-level analysis.

While many approaches target general vulnerability detection, specialized tools have also emerged. For instance, Divak [16] presents a robust and non-invasive dynamic analysis framework for detecting out-of-bounds (OOB) write

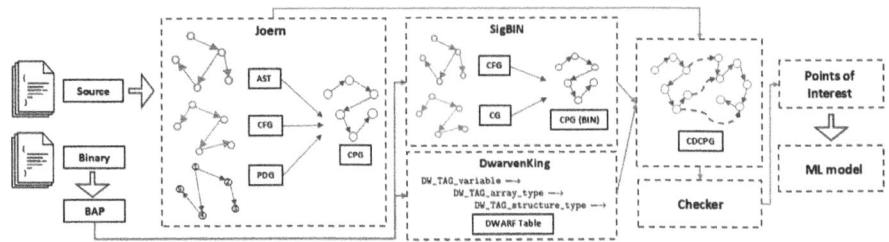

Fig. 1. CodeGrafter's pipeline. It converts source code into ASTs, CFGs, and PDGs using Joern to form source-level CPGs. SigBIN employs BAP to extract binary-level CFGs and Call Graphs (CGs), creating binary-level CPGs (BINs). DwarvenKing maps source lines to binary addresses via DWARF tables, generating CDCPGs. Checker then applies query templates to detect vulnerable POIs for training classification models

vulnerabilities, with a particular focus on preserving the integrity of real-world memory layouts. Similar to CodeGrafter, Divak leverages DWARF debug information to bridge source-level constructs with binary-level operations. However, Divak's source-to-binary mapping reconstructs only variable region, type, and lifetime information to support dynamic analysis of memory writes. In contrast, CodeGrafter reconstructs all available DWARF information for all instructions, such as source-binary correspondences for code blocks that contain no memory writes, as features for building a general-purpose, statically queriable CDCPG graph. The graph embeds static binary features alongside source-level semantics, allowing for CodeGrafter to reason about a broader class of vulnerabilities beyond those observable through Divak.

Graph Learning has also become a key method for automated vulnerability detection [42], using representations such as ASTs, CFGs, PDGs, and CPGs. CFGs model possible execution paths by representing sequences of program statements with nodes and execution order with edges [43], while ASTs capture the syntactic structure of the code, representing language constructs rather than the exact syntax [23]. PDGs extend CFGs and ASTs by modeling control and data dependencies in the program [15], originally developed for compiler optimization and now used for vulnerability detection in methods like IVDetect [20].

CPGs [39] further combine ASTs, CFGs, and PDGs, and are employed in techniques like Devign [44] and Reveal [8] for comprehensive vulnerability detection. CPG nodes can represent various program constructs (e.g., methods, variables), with labeled edges capturing relationships. Additionally, we incorporate binary-level CFGs (BIN), which model the logical flow of compiled code and offer different insights compared to source-level CFGs.

Our work addresses a research gap by combining source-level and binary-level graph information, thus improving the identification of vulnerabilities that are difficult to detect using either approach alone. The graph datasets we generate are formatted for use with PyTorch Geometric libraries and are organized similarly to other general graph datasets, such as MUTAG [17], making them suitable for graph learning tasks.

4 Design and Implementation

Figure 1 shows a high-level overview of CodeGrafter's architecture, which integrates three key parsers: (1) SigBIN, a binary parser; (2) DwarvenKing, a DWARF debug information parser; and (3) Joern, a C code parser [39]. Algorithm 1 outlines how these parsers work together to generate the CDCPG. Each parser produces data structures that feed into CodeGrafter's analysis component, Checker. While Joern is a third-party tool used as-is, SigBIN and DwarvenKing are custom-built for this framework.

Algorithm 1: CodeGrafter Pipeline

Input: Functions from source code, S, Functions from binary code, B

1 $P \leftarrow \emptyset$;
2 **foreach** $s \in S$ *and its corresponding* $b \in B$ **do**
3      $ast, cfg, cpg \leftarrow$ convert s to graphs using Joern;
4      $binary\_cfg \leftarrow$ convert b to graphs using BAP;
5      $cg \leftarrow$ extract call graphs for s;
6      $bin \leftarrow$ combine $binary\_cfg$ and cg;
7      $dwarf \leftarrow$ generate compiler metadata for s and b;
8      $cdcpg \leftarrow$ parse $dwarf$ to relate ast, cfg, cpg, and bin;
9      $poi \leftarrow$ extract POIs from $cdcpg$ using query templates in Checker;
10   append poi to P;
11 use P to train learning model;

4.1 SigBIN

Our binary parser, SigBIN, leverages the Binary Analysis Platform (BAP) [3] to extract assembly code semantics. BAP expresses these semantics in a platform-independent language called the BAP Intermediate Language (BIL)[7], which abstracts the details of various Instruction Set Architectures (ISAs). To generate control flow information, we parse the BIL representation of a binary to identify basic blocks–sequences of instructions with a single entry and exit point. The control flow relationships between these blocks, based on jumps, branches, and function calls, are used to construct a binary-level CFG, where basic blocks serve as nodes and control transfers form the edges (line 4 in Algorithm 1).

Each node in the CFG also contains a BIL code fragment that captures the instructions' effects on the processor state. CodeGrafter utilizes this information extensively to infer the source-level side effects of potential exploits. The process for achieving this through value set analysis is described in detail in Sect. 5.

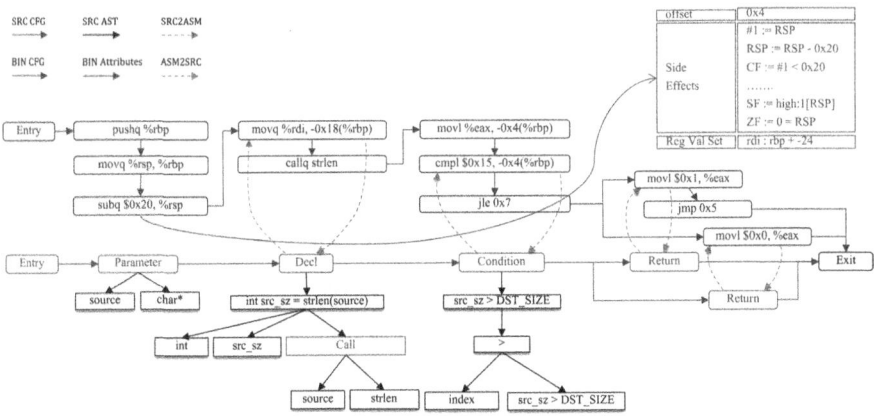

Fig. 2. CDCPG for code in Listing 1.5 showing binary CFG (purple), source CFG (blue), and AST components (black), BAP-lifted side-effect semantics (red), and source-binary mappings (orange/green dotted arrows), based on DWARF data (Color figure online)

However, CFGs constructed in this manner are inherently limited to intra-procedural paths, as they represent control flow within individual functions. To extend the analysis to inter-procedural paths, we integrate Call Graphs (CGs), which model the relationships between different function calls within a program (lines 5–6 in Algorithm 1). Instead of standard call graphs, we use enhanced causal call graphs. These graphs include additional nodes representing the call site basic blocks which connect caller and callee nodes.

For each call site in a basic block (in the caller's CFG), we add an edge from the call-site node to the entry basic block of the callee's CFG. A return edge is added from the exit point of the callee's CFG back to the basic block following the call site in the caller's CFG. This models the control flow returning to the caller after the function completes. The optimized combination of in-memory CFGs and CGs using this approach allows SigBIN to analyze inter-procedural paths that meet specific query constraints.

For storage and further analysis, these graphs are persisted in a Neo4j database and exported in the Graphviz DOT format (cf., [35]). Balancing the use of in-memory data structures with on-disk storage is crucial for analyzing large codebases efficiently. Performing path searches on complex CFGs with numerous branches can be computationally prohibitive due to path explosion. To optimize query performance, we maintain the intra-procedural CFGs and the full call graph in memory, passing them to the Checker module as needed, thereby avoiding excessive database reads and accelerating the analysis.

4.2   DwarvenKing

CPUs execute binaries by interpreting low-level instructions and their effects on machine state (e.g., registers and memory) without reference to the original source code. The transformation from high-level source code to machine instructions is performed by the compiler. During compilation, code may be relocated, merged, or even eliminated [40]. Compiler optimizations may also include techniques such as peephole, local, global, and loop optimizations [18, 24].

The compiler's primary goal is to produce a compact and efficient executable, often without documenting these transformations in the final binary. However, developers can generate binaries with additional debugging information for testing. For example, Executable and Linkable Format (ELF) binaries can include Debugging With Arbitrary Record Format (DWARF) data, a symbol table that contains names and addresses for functions, variables, and other program components. This metadata aids in linking the initial source code with its corresponding machine code addresses and is generated using compiler flag -g.

The Dwarvenking module parses DWARF information from ELF binaries to link executable code with source lines (line 7 in Algorithm 1). This mapping is crucial for debugging, as it reveals details not visible in the source code, such as side effects and low-level semantics. For instance, the actual layout and sizes of stack variables may differ from their source code representation. Understanding the binary-level layout helps eliminate false positives. Dwarvenking employs a two-pass algorithm to parse DWARF data. In the first pass, it records all Debugging Information Entries (DIEs). Each DIE contains attributes such as variable names, types, and source code line numbers. In the second pass, it extracts additional details from child and member DIEs, including locations, sizes, and base types, to determine the actual variable lengths at execution. This reveals the order and spacing of stack variables, which may differ from the source layout.

This information enables us to ascertain the effective length of the DIEs during runtime. It also provides the mappings between machine code addresses and the corresponding lines of source code. This line-to-address data sourced from DWARF serves as the basis for connecting source-level statements with their binary-level addresses (line 8 in Algorithm 1). In the special case of OOB-write detection, this yields variable storage region, type, and lifetime information (cf., [16]). In the more general case, applying this approach to all DWARF information in the program yields a rich collection of graph-minable features that characterize how the compiler has interpreted every code block in the program. For example, it reveals whether the compiler considers some source code to be unreachable (evidenced by dead code elimination), invariant (evidenced by invariant hoisting code-motion), static (evidenced by rerepresentation of a source expression as a binary constant), etc. This affords formulation of general-purpose queries capable of traversing between the program's two levels: source and binary. We can also effectively pinpoint disparities between corresponding pathways within the two program domains by using the domain-specific graphs.

4.3   Cross Domain Code Property Graphs

To construct Cross Domain Code Property Graphs (CDCPG), we begin by extracting source-level ASTs, CFGs, and PDGs using the Joern parser (line 3 in Algorithm 1), while obtaining binary-level CFGs and CGs from SigBIN.

```
1 int foo(char* source) {
2     int src_sz = strlen(source);
3     if (src_sz > DST_SIZE) return 1;
4     return 0;
5 }
```

Listing 1.5. Example code from which CDCPG graph in Figure 2 was generated

These graphs are then integrated using line-to-address mappings derived from DWARF tables provided by DwarvenKing. Figure 2 illustrates the structure of a CDCPG for the code example shown in Listing 1.5, where the purple nodes represent binary CFG elements connected by edges indicating fall-through, static branch, and jump relationships.

Indirect control transfers, such as method calls and function returns, pose a challenge because they are often unpredictable at compile-time and can be manipulated by adversaries. As a result, their edges are not explicitly defined in the CDCPG. Instead, these nodes are marked as *indirect*, with potential targets inferred using side-effect semantics lifted by BAP and encoded in the node details, as depicted in the red table in Fig. 2. Queries within the system use source-level CFG edges to estimate intended destinations while leveraging binary-level side-effect data to determine actual destinations, including those in exploits.

In the CDCPG, binary nodes contain detailed information such as address offsets, assembly syntax, register states, basic block identifiers, and function identifiers. The blue and black nodes in Fig. 2 correspond to their source-level counterparts, with blue representing CFG nodes and black representing AST nodes. The dotted orange and green arrows illustrate the links between source and binary representations, derived from DWARF data, allowing for bidirectional traversal across domains.

These source-to-binary and binary-to-source mappings are crucial for comprehensive analysis. Source-level mappings make it possible to translate binary-level analysis results into a source code context, which aids in human interpretation. Conversely, binary-level mappings enable the assessment of the exploitability and security impact of vulnerabilities identified in the source code.

Discrepancies between source and binary representations can reveal compiler optimizations, such as code elimination during compilation. Additionally, the binary semantics expose side-effects, including changes to registers, status flags, and memory, providing insights at both the function and basic block levels. To fully assess security implications, CodeGrafter performs symbolic evaluation of call graph paths to determine trace-level effects.

The combined analysis of source and binary domains facilitates the detection of security-relevant flaws, such as cases where binary-level instruction side-effects may compromise critical source-level variables. Overall, the CDCPG provides a powerful tool for discovering new vulnerabilities that cannot be identified using only source- or binary-level information, offering a more holistic view of the code's security landscape.

Algorithm 2: Insecure Dataflow Analysis

Input: *call_graph*, *cdcpg*

1 *ext_sources* ← [recv, gets, read, scanf, getenv, ...];
2 *ext_sinks* ← [system, write, send, ...];
3 **foreach** *src* ∈ *ext_sources* **do**
4   **foreach** *sink* ∈ *ext_sinks* **do**
5   **if** has_path(*callgraph*, *src*, *sink*) **then** **return** POI(*cdcpg*)

4.4 Checker

The Checker consists of reusable analysis query templates produced by experts. An analysis template is a predefined code explorer wrapped around CDCPG queries. We use the Gremlin query language to interface with the Neo4j database. For brevity, query examples are here expressed as pseudo-code algorithms that abstract the actual Gremlin and Python code for each query.

Each template performs an independent analysis to detect specific Points of Interest (POIs) (line 9 in Algorithm 1). CodeGrafter back-traces the CDCPG to find source locations, yielding high-level information to users, such as line number, function name, and class. This modular design allows new analyses to be added as needed without recreating the CDCPG.

Insecure Dataflows. Algorithm 2 sketches CodeGrafter's *insecure dataflow analysis*, which implements taint tracking (cf., [26]) on the CDCPG to discover exploitable inter-procedural paths between user-controlled program inputs and security-sensitive operations. The analysis identifies external sources (e.g., command-line parameters), sinks (e.g., system calls favored by attackers), and paths between them. Subsequent queries search for sanitation failures and data corruption vulnerabilities along the discovered paths.

The implementation in Algorithm 2 uses the call graph to analyze interprocedural flows. This affords early detection and deprioritization of POIs whose variables are not affected by attacker-controlled inputs—a major source of false positives in real-world audits. To efficiently infer the causal order of call sites, which is important for eliminating these false positives, the CFG is internally reduced to a relational graph between basic blocks. This allows the internal instruction nodes to be ignored to achieve better performance.

Insecure String Library Functions. Certain library functions that do not check sizes when copying data, such as `strcpy` and `memcpy`, are hotspots of attacker abuse. Many large C codes are cluttered with such functions, but most are typically not vulnerable; reporting them all would therefore raise too many false alarms. CodeGrafter therefore analyzes call sites for such functions at the binary level to obtain more precise bounds for the callee's memory references.

Algorithm 3: Insecure Library Call Analysis

   Input: *stack_layout, cdcpg*

1 *insec_calls* ← [recv, strcat, strcpy, memcpy, ...];

2 **foreach** *f* ∈ *insec_calls* **do**

3   *dst* ← *get_dst*(*f*, *stack_layout*);

4   *src* ← *get_src*(*f*, *stack_layout*);

5   **if** ¬bounded(*stack_layout*, *dst*, *src*) **then** **return** POI(*cdcpg*)

Algorithm 4: BOIL Analysis

   Input: *cdcpg*

1 *cfg* ← *get_cfg*(*cdcpg*);

2 *ptr_list* ← *get_ptr_arithmatic_assignments*();

3 **foreach** *p* ∈ *ptr_list* **do**

4   **if** *in_loop*(*cfg*, *p*) ∧ ¬*guarded*(*p*) **then** **return** POI(*cdcpg*)

The binary semantics can also reveal extenuating mitigations, such as the compiler's replacement of an unsafe library function with a safe one, or stack memory layouts that frustrate productive exploitation of a potential vulnerability.

Algorithm 3 sketches such a search. To effectively analyze these call sites, it is essential to understand the calling conventions used by the target architecture. The x86-64 System V AMD64 ABI calling conventions [25] specify a kernel interface that uses registers RDI, RSI, RDX, R10, R8, and R9 for passing parameters to functions. This information is crucial for our algorithm because it determines how values are passed and how they can be analyzed at the binary level.

Algorithm 3 implements *value set analysis* (VSA) [2] on the binary code property subgraph to derive symbolic expressions for the values of these registers at call sites identified in the source-level subgraph. The VSA queries the BIL side-effect information in the binary CDCPG nodes. The resulting register values refer to an offset relative to the RSP register, and the stack layout provides the size and type information related to that location. A candidate POI is returned if the inferred stack layout contains security-sensitive data within the bounds of the resulting expressions.

Buffer Overflow Induction Loop. While dangerous library functions with standard names are easy to identify at the source level, developers frequently implement the same risky functionality via custom code that comes in many forms, and is therefore more difficult to identify. CodeGrafter therefore implements *buffer overflow induction loop* (BOIL) [32] detection, which searches for code patterns that match these semantic behaviors. The detection searches for cases where data is moved from one pointer to another within a loop without considering destination buffer size.

Algorithm 4 sketches the analysis. Helper function $in\_loop(cfg, p)$ detects whether node p is within an intra-procedural CFG cycle, and $guarded(p)$ appeals to the VSA engine (Sect. 5) to search for a sufficient dominating bounds check.

5   Benchmark Experiments

The CDCPG analyses described in Sect. 4 are designed to serve as feature extractors for machine learning-based classification and ranking of the resulting POIs. Machine learning has emerged as a critical ingredient for many automated vulnerability detection approaches in recent years (e.g., [8,22,44]). We therefore evaluate our approach's effectiveness for improving these ML models, comparing CDCPGs to other graph representations, such as CPG (source only) and BIN (binary only), across multiple state-of-the-art (SOTA) learning algorithms. Our evaluation addresses the following research questions (RQs):

- **RQ1:** How effective is CDCPG in detecting software vulnerabilities compared to other graph types, such as CPG and BIN?
- **RQ2:** Does CDCPG's performance generalize well across different applications and CWE types, and how does this compare to other graph types?

For evaluation, we use novel graph learning techniques that leverage contrastive learning methods, including GraphCL [41], InfoGraph [36], and DGI [38], to classify graphs. These methods generate multiple augmentations of the original graph, labeling slight variations as positive instances and distinct ones as negative instances. During training, the contrastive loss function is optimized to bring positive instances closer and push negative instances apart. This approach is especially beneficial in our context, as the POIs collected from CodeGrafter may not provide enough data to train an effective supervised learning model.

To further assess dataset generalizability, we implement it with two SOTA vulnerability detection methods: Devign [44] and VulMAE [42]. Devign constructs a program-semantic-based graph from source code and learns node representations by aggregating neighboring node information through a new GNN module *Conv*, which extracts meaningful representations for graph-level predictions. VulMAE is a semi-supervised learning technique that improves graph autoencoding by incorporating masked feature reconstruction and scaled cosine error to overcome the limitations of traditional graph autoencoders and handle class imbalance. It also uses SMOTE [9] for enhanced robustness against imbalanced data. For fair comparison, we apply SMOTE across all the graph learning methods. We choose these algorithms because they currently represent the SOTA in this domain.

5.1   Dataset construction

The Common Vulnerabilities and Exposures (CVE) database [27] documents vulnerabilities across various applications. For our experiments, we select six widely used open-source projects written in C/C++, as shown in Table 2. These

Table 2. Description of applications in our CodeGrafter dataset

Application	Description
Sudo	Enables users to run programs with the security privileges of another user
LibTIFF	Library for reading and writing Tagged Image File Format (TIFF)
LibPNG	Library for reading, creating and manipulating Portable Network Graphics (PNG)
TinTin	A console telnet client for playing MUD (Multi-User Dimension)
TCPdump	Data network packet analyzer
OpenSSH	A secure networking utility works on Secure Shell protocol

applications have large codebases with numerous functions and lines of code, making manual exploration impractical.

CVE offers brief vulnerability descriptions with links to relevant sources and works in tandem with the National Vulnerability Database (NVD) [29]. It assigns severity scores and offers links to available solutions. Using these references, we analyzed the repositories of selected applications, extracting functions linked to these vulnerabilities. Additionally, we examined commit logs to identify further vulnerabilities addressed by patches. Some of the CVEs used in our evaluation include CVE-2018-15919, CVE-2016-10087, CVE-2023-3164, and CVE-2018-14879. To construct a binary dataset, we collected and labeled non-vulnerable functions from the same applications, resulting in 1,741 non-vulnerable functions. In contrast, we gathered approximately 180 vulnerable functions, comprising 10% of the non-vulnerable class.

5.2 Experimental Details

To quantify our evaluation, we reported three evaluation metrics for each experiment: (1) weighted F1-score, (2) weighted recall, and (3) average precision. We prioritize recall and F1 over accuracy since undetected vulnerabilities (false negatives) pose greater security risks than false positives, which only increase auditor workload. Weighted recall and F1 account for dataset imbalance and class instance distribution. Since both are threshold-dependent, we also report average precision (area under the Precision-Recall curve), which is more suitable for imbalanced datasets than AUC-ROC.

Devign was tested using the original model [44] with graph embedding, gated graph recurrent, and convolutional layers. Training employed Adam optimizer (learning rate: 0.0001, weight decay: 0.001), BCE loss, and a batch size of 8, with a 60%-20%-20% train-validation-test split. For InfoGraph, DGI, and GraphCL, we used Graph Attention Network (GAT) as the encoder, bilinear and inner product as discriminators, and Scaled Cosine Error (SCE) loss. Each experiment ran with an 80%-10%-10% data split, a learning rate of 0.01, and mean pooling for graph embeddings. VulMAE used a learning rate of 0.0005, batch size 16, hidden size 32, and masking rate 0.75. Data was split 80-20, with 30% additional training samples generated using SMOTE to address class imbalance.

Table 3. Comparison of BIN (binary only), CPG (source only) and CDCPG reporting scores for contrastive, generative, and supervised learning methods. Experiments were repeated five times with random seeds, and maximum scores across runs are reported.

Graph Type	Metrics	Contrastive			Supervised	Generative
		GraphCL [41]	DGI [38]	InfoGraph [36]	Devign [44]	VulMAE [42]
BIN	Wtd-F1	0.861	0.842	0.868	0.461	0.830
	Wtd-Recall	0.901	0.883	0.903	0.452	0.830
	Avg-Precision	0.845	0.823	0.856	0.472	0.882
CPG	Wtd-F1	0.858	0.852	0.862	0.901	0.932
	Wtd-Recall	0.902	0.893	0.902	0.852	0.942
	Avg-Precision	0.850	0.815	**0.867**	**0.960**	0.944
CDCPG	Wtd-F1	**0.872**	**0.858**	**0.868**	0.930	**0.937**
	Wtd-Recall	**0.908**	**0.900**	**0.907**	0.921	0.945
	Avg-Precision	**0.855**	**0.845**	0.854	0.939	**0.948**

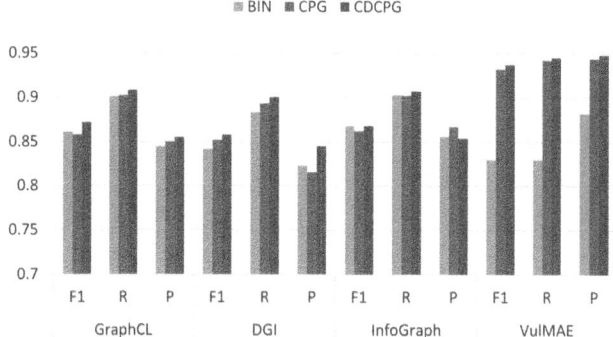

Fig. 3. Bar chart showing the differences in F1, Recall (R) and Precision (P) scores for the different graph types using various SOTA graph learning algorithms

5.3 Benchmark Results

RQ 1: To answer this research question, we quantify the effectiveness of CDCPG in detecting vulnerabilities compared to other existing graph types. For comparison, we report the results for CPG and BIN, which are graphs using information from only source code and binary code, respectively. Table 3 and Fig. 3 compare results from five state-of-the-art methods. Each experiment is repeated five times with random seeds and the best scores are reported in the table, with the highest score for each combination of method and metric highlighted. Results show that in almost all cases CDCPG outperforms existing graph types except for only two cases where the precision for CPG is slightly better.

While CDCPG did not identify previously unknown vulnerabilities in our dataset, it outperformed other graph types in detecting known ones, highlighting its ability to localize critical code regions. Once trained on CDCPG, our models generate softmax-based probability scores that allows us to rank POIs by likelihood of being vulnerable. This ranking aids in quantifying risk and streamlining audits by focusing attention on the most vulnerable areas.

BIN's usually worst performance corroborates our expectation that it is difficult to extract concept-relevant features from binary-only software models, making source-aware classification easier. However, the binary data enhances source-level data to improve vulnerability detection effectiveness beyond what is possible with source code alone. It allows the source level information to guide the model to learn better features for the binary code and vice versa. Also, the performance of our dataset generalizes well to existing vulnerability detection models as shown by the results of Devign and VulMAE.

Table 4. Quantitative analysis of the performance of CDCPG across different applications when run on VulMAE. Results are compared with existing graph types. Each experiment is run five times with random seeds and the highest scores are reported.

Graph Type	Metrics	Applications					
		LibPNG	LibTiff	TCPDump	Sudo	OpenSSH	TinTin
BIN	Wtd-F1	0.68	0.88	0.88	0.67	0.67	0.88
	Wtd-Recall	0.78	0.86	0.89	0.77	0.77	0.88
	Avg-Precision	**0.83**	0.87	0.89	0.61	0.59	0.90
CPG	Wtd-F1	0.67	0.90	0.85	0.93	0.70	0.67
	Wtd-Recall	0.77	0.91	0.86	0.93	0.77	0.78
	Avg-Precision	0.59	0.89	0.86	0.90	0.62	0.60
CDCPG	Wtd-F1	**0.76**	**0.96**	**0.89**	**0.94**	**0.80**	**0.97**
	Wtd-Recall	**0.80**	**0.96**	**0.90**	**0.94**	**0.82**	**0.97**
	Avg-Precision	0.78	**0.96**	**0.90**	**0.94**	**0.81**	**0.97**

Figure 4 shows that while Devign performs well on CPG graphs, it struggles with BIN graphs–unlike CDCPG, where Devign achieves the best results. This highlights the suitability of our dataset for existing vulnerability detection models and its potential for enabling future researchers to derive improved performance from their models. Also, among the array of models explored, the most favorable and comprehensive outcomes are achieved with VulMAE, which aligns with expectations given its compatibility with imbalanced datasets. That is why, based on this finding, we use VulMAE to address our next research question.

RQ 2: We next assess whether CDCPG's success in RQ1 stems from a narrow set of graphs tied to specific applications or reflects strong, consistent performance across all applications. Additionally, given the variability in coding styles, we also examine the robustness of our graph type against such differences. The scores presented in Table 4 correspond to the highest values among five separate runs using different random seed values. Similar to the prior experiment, the best scores for each application-metric combination are highlighted.

The outcomes demonstrate that the efficacy of CDCPG is consistent across all applications, with scores consistently surpassing 0.75. Conversely, the performance of the other two graph types lacks the same level of consistency. Specifically, in two out of the six applications under experimentation, both BIN and CPG yield scores below 0.75. This observation supports the notion that CDCPG boasts considerable reliability, making it a dependable option for diverse applications irrespective of the underlying coding styles and script formatting.

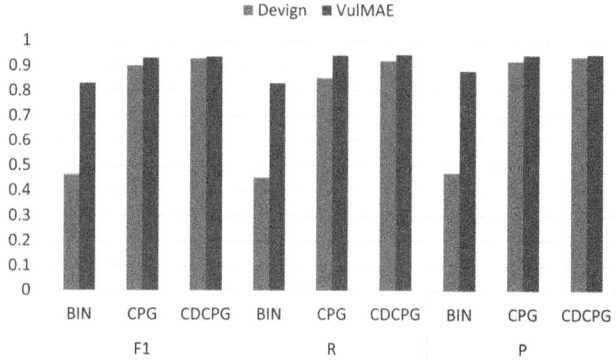

Fig. 4. Comparison of F1, Recall (R) and Precision (P) scores between VulMAE and Devign for the different graph types

We further extend our experiments to evaluate CDCPG's performance across different CWE types. Using the NVD database, we identify the CWE classification for vulnerable functions with a defined type. The distribution is highly imbalanced, with 80% of samples classified as CWE-119, followed by 6% as CWE-120, 4% as CWE-200, 2% as CWE-416 and 8% as CWE-476. CWE-119 records the highest F1 score of 0.97, primarily due to its larger data volume. Despite the imbalance, CodeGrafter attains fairly good F1 scores of 0.89, 0.86, 0.81 and 0.91 for CWE-120, CWE-200, CWE-416, and CWE-476 respectively, aided by SMOTE-based data augmentation. These consistently good results across diverse vulnerability types suggest that CDCPG generates informative features that support reliable performance, even with limited training data.

6   Limitations and Future Work

CodeGrafter uses compiler-generated DWARF information to create feature-rich graphs linking source- and binary-level code features. Prior work [1,12] has noted that DWARF data can be inaccurate–especially for optimized binaries lacking clear source-line mappings. Although such inaccuracies could potentially impair CodeGrafter's effectiveness, experimental results (Sect. 5.3) show that standard compiler optimizations still yield DWARF data accurate enough to improve vulnerability detection in many cases. However, future work should assess the effect of DWARF inaccuracies by evaluating CodeGrafter on corrupted inputs.

Moreover, our current approach focuses on only graph-level classification, since our graphs are labeled at the function level. This limits its ability to detect vulnerabilities at finer granularity such as line-level. Future improvements will enable subgraph-level analysis to refine the granularity. In addition, our dataset generation pipeline can struggle with very large programs due to high memory demands when handling large numbers of functions. Improvements to distributed storage and lazy graph traversal will improve scalability.

CodeGrafter currently performs binary classification but can be extended to multi-class classification using CWE labels for more precise vulnerability identification. Additionally, existing vulnerability databases do not differentiate between source-detectable and binary-only vulnerabilities. Curating a dataset that distinguishes these categories would highlight CodeGrafter's advantages over source- or binary-only approaches. Expanding the feature set with detailed binary-level indicators, such as high-risk gadgets, can further refine feature weights and complexity thresholds for better predictions. Finally, incorporating more semantic-based vulnerability templates (see Sect. 4.4) that target more CWE categories will enable CodeGrafter to detect a wider range of security flaws. For example, type-confusion attacks (CWE-843) can potentially be addressed by templates that compare binary-level dataflows against CDCPG edges to source-level types.

7   Conclusion

CodeGrafter integrates source- and binary-level semantics into a unified CPG for vulnerability detection in C/C++ programs. Unlike prior methods requiring expert input or focusing on general debugging, CodeGrafter enables low-skill auditors to identify high-impact vulnerabilities in large, unstructured codebases. Our new CDCPG graph model captures data flow, dependencies, and control flow from both source and binary code, allowing detection of vulnerabilities only identifiable when we consider both set of features, such as those that occur during compilation. Using a dataset curated from six open-source C/C++ programs, we evaluate various graph learning techniques. We found that graph autoencoders outperform other methods, significantly surpassing recent works.

Acknowledgments. The research herein was supported in part by DARPA awards N6600121C4024 and FA8750-19-C-0006, ARO awards W911NF-21-1-0032 and W911NF-24-2-0114, and DARPA subcontract 140D04-23-C-0070 from Immunant, Inc. Any opinions, recommendations, or conclusions expressed are those of the authors and not necessarily of the above supporters.

References

1. Assaiante, C., D'Elia, D.C., Di Luna, G.A., Querzoni, L.: Where did my variable go? Poking holes in incomplete debug information. In: Proceedings of an ACM International Conference Architectural Support for Programming Languages and Operating Systems, vol. 2, pp. 935–947 (2023)
2. Balakrishnan, G., Reps, T.: Analyzing memory accesses in x86 executables. In: Proceedings of International Conference Compiler Construction, pp. 5–23 (2004)
3. BAP: binary analysis platform. https://github.com/BinaryAnalysisPlatform/bap-python (2020), Accessed 19 Feb 2024
4. Boudjema, E.H., Verlan, S., Mokdad, L., Faure, C.: VYPER: vulnerability detection in binary code. IEEE Secur. Priv. **3**(2) (2020)
5. Brand, M.: In-the-wild series: Android exploits. Project Zero Blog, https://googleprojectzero.blogspot.com/2021/01/in-wild-series-android, January 2021
6. Brandom, R.: Google Rebuilt a Core Part of Android to Kill the Stagefright Vulnerability for Good. The Verge (2016)
7. Casinghino, C., Jamner, D., Gotovchits, I.: A formal specification for BIL: BIL instruction language. https://github.com/BinaryAnalysisPlatform/bil.pdf (2015)
8. Chakraborty, S., Krishna, R., Ding, Y., Ray, B.: Deep learning based vulnerability detection: are we there yet. IEEE Trans. Softw. Eng. (2022)
9. Chawla, N.V., Bowyer, K.W., Hall, L.O., Kegelmeyer, W.P.: SMOTE: synthetic minority over-sampling technique. J. Artif. Intell. Res. **16**(1) (2002)
10. CodeGrip: code review trends in 2020. https://assets.codegrip.tech/wp-content/uploads/2020/03/17142706/Code-Review-Trends-in-2020-By-Codegrip.pdf
11. Croft, R., Newlands, D., Chen, Z., Babar, M.A.: An empirical study of rule-based and learning-based approaches for static application security testing. In: Proceedings of the ACM/IEEE International Symposium Empirical Software Engineering and Measurement (2021)
12. Di Luna, G.A., Italiano, D., Massarelli, L., Österlund, S., Giuffrida, C., Querzoni, L.: Who's debugging the debuggers? Exposing debug information bugs in optimized binaries. In: Proceedings of an ACM International Conference Architectural Support for Programming Languages and Operating Systems, pp. 1034–1045 (2021)
13. Du, X., et al.: Leopard: identifying vulnerable code for vulnerability assessment through program metrics. In: Proceedings of International Conference Software Engineering, pp. 60–71 (2019)
14. Edgescan: 2022 vulnerability statistics report. https://www.edgescan.com/stats-report (2022)
15. Ferrante, J., Ottenstein, K.J., Warren, J.D.: The program dependence graph and its use in optimization. ACM Trans. Program. Lang. Syst. **9**(3), 319–349 (1987)
16. Hafkemeyer, L., Starink, J., Continella, A.: Divak: non-invasive characterization of out-of-bounds write vulnerabilities. In: Proceedings of International Conference Detection of Intrusions and Malware & Vulnerability Assessment, pp. 211–232 (2023)
17. Kazius, J., McGuire, R., Bursi, R.: Derivation and validation of toxicophores for mutagenicity prediction. J. Med. Chem. **48**(1), 312–320 (2005)
18. Keutzer, K., Wolf, W.: Anatomy of a hardware compiler. In: Proceedings of the ACM Conferences Programming Language Design and Implementation, pp. 95–104 (1988)

19. Li, X., Feng, B., Li, G., Li, T., He, M.: A vulnerability detection system based on fusion of assembly code and source code. Secur. Comm. Netw. (2021)
20. Li, Y., Wang, S., Nguyen, T.N.: Vulnerability detection with fine-grained interpretations. In: Proceedings of the ACM Joint Meeting European Software Engineering Conference and Symposium Foundations Software Engineering, pp. 292–303 (2021)
21. Li, Z., Zou, D., Xu, S., Jin, H., Qi, H., Hu, J.: Vulpecker: an automated vulnerability detection system based on code similarity analysis. In: Proceedings of the Annual Computer Security Applications Conferences, pp. 201–213 (2016)
22. Li, Z., Zou, D., Xu, S., Jin, H., Zhu, Y., Chen, Z.: SySeVR: a framework for using deep learning to detect software vulnerabilities. IEEE Trans. Dependable Secure Comput. **19**(4), 2244–2258 (2022)
23. Lin, G., Zhang, J., Luo, W., Pan, L., Xiang, Y.: POSTER: vulnerability discovery with function representation learning from unlabeled projects. In: Proceedings of the ACM Conferences Computer and Communications Security, pp. 2539–2541 (2017)
24. Liu, Y.A., Stoller, S.D., Li, N., Rothamel, T.: Optimizing aggregate array computations in loops. ACM Trans. Program. Lang. Syst. **27**(1), 91–125 (2005)
25. Lu, H., Matz, M., Girkar, M., Hubička, J., Jaeger, A., Mitchell, M.: System V application binary interface AMD64 architecture processor supplement version 1.0. AMD (2018)
26. Ming, J., Wu, D., Xiao, G., Wang, J., Liu, P.: TaintPipe: pipelined symbolic taint analysis. In: Proceedings of USENIX Security Symposium, pp. 65–80 (2015)
27. MITRE: common vulnerabilities and exposures. www.cve.org (2024), Accessed 19 Feb 2024
28. Morrison, P., Herzig, K., Murphy, B., Williams, L.: Challenges with applying vulnerability prediction models. In: Proceedings of the Symposium and Bootcamp Science of Security (2015)
29. NIST: national vulnerability dataset. nvd.nist.gov (2024), Accessed 19 Feb 2024
30. Pinconschi, E., Abreu, R., Adão, P.: A comparative study of automatic program repair techniques for security vulnerabilities. In: Proceedings of the IEEE International Symposium on Software Reliability Engineering, pp. 196–207 (2021)
31. Positive technologies: top cyberthreats on enterprise networks. https://ptsecurity.com/upload/corporate/ww-en/analytics/Top-cyberthreats.pdf (2020)
32. Rawat, S., Mounier, L.: Finding buffer overflow inducing loops in binary executables. In: Proceedings of the IEEE International Conference on Software Security and Reliability (2012)
33. Shen, D., Fang, J.: Rooting every Android. Technical Report, KEEN Security Lab (2016), blackHat Europe
34. Shimchik, N.V., Ignatyev, V.N., Belevantsev, A.A.: Improving accuracy and completeness of source code static taint analysis. In: Proceedings of the Ivannikov Ispras Open Conference, pp. 61–68 (2021)
35. Song, D., et al.: BitBlaze: a new approach to computer security via binary analysis. In: Proceedings of the International Conference Information Systems Security (2008)
36. Sun, F.Y., Hoffmann, J., Verma, V., Tang, J.: InfoGraph: unsupervised and semi-supervised graph-level representation learning via mutual information maximization. In: Proceedings of International Conference Learning Representations (2020)
37. Synopsys: open source security risk analysis. https://synopsys.com/software-integrity/resources/analyst-reports/security.html (2024), Accessed 19 Feb 2024

38. Veličković, P., Fedus, W., Hamilton, W.L., Liò, P., Bengio, Y., Hjelm, R.D.: Deep graph infomax. In: Proceedings of the International Conference Learning Representations (2019)
39. Yamaguchi, F., Golde, N., Arp, D., Rieck, K.: Modeling and discovering vulnerabilities with code property graphs. In: Proceeding of the IEEE Symposium on Security and Privacy, pp. 590–604 (2014)
40. Yang, Z., Johannesmeyer, B., Olesen, A.T., Lerner, S., Levchenko, K.: Dead store elimination (still) considered harmful. In: Proceeding of the USENIX Security Symposium (2017)
41. You, Y., Chen, T., Sui, Y., Chen, T., Wang, Z., Shen, Y.: Graph contrastive learning with augmentations. In: Proceeding of the Conference on Advances in Neural Information Processing Systems, pp. 5812–5823 (2020)
42. Zamani, M., Irtiza, S., Khan, L., Hamlen, K.W.: VulMAE: graph masked autoencoders for vulnerability detection from source and binary codes. In: Proceedings of the International Symposium on Foundations and Practice Security, pp. 191–207 (2023)
43. Zhou, M., et al.: A method for software vulnerability detection based on improved control flow graph. Wuhan Univ. J. Nat. Sci. **24**(2), 149–160 (2019)
44. Zhou, Y., Liu, S., Siow, J., Du, X., Liu, Y.: Devign: effective vulnerability identification by learning comprehensive program semantics via graph neural networks. In: Proceedings of the Conference on Neural Information Processing Systems (2019)

SyzForge: An Automated System Call Specification Generation Process for Efficient Kernel Fuzzing

ZhiZhuo Tang, Jian Lin$^{(\boxtimes)}$, Weiyu Dong, Hang Ma, and Tieming Liu

Information Engineering University, Zhengzhou, China
ling_pro@163.com

Abstract. The Linux kernel, a cornerstone of modern computing across servers, mobile devices, and embedded systems, is increasingly vulnerable due to its vast complexity and continuous evolution. Fuzz testing has emerged as a critical technique for identifying kernel vulnerabilities, with tools like Syzkaller uncovering thousands of bugs through system call (syscall) fuzzing. However, the effectiveness of such tools relies heavily on manually crafted syscall specifications, a process that struggles to keep pace with the kernels dynamic nature and intricate semantics. This paper presents SyzForge, a novel automated framework to generate precise syscall specifications for Linux kernel drivers. SyzForge integrates four stages: static analysis to distill kernel semantics from source code, symbolic execution for dynamic constraint-based parameter solving, fuzzing with Syzkaller to assess coverage, and large language model (LLM)-driven refinement to correct specification errors. Evaluated on Linux kernel version 6.12, SyzForge achieves a 13.3% increase in code coverage compared to default Syzkaller specifications, outperforming KernelGPT by 4.3%, SyzDescribe by 5.5%, and DIFUZE by 24.3%. Furthermore, it identifies 19 previously unreported vulnerabilities, demonstrating its practical impact. By automating a traditionally manual process, SyzForge enhances fuzzing efficiency, improves vulnerability detection, and strengthens kernel security. This work addresses key limitations in existing specification generation methods, offering a scalable and adaptable solution to safeguard the Linux ecosystem amid its ongoing development.

Keywords: LLMs · Fuzz · Kernel · Symbolic execution

1 Introduction

The Linux kernel [1], fundamental to modern computing across diverse platforms, faces significant security challenges due to its complexity and vast codebase. Vulnerabilities [2], such as privilege escalation or system crashes, threaten stability and privacy, making their detection and mitigation a critical research priority.

M. Egele et al. (Eds.): DIMVA 2025, LNCS 15747, pp. 118–139, 2025.
https://doi.org/10.1007/978-3-031-97620-9_7

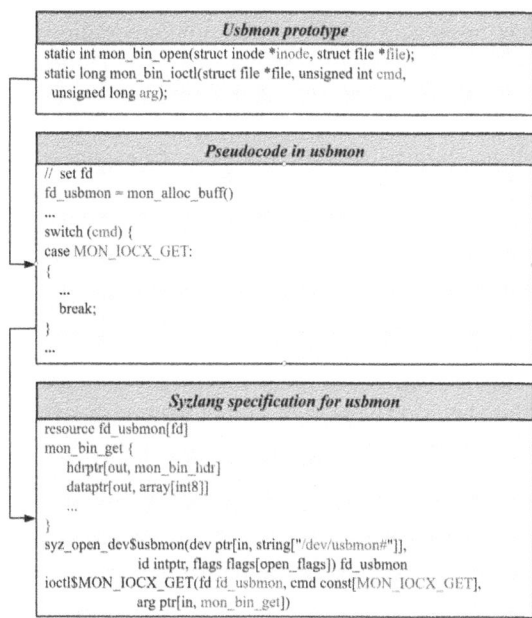

Fig. 1. Syzlang Generation Challenges

Fuzz testing [4,5] has proven to be one of the most effective methods to uncover vulnerabilities in the Linux kernel. Tools like Syzkaller [3], a highly successful system call fuzzer, have identified thousands of kernel bugs, demonstrating the efficacy of automated testing techniques. However, the process of fuzz testing the Linux kernel involves significant challenges, particularly in generating syscall sequences in user space that conform to the complex rules and semantics of the kernel. These sequences must not only meet strict syntax and logic requirements, but also effectively target deep-seated vulnerabilities within intricate kernel subsystems. Addressing these challenges requires advanced automation techniques and precise modeling of syscall behaviors, which are essential to enhance the fuzzing coverage and improve the overall security of the Linux kernel.

As shown in Fig. 1, the prototype of the usbmon driver system call in Linux, written in C, provides limited information for its execution in user space. For example, the kernel function mon_bin_open requires two parameters. However, executing the corresponding system call in user space requires an additional third parameter to specify the file path. This path is resolved by the Virtual File System (VFS) [6] to locate the character device associated with usbmon [7], which then triggers the mon_bin_open function in the kernel to handle the device opening operation.

Similarly, for ioctl-related functions, it is essential to thoroughly analyze the cmd and arg parameters to help test cases reach deeper kernel logic and bypass barriers during fuzz testing. This process is not only time-consuming, but also prone to human error, as manually crafting system call specifications can overlook critical dependencies and system call interfaces. Moreover, as the Linux kernel continues to evolve, the corresponding system call specifications must be updated to reflect changes in the codebase. Consequently, automating the generation of system call specifications that adhere to Linux kernel rules and can effectively explore deep kernel logic is a critical challenge in advancing fuzz testing techniques.

Current automated specification generation approaches [15] face limitations. Static analysis methods like DIFUZE [16] and SyzDescribe [18] can lack precision or generalizability, while dynamic analysis techniques such as KSG [17] may incur high overhead or require complex setup. As detailed in Sect. 2.2, these shortcomings hinder the effectiveness of kernel fuzzing, particularly in exploring

In summary, the current state of system call specification generation for the Linux kernel requires significant improvement to enhance fuzzing efficiency and coverage. Despite advances, key challenges remain, particularly in accurately capturing complex semantic constraints (such as intricate parameter dependencies or specific 'magic number' values) and systematically ensuring the fuzzing effectiveness of automatically generated specifications. Existing methods often lack mechanisms to precisely derive these deep constraints or to validate and refine specifications based on their actual performance in guiding fuzzers, highlighting the need for more robust and adaptable solutions.

To bridge these gaps, this paper presents SyzForge, a systematic framework focused on Linux kernel drivers that uniquely integrates deep program analysis with a closed-loop validation and refinement process. Our core innovations lie in: (1) employing symbolic execution specifically to extract fine-grained argument constraints, including crucial 'magic numbers' often missed by static analysis (Sect. 4.3), and (2) implementing a novel 'Generate-Validate-Correct' loop where specifications are evaluated based on their fuzzing performance (coverage metrics, Sect. 4.4), feeding this dynamic feedback into an LLM-driven correction stage [8] (Sect. 4.5) for targeted refinement.

SyzForge tackles key challenges in kernel fuzzing by automating the creation of high-quality system call specifications, enabling tools like Syzkaller to explore deeper kernel logic and uncover vulnerabilities more effectively. The evaluation demonstrates notable success: SyzForge achieves a 13.3% increase in code coverage compared to default Syzkaller specifications, outperforming KernelGPT by 4.3%, SyzDescribe by 5.5%, and DIFUZE by 24.3%. Furthermore, our fuzzing campaign using these specifications on Linux kernel version 6.12 [11] yielded 73 crash reports; subsequent analysis and deduplication led to the identification of 19 previously unknown vulnerabilities, underscoring SyzForge's practical utility in enhancing kernel security.

Our contributions are summarized as follows:

- **Dynamic Constraint-Based Argument Range Solving:** We propose a symbolic execution-based approach to extract finer-grained parameter relationships and crucial constant values ('magic numbers'), aiding fuzz testing in bypassing constraints (Sect. 4.3).

- **Fuzzing-Driven Specification Evaluation and Refinement:** We introduce a closed-loop system that evaluates generated specifications based on dynamic fuzzing metrics (coverage, path discovery) and uses this feedback to guide targeted LLM-based correction, ensuring practical effectiveness (Sect. 4.4).
- **Large Language Model (LLM) Assisted Correction:** We leverage advanced LLMs within our feedback loop to analyze and correct specification errors identified during validation, utilizing contextual information to generate optimized specifications (Sect. 4.5).
- **Integrated Framework Implementation:** We design and implement SyzForge, demonstrating the practical feasibility and effectiveness of combining symbolic execution, fuzzing feedback, and LLM refinement for automated specification generation, achieving state-of-the-art results (Sect. 5).

2 Background and Related Work

2.1 Kernel and Device Drivers

The operating system kernel provides essential services to user-space applications, including process management, memory management, file system operations, network communication, and device access. To ensure the overall security of applications and users, interactions between user space and the kernel are strictly governed by system call (syscall) interfaces, which typically follow standardized definitions like POSIX. However, if a system call triggered by a user-space application exploits a kernel vulnerability or causes a kernel crash, it can have severe consequences. Such issues not only threaten all applications relying on the kernel but can also allow attackers to bypass security mechanisms enforced by the kernel, potentially escalating their privileges. Therefore, detecting kernel vulnerabilities through system call interfaces is crucial to maintaining system security.

Device drivers are critical modules within the operating system kernel responsible for interacting with hardware. From the perspective of device drivers, their primary role is to act as a bridge between the hardware and the operating system, providing a unified interface for hardware access while abstracting the complexities of underlying hardware. This enables user-space applications and other kernel subsystems to interact with various hardware devices transparently. To facilitate communication between user space and drivers from different vendors, modern operating systems define specific system calls, such as read, write, open, and ioctl. These system calls are key interfaces for communication between device drivers and user space, offering a standardized way to interact with devices. They can accept data structures specified by the driver as input, allowing for flexible and efficient device management.

2.2   Kernel Fuzzing and Specification Generation

Fuzzing is a highly effective automated technique for vulnerability discovery, generating varied inputs to uncover anomalous program behaviors. While successful in user-space, kernel fuzzing faces unique challenges like complex system call (syscall) interfaces, context-switching overhead, and the risk of system crashes, demanding domain-specific knowledge for effective input generation.

Syzkaller is the state-of-the-art kernel fuzzer, but its effectiveness heavily relies on Syzlang [10] specifications, which describe syscall interfaces, argument types, and basic constraints. Manually creating and maintaining these specifications for the rapidly evolving Linux kernel is a significant bottleneck, demanding considerable expert effort despite Syzkaller's proven success in finding thousands of bugs.

To address this specification bottleneck, various automated or semi-automated approaches have emerged, often analyzing kernel code either statically or dynamically.

Static analysis tools like DIFUZE [16] (using LLVM IR) and SyzDescribe [18] (using programming conventions) automatically generate specifications by analyzing Linux source code. However, these methods can suffer from lack of precision (generating overly broad constraints) or limited generalizability (if specific code conventions are not universal), potentially hindering deep fuzzing exploration.

Alternatively, dynamic analysis, exemplified by KSG [17] for Linux, recovers driver behavior via runtime tracing to generate specifications. While achieving partial automation, this approach often requires complex setup (instrumentation and recompilation) and incurs high runtime overhead, limiting its practical adoption, especially for fuzzing at scale.

Other techniques focus on binary analysis, particularly for complex driver interactions or specific environments. Syzgen++ [20] infers syscall dependencies by applying symbolic execution pattern matching directly on driver binaries, primarily targeting dependency relationships rather than complete specifications or specific value constraints like 'magic numbers'. Tools like Syzgen [15] also employ dynamic and binary analysis but target closed-source macOS drivers, limiting their generalizability to Linux and potentially facing precision challenges compared to source-aware methods like SyzForge.

More recently, Large Language Models (LLMs) have been explored. Kernel-GPT [19] uses LLMs [9] to generate Syzlang by interpreting source code context. While promising for automation, it faces potential LLM inaccuracies [25] and relies on static validation checks, lacking SyzForge's dynamic feedback loop that uses fuzzing coverage to refine specifications for improved effectiveness.

A fundamentally different approach is taken by FuzzNG [21], which avoids explicit specifications altogether via runtime input reshaping using kernel hooks. This significantly reduces upfront effort but inherently limits the ability to satisfy deep, complex semantic constraints that specification-guided fuzzers aim to explore.

Despite these diverse advances, generating specifications that are simultaneously comprehensive, precise, and demonstrably effective at guiding fuzzers deep into kernel code remains challenging. SyzForge aims to bridge these gaps by integrating deep program analysis with a unique fuzzing-driven feedback loop and LLM-based refinement.

3 Motivation

3.1 Overcoming Barriers in Kernel Fuzzing

Figure 2 illustrates the implementation of a simplified kernel driver ioctl function (adapted from real-world code), highlighting limitations inherent in current automated specification generation tools. While state-of-the-art tools for specification generation commonly focus on the cmd parameter–analyzing switch-case statements to infer the logic of functions or sub-functions associated with different cmd values–they frequently neglect the constraints imposed on the arg parameter. In this context, the cmd parameter corresponds to the second argument of ioctl, while the arg parameter refers to the third.

```
1  struct example_object {
2      int A;
3      long long B;
4      unsigned short C;
5  };
6  static long example_ioctl(struct file *file, unsigned int cmd, unsigned long
       arg) {
7      int __user *user_arg = (int __user *)arg;
8      switch (cmd) {
9      case CMD_1: {
10         struct example_object obj;
11         if (copy_from_user(&obj, user_arg, sizeof(struct example_object))) {
12             return -EFAULT;
13         }
14         if (obj.A == MAGIC_NUMBER_1 && obj.B == MAGIC_NUMBER_2) {
15             // deep logic
16         } else {
17             return -EINVAL;
18         }
19         break;
20     }
21     default:
22         printk(KERN_ERR "Unsupported command: %u\n", cmd);
23         return -ENOTTY;
24     }
25     return 0;
26 }
```

Fig. 2. Example of Barriers in Kernel Code

As depicted in Fig. 2, the function copy_from_user copies data from the memory location pointed to by the arg pointer into the obj structure (line 11). This structure comprises three fields: A, B, and C. Function execution proceeds only if the conditions `obj.A == MAGIC_NUMBER_1` and `obj.B == MAGIC_NUMBER_2` are satisfied (line 14). In user-mode fuzzing, such conditional checks are termed "barriers," which demand substantial computational resources to overcome.

Although contemporary user-mode fuzzing tools typically employ symbolic execution to navigate these barriers, Syzkaller, in its current implementation, does not utilize symbolic execution to enhance fuzzing efficiency. Integrating symbolic execution into Syzkaller, which is written in Go, presents considerable technical challenges, and the symbolic execution method itself is often resource-intensive. Consequently, fuzzers like Syzkaller relying on existing specification formats often struggle to effectively handle such barriers.

To overcome this limitation, we propose a novel approach designed to enable Syzkaller to efficiently bypass these barriers in subsequent fuzzing iterations. This method is intended to significantly improve fuzzing efficiency and address the shortcomings of current techniques.

3.2   Leveraging Large Language Models for Specification Optimization

In the system call specification generation process, although existing methods can deeply analyze program logic, we have observed that specifications generated based on simple rules often contain errors or fail to effectively enhance fuzzing tool coverage. To optimize this process, we propose utilizing large-scale language models (LLMs) to refine the generated specifications. Ultimately, we aim to construct a standardized system that improves both specification accuracy and testing efficiency.

Although KernelGPT has used large models to address some related issues, its model lacked deep reasoning capabilities, and its prompt design was insufficiently comprehensive. More critically, the method was unable to assess the quality of generated specifications, preventing effective optimization for specific specifications. Thus, current methods face limitations in specification optimization.

To overcome these challenges, we introduce a new approach that leverages the state-of-the-art open-source model, DeepSeek-r1, to automate the optimization of system call specifications. Furthermore, we employ the Syzkaller fuzzing tool to monitor the coverage of the initial system call specifications, allowing us to identify specifications that require further optimization. This method aims to significantly improve specification quality and coverage, providing more efficient and precise support for subsequent fuzzing efforts.

4   Design and Implement

SyzForge aims to automate the generation of accurate syscall specifications for Linux kernel drivers, which can be directly loaded and used by Syzkaller. In this section, we provide a detailed description of the four stages employed by SyzForge and the key techniques utilized in each. We outline the overall workflow of SyzForge, followed by an in-depth exploration of the specific components it includes.

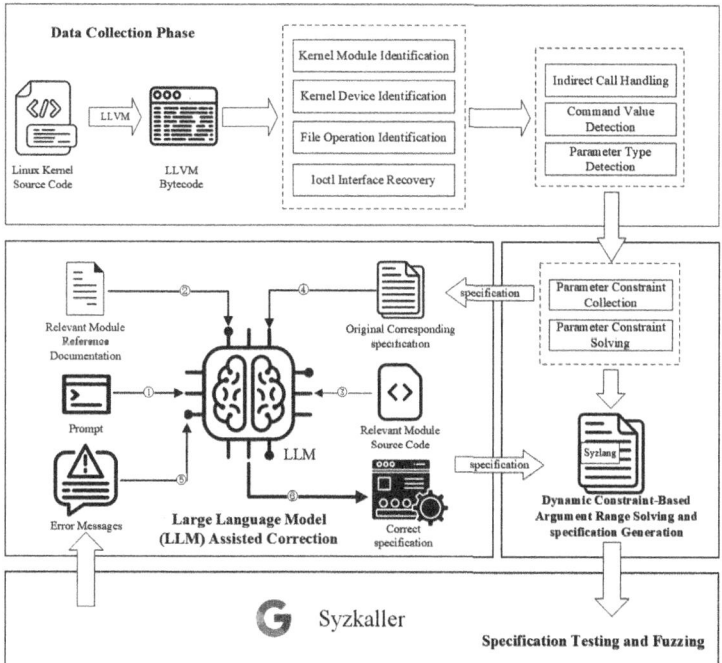

Fig. 3. Automatic Generation System of Linux System Call Specification

4.1   Overview

The SyzForge automated system call specification generation process, as shown in Fig. 3, takes Linux kernel source code as input and incrementally refines it to produce high-quality specifications compliant with Syzlang grammar. These specifications are designed to drive kernel fuzzing effectively. The process comprises four core stages integrated within a feedback loop:

(1) **Data Collection Stage:** Performs static analysis on kernel source code (via LLVM IR [12]) to extract essential metadata, including system call interfaces, parameter types, and device dependencies. This stage establishes the initial static semantic context.
(2) **Symbolic Execution and Specification Generation Stage:** Employs symbolic execution (KLEE [22]) to explore kernel execution paths based on the static context, dynamically inferring constraints and valid values for system call parameters. It generates initial Syzlang specifications from this analysis.

(3) **Specification Testing and Validation Stage:** Validates the generated specifications through dynamic testing, assessing compilability, runtime behavior, and fuzzing coverage achieved when used with Syzkaller. Specifications failing validation are flagged for correction.

(4) **Large Language Model-Assisted Correction Stage:** Leverages an LLM to automatically refine or correct the flagged specifications. It uses the initial specification, error messages, fuzzing feedback, and related code context to generate improved versions, completing the feedback loop.

The core innovation of SyzForge lies in this "Generate - Validate - Correct" feedback loop, which deeply integrates static analysis, symbolic execution, dynamic fuzzing feedback, and large language model reasoning. Unlike traditional approaches that rely on manual expertise to write specifications, SyzForge automates and iteratively optimizes specification generation, aiming to significantly enhance fuzz testing effectiveness and vulnerability discovery. The subsequent sections detail each of these stages.

4.2 Data Collection

In this phase, we adopt the Kernel Development Contracts proposed in the SyzDescribe paper as the core guiding principle to statically analyze LLVM bytecode and reconstruct system call interfaces. Consistent with SyzDescribe, we assume that kernel driver development adheres to the following stable contracts:

1. Module Initialization Order: The loading sequence of modules is determined by the priority defined by macros such as module_init, with a priority of 6 corresponding to loadable modules.
2. Device Object Registration Rules: Driver objects (e.g., struct cdev) and device objects (e.g., struct device) are uniquely associated through device numbers (Major/Minor).
3. File Operation Interface Binding: System call handler functions (e.g., ioctl) are registered to driver objects via the file_operations structure.

Initially, we identify relevant kernel modules based on the module initialization order contract. Next, using the device object registration rules, we employ hash tables to identify associated kernel devices. Following this, we recognize relevant file operations based on the file operation interface binding contract. Finally, utilizing the information obtained from the above steps, we reconstruct the ioctl interfaces.

Subsequently, we apply Multi-Layer Type Analysis (MLTA) to identify indirect calls, thereby assisting in constructing a more comprehensive control flow graph. By analyzing command values, we determine that different cmd values correspond to distinct internal logics. Lastly, we identify the types of arg parameters, as different parameter types have varying valid values, which aids in the subsequent parameter solving process.

The code for the data collection phase is primarily inherited from SyzDescribe, with modifications such as the incorporation of the MLTA [24] method to identify indirect calls. Through debugging the source code, we discovered that the original MLTA method was not utilized. Building upon this, we added the relevant MLTA methods. Additionally, other related articles employing data flow analysis can assist in identifying indirect calls.

This approach enhances the accuracy and efficiency of system call interface reconstruction, providing a solid foundation for subsequent fuzz testing and vulnerability detection.

4.3 Dynamic Constraint-Based Argument Range Solving and Specification Generation

Following the data collection phase, SyzForge determines valid value ranges for the arg parameters of system calls, utilizing extracted data to circumvent obstacles in subsequent fuzzing stages. A primary challenge lies in the intricate nature of kernel parameters, often characterized by complex structures such as nested unions, with valid ranges tightly coupled to execution paths. Traditional static analysis struggles to derive precise solutions in these scenarios, prompting SyzForge to adopt a dynamic constraint-solving approach based on symbolic execution, as detailed in Algorithm 1.

Algorithm 1. Constraint Extraction and Solving Algorithm

Require: $TargetFunction, ValueType, cmd$
Ensure: $ValueRes$
1: **Initialize** $Constraint \leftarrow$ an empty set;
2: $Constraint \leftarrow$ **CollectConstraints**(TargetFunction);
3: $ValueRes \leftarrow$ **Solve**($Constraint, ValueType$);
4: **function CollectConstraints**($TargetFunction$)
5:     **Initialize** $ConstraintSet \leftarrow$ an empty set;
6:     **Initialize** $ValueAffectedIns \leftarrow$ an empty set;
7:     **Identify** relevant $Instructions$ for cmd in $Function$;
8:     $ValueAffectedIns \leftarrow$ **CollectInstructions**($Instructions[cmd]$);
9:     **for** Inst **in** $ValueAffectedIns$ **do**
10:         **if** $IsCmpInst(Inst)$ **then**
11:             $Constraint \leftarrow Constraint+$**ExtractConstraint**($Inst$);
12:         **end if**
13:         **if** $IsCallInst(Inst)$ **then**
14:             $CalleeFunction \leftarrow$ **GetCalleeFunction**($Inst$);
15:             $Constraint \leftarrow Constraint+$**CollectConstraints**($CalleeFunction,$ cmd);
16:         **end if**
17:     **end for**
18:     **return** Constraint
19: **end function**

A key feature of ioctl system calls in kernel modules is the rigid dependency between the command value (cmd) and the argument (arg). Different cmd values trigger distinct control flow branches, each linked to specific arg value ranges. SyzForge exploits this by establishing a dynamic constraint mapping from cmd to arg, enhanced by pre-processed data from the collection phase, which accelerates constraint resolution.

Algorithm 1 outlines the framework, accepting a target function, parameter type, and cmd value as inputs (lines 2-3). It employs recursive constraint collection and resolution to produce effective value domains. The collection phase uses a depth-first search to traverse the target function's control flow graph (CFG) (lines 5–14), leveraging LLVM's Static Single Assignment (SSA) form for cross-procedural data flow tracking. This involves: (1) identifying conditional branches tied to the current cmd and extracting comparison operations (e.g., CMP, TEST) that define arg ranges; (2) conducting context-sensitive analysis of function calls to gather constraints from callees recursively; and (3) applying type inference to resolve ambiguities from pointer aliasing via LLVM cast operations. Constraints from caller and callee functions are merged into a unified set, which the Z3 [23] solver then processes to yield arg value ranges (line 3).

This methodology enables SyzForge to derive path-specific parameter ranges within kernel submodules. For specification generation, it builds on prior work, adhering to the Syzlang grammar to produce initial specifications that adeptly navigate fuzzing barriers. Implementation relies on KLEE for symbolic execution and C++ for framework development and specification creation.

4.4   Specification Testing and Fuzzing

After the initial generation of specifications, they must be seamlessly integrated into the testing process to assess their effectiveness. However, as these specifications are derived from simple rule-based methods, their quality cannot be guaranteed without thorough evaluation. To address this, a rigorous testing phase is essential to validate their suitability for practical use. Specifications that pass the evaluation can be deployed for regular vulnerability discovery, while those that fail are flagged for further refinement. In such cases, the process advances to the next stage, where corrections are made using a large language model to enhance their quality.

To ensure that only high-quality specifications are deployed, establishing precise evaluation criteria is critical. These criteria serve as benchmarks to determine whether a generated specification is viable for testing and deployment. To this end, we have developed a comprehensive set of standards designed to assess the quality and effectiveness of the generated specifications. To formalize the assessment process, we define a specification quality evaluation function, denoted as $Q(Spec)$. This function evaluates specifications based on two key dimensions: static validation and dynamic validation scores. The quality function $Q(Spec)$ is formally expressed as:

$$Q(Spec) = \begin{cases} 1 & \text{if } S(Spec) \geq \theta_s \land D(Spec) \geq \theta_d \\ 0 & \text{otherwise} \end{cases} \tag{1}$$

where:
- $S(Spec)$ is the static validation score.
- $D(Spec)$ is the dynamic validation score.
- θ_s and θ_d are predefined thresholds for static and dynamic scores, respectively, representing minimum acceptable quality levels.

Static Validation Score. The static validation score, $S(Spec)$, is a Boolean function derived from the logical conjunction of two binary criteria:

$$S(Spec) = \prod_{i=1}^{2} s_i(Spec) \tag{2}$$

Each component $s_i(Spec)$ evaluates a specific static property of the specification $Spec$:

$$s_1(Spec) = \begin{cases} 1 & \text{if } Spec \text{ compiles successfully in Syzkaller} \\ 0 & \text{if compilation fails} \end{cases} \tag{3}$$

$$s_2(Spec) = \begin{cases} 1 & \text{if ABI compatibility check passes} \\ 0 & \text{if check fails (type/struct conflicts)} \end{cases} \tag{4}$$

Static validation ensures fundamental specification integrity and compatibility. A specification must satisfy both criteria to achieve $S(Spec) = 1$.

Dynamic Validation Score. The dynamic validation score, $D(Spec)$, provides a quantitative measure of specification effectiveness during runtime. It is defined as a weighted composite evaluation of two key metrics:

$$D(Spec) = \alpha \cdot C(Spec) + \beta \cdot P(Spec) \tag{5}$$

where:

- **Basic Block Coverage** ($C(Spec)$): Quantifies code path exercise, calculated as the ratio of executed basic blocks to the total basic blocks:

$$C(Spec) = \frac{\sum_{b \in B} \delta(b)}{|B|}, \quad \text{where } \delta(b) = \begin{cases} 1 & \text{if basic block } b \text{ is executed} \\ & \text{otherwise} \end{cases} \tag{6}$$

B is the set of all basic blocks, and $|B|$ is the total count. Higher coverage suggests more comprehensive code traversal.
- **New Path Discovery Rate** ($P(Spec)$): Assesses the specification's ability to explore novel execution paths relative to a baseline:

$$P(Spec) = \frac{|Path_{\text{new}}|}{|Path_{\text{new}}| + |Path_{\text{base}}| + \epsilon} \tag{7}$$

$|Path_{\text{new}}|$ is the number of new paths, $|Path_{\text{base}}|$ is the number of baseline paths, and $\epsilon = 10^{-5}$ prevents division by zero. A higher rate indicates stronger exploratory capacity.

In Eq. 6, weights are set to $\alpha = 0.67$ and $\beta = 0.33$, prioritizing coverage and path discovery. Thresholds are $\theta_s = 1$ and $\theta_d = 0.75$. A high-quality specification for fuzzing achieves $S(Spec) = 1$ and $D(Spec) \geq 0.75$. These metrics ensure specifications are syntactically and semantically sound, and effective in kernel code exploration and vulnerability discovery.

4.5   Large Language Model (LLM) Assisted Correction

During the initial phase, system call specifications were generated based on pre-defined rules, which could result in some flawed or even incorrect specifications. Manually correcting these specifications is a time-consuming and labor-intensive process.

With the continuous development of large language models (LLMs), they have gained the ability to perform logical reasoning. By providing appropriate prompts and related information, LLMs can assist in generating corresponding specifications. Based on this, we propose an Algorithm 2 that utilizes large language models to help correct defective specifications, as detailed below.

Algorithm 2. Algorithm for LLM processing data and generating specifications

---

Require: error_message, original_specification
Ensure: LLM_specification
  1: $prompt \leftarrow$ **GenPrompt**($usage\_info, related\_example$);
  2: $related\_document \leftarrow$ **WebScrape**($related\_info, module\_info$);
  3: $related\_code \leftarrow$ **FindCodeInModule**($module\_info$);
  4: $Related \leftarrow related\_document, related\_code$;
  5: $Info \leftarrow Related, error\_message, original\_specification$;
  6: $LLM\_specification \leftarrow$ **QueryLLM**($prompt, Info$);
  7: $Step \leftarrow 0$;
  8: **while** $True$ **do**
  9:     **if** $step > max\_iter$ **then**
 10:         **return** NULL;
 11:     **end if**
 12:     $error\_message \leftarrow$ **SyzCHECK**($LLM\_specification$);
 13:     **if** $error\_message ==$ None **then**
 14:         **return** $LLM\_specification$;
 15:     **end if**
 16:     $LLM\_specification \leftarrow$ **QueryLLM**($Info, error\_message$);
 17: **end while**

The algorithm accepts an error message and an original specification as inputs and produces a corrected specification using the LLM. Initially, it generates a prompt using system call specifications and related examples to guide subsequent processing. Web scraping then extracts relevant system call descriptions and usage guidelines from the Linux Kernel documentation, which the LLM references to generate accurate specifications. Although the DeepSeek LLM now supports real-time web searches, this feature was unavailable during the drafting of this paper. Next, a function retrieves pertinent code from Linux kernel modules using module information, enhancing the dataset.

Subsequently, the algorithm consolidates the related documentation and code into a single set, combining it with the error message and original specification to form a comprehensive input. A query function processes the prompt and this input to generate the corrected specification. The algorithm iterates to ensure the specification's syntactic accuracy, initializing a step counter to 0. In each iteration, it checks whether the step count exceeds the maximum limit, terminating the process if it does. A validation function checks the syntactic correctness of the corrected specification, producing a new error message if syntactic errors persist. If no error message is produced, the process concludes, returning the corrected specification. Otherwise, the updated error message and input refine the specification via the query function, continuing until it passes validation without errors.

The implementation uses Python for web scraping and employs the DeepSeek-r1 LLM via the OpenAI library. This approach automates the correction of flawed system call specifications, significantly enhancing the efficiency and accuracy of specification generation for kernel fuzz testing.

5 Evaluation

To assess the effectiveness of our approach, we designed experiments to investigate the following research questions:

RQ1: Can SyzForge generate high-quality and effective specifications?
RQ2: How effective are the generated specifications in improving the coverage of kernel fuzzers?
RQ3: Can the specifications generated by SyzForge detect real-world vulnerabilities in the kernel?

To address these questions, we conducted evaluations using the following setup. The testing environment consisted of a workstation equipped with a 64-core processor and 512 GB of memory, running Ubuntu 22.04.5 LTS. As our target, we selected a Linux kernel development snapshot based on commit 'adc2186' (from February 2024, during the v6.12 release candidate cycle). The version of Syzkaller used was based on the 'cfe3a04' commit from the Git repository. During the specification generation and evaluation process, we utilized the Syzbot configuration tailored for the Linux kernel. This configuration reflects Google's approach when performing kernel fuzz testing using QEMU. Our fuzz testing configuration employed Syzkaller's default settings, which include 4 QEMU instances, each utilizing 2 CPU cores. To ensure a fair comparison, we disabled the reproduction feature when collecting and comparing coverage results.

To demonstrate the quality and effectiveness of the specifications generated by SyzForge, we compare them against several baselines that, like SyzForge, generate Syzlang specifications for use with Syzkaller: state-of-the-art tools SyzDescribe [18] and DIFUZE [16], the LLM-based approach KernelGPT [19], and the default specifications meticulously crafted for Syzkaller [3]. Comparisons with FuzzNG [21] and Syzgen++ [20] are omitted due to fundamental differences in their approaches, making them less suitable for direct comparison within our evaluation framework focused on Syzkaller specification effectiveness. Specifically, FuzzNG operates on a distinct paradigm avoiding Syzlang specifications via runtime input reshaping, while Syzgen++ primarily focuses on inferring explicit dependencies from binaries using pattern matching.

5.1 Overall Results

This section evaluates the overall quality of system call specifications generated by SyzForge by comparing them against three baselines: SyzDescribe, DIFUZE, and the official Syzkaller specifications (addressing RQ1). Our analysis targets Linux kernel v6.12 (commit adc2186) using the syzbot configuration. Table 1 summarizes the key quantitative metrics for this comparison.

As shown in Table 1, we measure the following key metrics to assess the quality of the generated specifications: (1) #HANDLER represents the number of identified system call handler structures, reflecting the coverage of potential entry points. (2) #NAME denotes the number of extracted device file names associated with interactable device nodes. (3) #CMD&TYPE quantifies the number of valid command and parameter type combinations, directly indicating the modeling depth for diverse operations within interfaces like ioctl. (4) #ARG-NUM counts the number of 'magic numbers' (specific constant values) identified, which are crucial for bypassing specific checks ('fuzzing barriers') in kernel execution paths.

Comparing the #HANDLER metric (Table 1), while DIFUZE reports a high handler count (403), its significantly lower counts for associated device files (#NAME, 33) and valid command-type combinations (#CMD&TYPE, 97) suggest potential false positives. Excluding DIFUZE, SyzForge identifies the most handlers (211), surpassing SyzDescribe (194) and the official Syzkaller set (163).

Next, we examine #NAME (device file names) and #CMD&TYPE (valid command-type pairs), which reflect the breadth of interface detail captured. Table 1 shows SyzForge leads in both metrics (263 names, 2,437 combinations), indicating its ability to generate a more comprehensive set of specifications that model diverse kernel interactions compared to SyzDescribe (235, 2,237) and Syzkaller (241, 1,867).

Finally, we evaluate #ARG-NUM, the count of specific constant values ('magic numbers') identified via symbolic execution, crucial for bypassing fuzzing barriers. SyzForge identified 17 such constants (Table 1), whereas SyzDescribe and DIFUZE found none, and the official specifications contain only 5. Although 17 might seem small relative to the kernel's scale, these constants are highly valuable. Identifying even one allows fuzzers like Syzkaller to overcome conditional checks that often block exploration. Syzkaller can leverage these constants provided in the specification, for instance, by incorporating them into its mutation strategies or directly assigning them to relevant structure fields. This capability can lead to uncovering deeper bugs (further discussed in Sect. 5.2).

Beyond these quantitative comparisons against baselines, we also evaluated the effectiveness of SyzForge's internal stages, particularly the LLM-assisted repair mechanism, when generating specifications for previously undescribed drivers. This evaluation focused on 67 selected drivers for which SyzForge successfully generated 48 initial specifications. As summarized in Table 2, automated checks found 24 of these to be immediately valid according to the quality standards defined in Sect. 4.4. The LLM repair module was then applied to the remaining 24 invalid specifications, successfully repairing 8. This increased the total number of valid specifications to 32 (24 + 8). These results demonstrate SyzForge's capability to handle new drivers and confirm the practical utility of the integrated LLM repair stage in improving specification quality, although the 33% repair success rate (8/24) also points to areas for future improvement in LLM reliability (discussed further in Sect. 6.1).

In conclusion, the overall results highlight SyzForge's advantages. The quantitative metrics (Table 1) indicate that it generates more comprehensive specifications than baseline methods, particularly by identifying barrier-bypassing constants through symbolic execution. Furthermore, the evaluation involving LLM repair (Table 2) demonstrates the practical benefit of this stage in enhancing specification quality, contributing to the overall effectiveness of the SyzForge framework.

Table 1. SyzForge Syscall Description Comparison

Name	Config	#HANDLER	#NAME	#CMD & TYPE	#ARG-NUM
SyzForge	syzbot	211	263	2,437	17
SyzDescribe	syzbot	194	235	2,237	0
DIFUZE	syzbot	403	33	97	0
syzkaller	cfe3a04	163	241	1,867	5

Table 2. Specification generation for the undescribed drivers

Generation		Evaluation and Repair		LLMs Repair
#Total	#Success	#valid	#Valid after repair	#LLMs Success
67	48/67	24/48	32/48	8/24

5.2 Coverage Improvement

To address **RQ2** concerning the effectiveness of generated specifications in improving fuzzer coverage, we conducted comparative experiments. Syzkaller was employed as the fuzzing engine, utilizing specifications generated by Syz-Forge alongside several baselines: KernelGPT, SyzDescribe and DIFUZE, and the default specifications distributed with Syzkaller. These experiments ran on Linux kernel v6.12 (commit adc2186) for 168 h under identical configurations, with results averaged over three runs for reliability. Figure 4 plots the resulting code coverage growth over time.

The coverage trajectories presented in Fig. 4 reveal distinct performance characteristics among the specification sources. SyzForge consistently yielded the highest code coverage throughout the 168-hour period. Specifically, SyzForge achieved a 13.3% higher coverage than the default Syzkaller specifications (yellow curve). It also surpassed KernelGPT (purple curve) by 4.3%, SyzDescribe (red curve) by 5.5%, and DIFUZE (green curve) by a significant 24.3%.

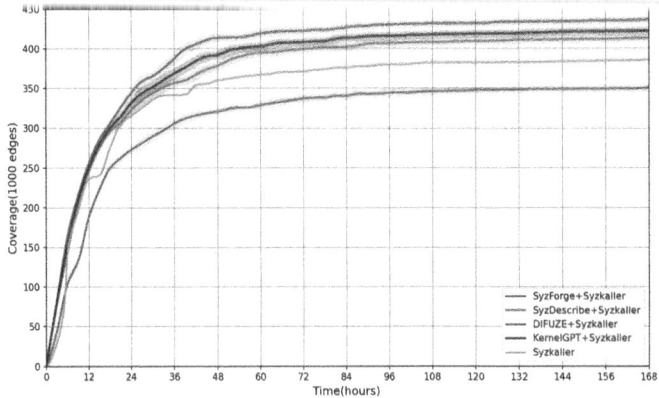

Fig. 4. Comparative code coverage growth achieved by Syzkaller when using specifications generated by different tools over 168 h (Color figure online)

While initial coverage gains were significant for all methods within the first 12 h, their long-term trajectories diverged. DIFUZE consistently showed the lowest performance, likely due to limitations in specification quantity and quality inherent in its static approach (see Table 1). SyzDescribe performed slightly better than the Syzkaller baseline, indicating some value in its automated, convention-based generation, although its effectiveness plateaued compared to top performers.

KernelGPT, utilizing LLMs, notably outperformed both SyzDescribe and the default specifications, demonstrating the potential of LLMs for generating more effective specifications than rule-based or manual methods alone. However, its coverage fell short of SyzForge's. We attribute this to SyzForge's distinct validation and refinement process. Unlike SyzForge's dynamic evaluation loop incorporating fuzzing feedback, KernelGPT appears to lack this mechanism. Consequently, while KernelGPT may generate syntactically valid specifications, some might be less effective in driving fuzzing exploration compared to SyzForge's specifications, which are actively refined based on observed performance.

Indeed, SyzForge achieved the fastest initial growth and maintained the highest coverage throughout the 168 h. While the rate of new coverage discovery slowed for all methods in the latter half of the experiment, SyzForge's sustained lead underscores the efficacy of its integrated methodology. By combining program analysis with a crucial dynamic testing feedback loop and LLM refinement, SyzForge produces specifications demonstrably better optimized for deep kernel exploration.

In summary, these coverage experiments suggest that while LLM-based generation like KernelGPT improves upon prior methods, SyzForge's synergistic approach, featuring dynamic evaluation and feedback, yields the most significant coverage gains in extended kernel fuzzing.

5.3   Kernel Bug Detection

To answer **RQ3**, we conducted a two-week fuzzing experiment on the Linux kernel using Syzkaller equipped with SyzForge-generated specifications. This resulted in the discovery of 73 crash reports. After careful analysis and deduplication against public Syzbot records, we identified 19 potential vulnerabilities that were previously unreported at the time of our experiment. Table 3 provides details of these potential vulnerabilities, including the affected modules, related operations, and their types. These findings were reported privately to the relevant maintainers following responsible disclosure guidelines, several of which have since been confirmed or addressed. Despite Syzkaller's extensive computational resources dedicated to testing the Linux kernel, these 19 potential vulnerabilities had not been captured by prior public fuzzing efforts. The reason SyzForge enabled Syzkaller to uncover these previously undetected issues likely lies in the higher quality and semantic richness of the specifications it generates. By providing better domain knowledge, Syzkaller was able to explore deeper into the kernel's codebase. This result clearly demonstrates that specifications automatically generated by SyzForge can significantly improve the vulnerability detection capabilities of fuzzers like Syzkaller.

Case Study: SyzForge assisted fuzzers in discovering a vulnerability within the HFS+ [14] filesystem driver, a hierarchical filesystem implementation in the Linux kernel. The bug is triggered during metadata cleanup operations, such as file deletion or renaming, and results in slab-out-of-bounds memory accesses due to improper management of BTree nodes. The root cause lies in the HFS+ driver's failure to validate BTree node indices and memory boundaries when reading node data. Specifically, during metadata cleanup (e.g., deleting file attributes or catalog entries), the driver attempts to access a non-existent B*Tree node, which occurs when the node index is corrupted or miscalculated during tree traversal. Additionally, the hfsplus_bnod_read function reads data from node->page[] without verifying whether the requested offset (off) and length (len) exceed the node's allocated memory (node->total_len). As a result, a read

Table 3. Bugs detected by SyzForge

Module	Operation	Type	Unique	Confirmed	Fixed
Z3fold	do_compact_page	Memory Corruption	✓	✓	✓
Em28xx	em28xx_init_extension	Memory Corruption	✓		
Netdevsim	nsim_fib_event_work	Memory Corruption	✓		
Bcachefs	crypto_skcipher_encrypt	NULL pointer dereference	✓		
Btrfs	lo_complete_rq	Memory Corruption	✓		
HFS	hfs_find_init	NULL pointer dereference	✓		
HFS	hfsplus_rename	out-of-bounds Read	✓		
JFS	dbAllocBits	Read Out of bounds	✓		
JFS	release_metapage	Use-After-Free	✓		
JFS	txEnd	Use-After-Free	✓	✓	
InhniBand	__ethtool_get_link_ksettings	Use-After-Free	✓		
list	__bpf_lru_node_move	Memory Corruption	✓	✓	✓
Gfs2	gfs2_invalidate_folio	Use-After-Free	✓		
HDM	hdm_disconnect	Use-After-Free	✓	✓	✓
Ocfs2	ocfs2_write_begin_nolock	Use-After-Free	✓		
OverlayFS	ovl_lookup_positive_unlocked	NULL pointer dereference	✓		
DRM	drm_mode_vrefresh	divide error	✓		
IP_tunnel	sk_skb_reason_drop	NULL pointer dereference	✓		
NTFS3	chrdev_open	Use-After-Free	✓		

operation targeting an invalid node may access 40 bytes beyond the allocated 152-byte kmalloc-192 region, triggering a KASAN error.

6   Limitation and Future Work

6.1   Limitations

The design and evaluation of SyzForge reveal several inherent limitations that require attention:

Dependency on Kernel Source Code: SyzForge relies on access to kernel source code for static analysis and symbolic execution. This dependency limits its applicability to open-source kernels, such as Linux, and excludes proprietary or closed-source systems where the source code is unavailable.

LLM Reliability: The correction phase depends on the DeepSeek-r1 large language model (LLM), which may occasionally produce incorrect or suboptimal suggestions due to limitations in reasoning or training data. For example, the paper reports that only 8 out of 24 invalid specifications were successfully repaired, suggesting potential inconsistencies in LLM performance. Improving model reliability through fine-tuning or additional validation mechanisms remains an ongoing challenge.

Limited Evaluation Scope: The experiments were limited to the Linux kernel, specifically version 6.12. While this version is a critical target, the exclusion of other operating systems (e.g., FreeBSD, Windows) and kernel variants restricts the insights into SyzForge's generalizability. Furthermore, the 168-hour coverage tests and two-week fuzzing period may not fully capture long-term performance or edge cases.

6.2   Future Work

Future research could address these limitations and further develop SyzForge:

Scalability Enhancements: Optimizing resource usage, for example, by replacing KLEE with a lighter symbolic execution tool or streamlining LLM queries, could enhance SyzForge's applicability.

Broader Testing: Evaluating SyzForge on additional kernels (e.g., FreeBSD) and configurations would validate its versatility and uncover new use cases.

Improved LLM Integration: Enhancing LLM reliability through domain-specific fine-tuning or ensemble methods could improve the success rate of corrections, reducing the 16 out of 24 unaddressed invalid specifications.

Dynamic Adaptation: Incorporating runtime feedback or hybrid static-dynamic analysis could improve SyzForge's ability to handle complex, context-dependent barriers.

7   Conclusion

In this paper, we propose SyzForge, a system that synthesizes the strengths of existing methodologies to construct an automated specification generation framework based on a closed-loop generate-verify-revise approach. SyzForge utilizes symbolic execution to help bypass barriers and employs the latest deepseek-r1 large language model for inference and revision. We conduct a comprehensive evaluation of SyzForge and analyze the root causes of discrepancies between SyzForge's results and the ground truth, providing insights for future improvements in automated syscall description generation. Results indicate that SyzForge enhances Syzkaller's coverage and detects 19 previously unknown bugs in the 6.12 Linux kernel. We plan to make SyzForge publicly available as opensource software in the future to encourage further development and adoption by the community.

Acknowledgments. We sincerely appreciate the guidance from our shepherd and the valuable comments from the anonymous reviewers which helped improve this paper significantly. This work was supported by the Natural Science Foundation of Henan Province of China (Grant No. 242300420698) and the Key R&D Project in Henan Province of China (Grant No. 221111210300). We sincerely thank the Syzkaller development team for their fundamental contributions to kernel fuzz testing. We also thank the research team members at the University of Information Engineering for their technical support and Prof. Dong for providing constructive feedback on the experimental validation. Finally, we appreciate the advances in large-scale language modeling that inspired our mutation strategy.

References

1. The Linux Kernel Archives. https://www.kernel.org/
2. Syzbot. https://syzkaller.appspot.com/upstream/
3. Google: Syzkaller. GitHub Repository. https://github.com/google/syzkaller/
4. NCC Group: TriforceLinuxSyscallFuzzer. GitHub repository (2016). https://github.com/nccgroup/TriforceLinuxSyscallFuzzer
5. Oracle: Kernel-Fuzzing. https://github.com/oracle/kernel-fuzzing
6. Wikipedia contributors. Virtual file system. https://en.wikipedia.org/wiki/Virtual_file_system/
7. The Linux Kernel Archives: usbmon documentation. https://docs.kernel.org/usb/usbmon.html
8. DeepSeek: DeepSeek-R1. GitHub Repository. https://github.com/deepseek-ai/DeepSeek-R1
9. OpenAI: GPT-4 Documentation. https://platform.openai.com/docs/models/gpt-4-and-gpt-4-turbo
10. Syzlang Syntax Documentation. https://github.com/google/syzkaller/blob/master/docs/syscall_descriptions_syntax.md
11. The Linux Kernel Archives: Linux Kernel v6.12 Source Code. https://elixir.bootlin.com/linux/v6.12/source
12. LLVM Developer Group: Clang Static Analyzer. https://clang-analyzer.llvm.org/
13. The FreeBSD Project. https://www.freebsd.org/
14. Wikipedia contributors: HFS Plus. https://en.wikipedia.org/wiki/HFS_Plus
15. Chen, W., Wang, Y., Zhang, Z., et al.: SyzGen: automated generation of Syscall specification of closed-source macos drivers. In: Proceedings of the 2021 ACM SIGSAC Conference on Computer and Communications Security (CCS), Virtual Event, South Korea, pp. 749–763. ACM (2021)
16. Corina, J., Machiry, A., Salls, C., et al.: DIFUZE: interface aware fuzzing for kernel drivers. In: Proceedings of the 2017 ACM SIGSAC Conference on Computer and Communications Security (CCS), Dallas, TX, USA, pp. 2123–2138. ACM (2017)
17. Sun, H., Shen, Y., Liu, J., et al.: KSG: augmenting kernel fuzzing with system call specification generation. In: Proceedings of the 2022 USENIX Annual Technical Conference (ATC), Carlsbad, CA, USA, pp. 351–366. USENIX Association (2022)
18. Hao, Y., Li, G., Zou, X., et al.: SyzDescribe: principled, automated, static generation of syscall descriptions for kernel drivers. In: Proceedings of the 2023 IEEE Symposium on Security and Privacy (SP), San Francisco, CA, USA, pp. 3262–3278. IEEE (2023)
19. Yang, C., Zhao, Z., Zhang, L.: KernelGPT: enhanced kernel fuzzing via large language models. arXiv preprint arXiv:2401.00563 (2023)
20. Chen, W., Hao, Y., Zhang, Z., et al.: SyzGen++: dependency inference for augmenting kernel driver fuzzing. In: Proceedings of the 2024 IEEE Symposium on Security and Privacy (SP), San Francisco, CA, USA, pp. 4661–4677. IEEE (2024)

21. Bulekov, A., Das, B., Hajnoczi, S., et al.: No grammar, no problem: towards fuzzing the linux kernel without system-call descriptions. In: Proceedings of the Network and Distributed System Security Symposium (NDSS), San Diego, CA, USA (2023)

22. Cadar, C., Dunbar, D., Engler, D.: KLEE: unassisted and automatic generation of high-coverage tests for complex systems programs. In: 8th USENIX Symposium on Operating Systems Design and Implementation (OSDI 2008), San Diego, CA, pp. 209–224. USENIX Association (2008)

23. de Moura, L., Bjørner, N.: Z3: an efficient SMT solver. In: Ramakrishnan, C.R., Rehof, J. (eds.) Tools and Algorithms for the Construction and Analysis of Systems (TACAS 2008). LNCS, vol. 4963, pp. 337–340. Springer, Heidelberg (2008). https://doi.org/10.1007/978-3-540-78800-3_24

24. Lu, K., Hu, H.: Where does it go? Refining indirect-call targets with multi-layer type analysis. In: Proceedings of the 2019 ACM SIGSAC Conference on Computer and Communications Security (2019)

25. Bang, Y., Cahyawijaya, S., Lee, N., et al.: A multitask, multilingual, multimodal evaluation of chatGPT on reasoning, hallucination, and interactivity. arXiv preprint arXiv:2302.04023 (2023)

Poster: Machine Learning for Vulnerability Detection as Target Oracle in Automated Fuzz Driver Generation

Gianpietro Castiglione$^{(\boxtimes)}$, Marcello Maugeri, and Giampaolo Bella

University of Catania, Catania, Italy
{gianpietro.castiglione,marcello.maugeri}@phd.unict.it,
giampaolo.bella@unict.it

Abstract. In vulnerability detection, machine learning has been used as an effective static analysis technique, although it suffers from a significant rate of false positives. Contextually, in vulnerability discovery, fuzzing has been used as an effective dynamic analysis technique, although it requires manually writing fuzz drivers. Fuzz drivers usually target a limited subset of functions in a library that must be chosen according to certain criteria, e.g., the depth of a function, the number of paths. These criteria are verified by components called *target oracles*. In this work, we propose an automated fuzz driver generation workflow composed of: (1) identifying a likely vulnerable function by leveraging a machine learning for vulnerability detection model as a target oracle, (2) automatically generating fuzz drivers, (3) fuzzing the target function to find bugs which could confirm the vulnerability inferred by the target oracle. We show our method on an existing vulnerability in LIBGD, with a plan for large-scale evaluation.

Keywords: Vulnerability Detection · Vulnerability Discovery · Fuzzing · Machine Learning · Automated Security Testing · Large Language Models

1 Introduction

In recent years, two automated techniques have emerged for finding 0-day vulnerabilities in functions: *Machine Learning for Vulnerability Detection (ML4VD)* [4] and *Fuzzing* for vulnerability discovery.

ML4VD models, when trained on large datasets, are employed to determine whether a given set of functions may exhibit specific vulnerabilities. However, the method involves static analysis, which cannot verify the vulnerability at runtime, may suffer from a significant false-positive rate [2], and ultimately, top-performing models may not be able to differentiate between vulnerable functions and patched functions [4].

G. Castiglione and M. Maugeri—These authors contributed equally.

M. Egele et al. (Eds.): DIMVA 2025, LNCS 15747, pp. 140–145, 2025.
https://doi.org/10.1007/978-3-031-97620-9_8

On the other hand, fuzzing employs dynamic analysis, reducing false positives from static analysis. However, no push-button fuzzing technique exists, as a *fuzzer* requires a *fuzz driver*, a test harness for parsing inputs and invoking the target function that, in turn, requires deep knowledge about the target function and the corresponding library and extensive manual work [1]. *Automated Fuzz Driver Generation (AFDG)*, despite its challenges, solves the burden, with OSS-FUZZ-GEN [3] standing on top due to the use of Large Language Models (LLMs). Contextually, a library could include several functions, and *target oracles* are employed to identify *interesting* functions [5], i.e. functions determined to be likely interesting targets. For example, OSS-FUZZ-GEN prioritises functions relying on FUZZ INTROSPECTOR's[1] heuristics, which mainly consider the cyclomatic complexity of the target function or the simplicity to generate the fuzz driver. Nevertheless, these heuristics do not account for the likelihood of a function being vulnerable. Hence, we propose a combined method employing ML4VD models as a target oracle to prioritise relevant functions for the AFDG.

Considering these assumptions, this study is based on the following research questions:

RQ1 Can machine learning for vulnerability detection be an effective target oracle for automated fuzz driver generation?
RQ2 To what extent can such a combined method confirm the true positives, and/or reduce the number of false positives?

In this work, we present the design and workflow that combines the two techniques in Sect. 2. We validate the method by selecting a confirmed vulnerable function from the DIVERSEVUL dataset and successfully applying OSS-FUZZ-GEN to generate a fuzz driver that triggers the vulnerability. The target function originates from a project already included in the OSS-FUZZ infrastructure[2], ensuring compatibility with OSS-FUZZ-GEN, but is not currently covered by an existing fuzz target. This allows us to generate a novel fuzz driver and achieve previously unreached code coverage. The complete experimental setup is detailed in Sect. 3. Subsequently, we examine the state-of-the-art, discussing insights or differences from our technique in Sect. 4. Finally, we discuss our practical contribution and propose a research plan for an in-depth evaluation in Sect. 5.

2 Design

To address the research questions, the proposed design employs two main techniques: a ML4VD model as the vulnerability detection component (static analysis of the target function code), and AFDG, for ultimately applying fuzzing as the discovery component (dynamic analysis that uncovers vulnerabilities during execution of the target function).

The overall workflow of the proposed method, illustrated in Fig. 1, comprises three steps. It represents a generalised pipeline for AFDG and emphasises the

---

[1] https://github.com/ossf/fuzz-introspector.
[2] https://github.com/google/oss-fuzz.

main contribution of this work. Namely, the use of the ML4VD model as a target oracle.

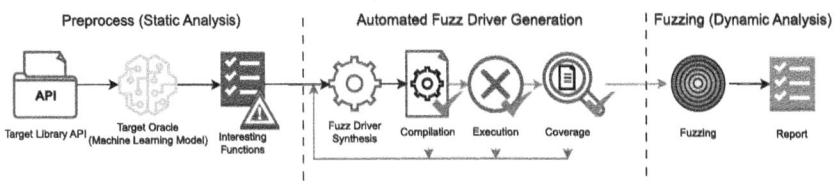

Fig. 1. Workflow of the proposed method

Preprocess (Static Analysis). Our method first identifies functions in the target library API that are likely to be vulnerable. The objective is accomplished by an ML4VD model, which performs a static analysis of each function and flags those that may present one or more weaknesses according to the *CWE* framework by MITRE. Ultimately, the flagged functions are considered potentially interesting functions for further analysis by fuzzing.

Automated Fuzz Driver Generation. Subsequently, the interesting functions are selected for the AFDG process to begin. This is an iterative process in which the fuzz driver synthesis produces a candidate fuzz driver. The candidate must (1) compile, (2) execute without immediate logical failure, and (3) gather sufficient coverage to ensure the input is correctly injected. Once these steps are met, the candidate proceeds to the next stage of the workflow.

Fuzzing (Dynamic Analysis). At this stage, the fuzzing process begins, leveraging the previously generated fuzz driver to inject the target function with a large volume of malformed inputs. At the expiration of the time budget, the fuzzer either discovers a crashing input that confirms the vulnerability or does not.

3   Case Study: LIBGD Library

Target Selection. To evaluate the design of the proposed method, we conducted initial experiments on the LIBGD[3] library. In particular, we assumed to have already a ML4VD model trained on the DIVERSEVUL dataset [2], which is considered the best collection of vulnerable functions in C/C++.

Subsequently, we selected among the projects in DIVERSEVUL one already included in the OSS-FUZZ infrastructure and with FUZZ INTROSPECTOR reports available. We used such requirements to identify functions currently not covered by existing fuzz drivers in such a project. Consequently, we directly take an uncovered function labeled with CWEs as most likely classified as vulnerable, belonging to the dataset itself.

---

[3] https://github.com/libgd/libgd.

In particular, we chose the function *gdImageWebpPtr*, labelled as weak to **CWE-415: Double Free**[4], which is confirmed by **CVE-2016-6912**[5].

Experimental Setting. Subsequently, we employed OSS-FUZZ-GEN to generate a fuzz driver. Initially, OSS-FUZZ-GEN relies on FUZZ INTROSPECTOR to retrieve the target function signature and its corresponding arguments, which are provided in YAML format, as illustrated in Fig. 2. Since the target function is selected by the target oracle, we assume that our ML4VD model has flagged the *gdImageWebpPtr* function as potentially vulnerable.

```
1    functions:
2       - name: "gdImageWebpPtr"
3    params:
4       - name: "im"
5           type: "gdImagePtr"
6       - name: "size"
7           type: "int *"
8    return_type: "void *"
9    signature: "BGD_DECLARE(void *) gdImageWebpPtr (gdImagePtr im, int *size)"
10   language: "c++"
11   project: "libgd"
12   target_name: "gd_webp_fuzzer"
13   target_path: "gd_webp_fuzzer.cc"
```

Fig. 2. Function Signature of *gdImageWebpPtr*

After this minimal setup, the AFDG process starts with the creation of a prompt from a template to be provided to a Large Language Model (LLM). The prompt template is shown in Fig. 3. In particular, we used the standard prompt template provided in the official repository of OSS-FUZZ-GEN; the only difference is the embedded information about the target's weaknesses we added in the prompt, which is shown in red.

In our initial experiments, we employed *GPT-4* with a temperature setting of 0. Although the model successfully generated valid fuzz drivers on first attempts, the fuzz drivers struggled to find a bug and ultimately confirm the vulnerability. This is because the specific vulnerability can be detected whether *gdImageWebpPtr* is invoked by passing a sufficiently large image. Instead, the fuzz drivers presented a hard-coded limit cap on the size of the image, likely to prevent out-of-memory errors during the fuzzing campaign. To solve the issue, we instructed the model to allow large images without setting a cap. From the initial experiments, we can conclude that if a valid fuzz driver fails to find a bug, this does not rule out a vulnerability, and further instructions based on the expected vulnerability could be needed. Ultimately, our contribution aims to (1) prioritise functions likely to expose a weakness, (2) confirm a vulnerability whenever a critical bug is found.

---

[4] https://cwe.mitre.org/data/definitions/415.html.

[5] https://nvd.nist.gov/vuln/detail/CVE-2016-6912.

OSS-Fuzz-Gen Prompt

1 - System Prompt

You are a security testing engineer who wants to write a C++ program to discover memory corruption vulnerabilities in a given function-under-test [...]

2 - C++ Specific Instructions

Use <code>FuzzedDataProvider</code> to generate these inputs [...]

3 - Instructions and Examples

[...] Do not create new variables with the same names as existing variables. WRONG:
<code>
int LLVMFuzzerTestOneInput(const uint8_t *data, size_t size) {
void* data = Foo();
} </code> [...]

4 - Problem Statement

Your goal is to write a fuzzing harness for the provided function-under-test signature [...] <function signature>
Note: The function is a candidate for the vulnerability CWE-415 (Double Free)

Fig. 3. OSS-Fuzz-Gen prompt (Color figure online)

4   Related Works

Risse *et al.* [4] have shown that top-performing ML4VD models are unable to distinguish between functions that contain a vulnerability and functions where the vulnerability is patched. Consequently, without a definitive solution, we expect an increase in false positives over time, which could be mitigated by our method. The state-of-the-art employs *Directed Greybox Fuzzing (DGF)* to steer the generation of inputs that can reach a specific program location. Zhu *et al.* [7], Yu *et al.*. [6], already employed ML4VD as target oracle for DFG. However, DFG itself does not ensure that the target function can be reached. For example, a compiled binary could never call a library function. Our work employs automated fuzz driver generation to generate fuzz drivers which call the target function.

5   Considerations and Future Works

Contributions. FUZZ INTROSPECTOR uses two heuristics to determine functions likely to be interesting targets. However, the heuristics fall short of considering interesting only functions having high cyclomatic complexity and containing *parse* in their name (Heuristic 1), and accepting the same argument types as the fuzzing interface *LLVMFuzzerTestOneInput* (Heuristic 2). Our ultimate aim is to propose a third heuristic, in which an ML4VD model identifies potentially vulnerable functions.

Threats to Validity. As the case study involves a CVE from 2015, this may raise a question about whether the code to reproduce such vulnerability is memorised by the model from the training data. Future studies will focus on evaluating the effectiveness of novel candidate functions.

Future Work. Future work primarily focuses on applying the implemented method to a broader range of functions. Our plan encompasses the integration of an ML4VD model as a target oracle (third heuristic) into FUZZ INTROSPECTOR and evaluating its effectiveness on at least ten projects from both OSS-FUZZ and DIVERSEVUL, relying on OSS-FUZZ-GEN for AFDG.

In particular, to answer **RQ1**, we plan to evaluate our target oracle in novel interesting functions, i.e. not included in the dataset.

To answer **RQ2**, we will focus on functions in a test set, measuring the precision of both the target oracle and, ultimately, the AFDG process.

References

1. Babi, D., et al.: Fudge: fuzz driver generation at scale. In: Proceedings of the 2019 27th ACM Joint Meeting on European Software Engineering Conference and Symposium on the Foundations of Software Engineering, pp. 975–985. Association for Computing Machinery (2019). https://doi.org/10.1145/3338906.3340456
2. Chen, Y., Ding, Z., Alowain, L., Chen, X., Wagner, D.: DiverseVul: a new vulnerable source code dataset for deep learning based vulnerability detection. In: Proceedings of the 26th International Symposium on Research in Attacks, Intrusions and Defenses, RAID 2023, pp. 654–668. Association for Computing Machinery, New York (2023). https://doi.org/10.1145/3607199.3607242
3. Liu, D., Chang, O., Metzman, J., Sablotny, M., Maruseac, M.: OSS-Fuzz-Gen: Automated Fuzz Target Generation (2024). https://github.com/google/oss-fuzz-gen
4. Risse, N., Böhme, M.: Uncovering the limits of machine learning for automatic vulnerability detection. In: Proceedings of the 33rd USENIX Conference on Security Symposium, SEC 2024. USENIX Association (2024)
5. Weissberg, F., et al.: SoK: Where to fuzz? Assessing target selection methods in directed fuzzing. In: Proceedings of the 19th ACM Asia Conference on Computer and Communications Security, ASIA CCS 2024, pp. 1539 1553. Association for Computing Machinery, New York (2024). https://doi.org/10.1145/3634737.3661141
6. Yu, L., Lu, Y., Shen, Y., Li, Y., Pan, Z.: Vulnerability-oriented directed fuzzing for binary programs. Sci. Rep. **12** (2022). https://api.semanticscholar.org/CorpusID: 247407573
7. Zhu, X., et al.: DeFuzz: deep learning guided directed fuzzing. CoRR **abs/2010.12149** (2020). https://arxiv.org/abs/2010.12149

Side Channels

Reverse-Engineering the Address Translation Caches

Philipp Ertmer, Robert Dumitru[(⊠)], and Yuval Yarom

Ruhr University Bochum, Bochum, Germany
{philipp.ertmer,robert.dumitru,yuval.yarom}@rub.de

Abstract. The address translation process and the responsible memory management unit (MMU) in modern CPUs have been the subject of multiple recent microarchitectural side-channel attacks. A precondition to many of these attacks is familiarity with the intimate details of the microarchitectural implementation of the process. However, because vendors do not typically publish extensive information on this, attackers must resort to reverse engineering techniques. Indeed, past works have investigated such techniques, providing insights and novel understanding on the implementation of components used in the address translation process.

In this work, we improve this understanding. We extend the cache desynchronization technique of Tatar et al., and apply it to the page translation caches, which store partial address translation information. We develop automated tooling for investigating five generations of Intel processors, ranging from Haswell to Alder Lake. Our investigations correct mistakes in prior publications, identify a cache level that was missed so far, and discover two hitherto unknown replacement policies. This new understanding of address translation can increase attack precision and facilitate better address-translation-based attacks.

1 Introduction

Microarchitectural side-channel attacks have emerged as a significant threat in the field of cyber security. Such attacks have been demonstrated to leak cryptographic keys [11,12,24,34,38], break address space layout randomization (ASLR) [10,13,15,17,41], and even compromise entire systems [7,19,20, 29,33,39]. As many of these techniques exploit the CPU data and instruction caches to infer sensitive information, many software- and hardware-based countermeasures have been proposed to protect these components from side-channel attacks [8,14,21,31,42]. In turn, several follow-up works have demonstrated limitations in these countermeasures, because they fail to consider microarchitectural components other than memory caches [3,12,25,27,34,37,40].

To stay ahead in this ongoing challenge, attackers and defenders need to inform themselves about the underlying microarchitecture. One possibility is to consult the official resources provided by CPU vendors [6,18]. However, these are often vague or incomplete due to their proprietary nature. Thus, researchers

© The Author(s), under exclusive license to Springer Nature Switzerland AG 2025
M. Egele et al. (Eds.): DIMVA 2025, LNCS 15747, pp. 149–168, 2025.
https://doi.org/10.1007/978-3-031-97620-9_9

often must resort to reverse engineering in order to recover the implementation details of targeted components.

One of the components that has recently garnered attention is the memory management unit (MMU), which is responsible for translating virtual memory addresses to the corresponding physical memory addresses. Two notable works in this direction are those of Gras et al. [12] and Tatar et al. [32], which reverse-engineer the translation lookaside buffer (TLB), a cache implemented by the MMU. Their insights enable attackers to bypass countermeasures against microarchitectural attacks [12,34,39] and efficiently execute attacks targeting the MMU [32].

However, several other components that implement address translation in modern CPUs have attracted much less attention. In particular, the MMU includes *translation caches*, which cache partial results of address translations. While these play a critical role in the address translation process and for many attacks targeting the MMU [13,34,41], their pertinent implementation details are still largely undocumented. Van Schaik et al. [35] investigated these structures, but the information they provide is partial, and has not been corroborated by independent studies. Understanding the exact structure and behavior of these components is crucial for finding potential vulnerabilities as well as for implementing effective defenses.

Contributions

In this work we close the gap and improve the overall understanding of the MMU. We adapt the TLB desynchronization approach of Tatar et al. [32] and apply it to reverse-engineer translation caches. We build an automated tool called *Talbot*, for reverse engineering translation caches,[1] which we use to evaluate six different Intel microarchitectures. Talbot recovers multiple properties of translation caches, including their size, hash function, replacement policy, and more.

Our adapted desynchronization approach allows us to reduce noise and produce more precise and consistent results than were previously possible. With this increased precision, we correct some minor mistakes in the information published in past works and identify that, contrary to past publications, the Skylake microarchitecture does feature four levels of translation caches.

Additionally, our reverse-engineering effort identifies two unpublished replacement policies used on Intel processors. The first, which we term *hit-updated PLRU*, is a variant of tree-based PLRU where the tree structure is only updated on cache hits but not on cache misses. Consequently, entries that are not used tend to be quickly replaced. The second, which we term *most-recently hit*, is a variant of the most-recently used policy where the entry that experienced the most recent hit is the first candidate for replacement.

In summary, this work makes the following contributions:

- We adapt TLB desynchronization for the purpose of reverse-engineering translation caches (Sect. 3).

---

[1] Talbot is open-source, available at https://github.com/0xADE1A1DE/Talbot.

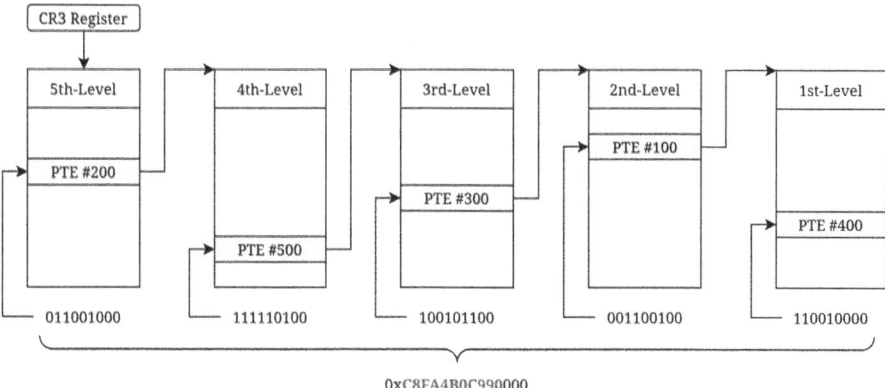

Fig. 1. The page table walk performed by a MMU to translate virtual address 0xC8FA4B0C990000 to its corresponding physical page. The most-significant 9-bit slice is used to index PTE number 200 in the fifth-level page table, which points to the fourth-level page table. The next 9-bit slice is then used to index entry 500 in the fourth-level page table, and so forth.

- We design and implement Talbot, an automated tool for investigating translation caches (Sect. 5).
- Using Talbot, we reverse-engineer the translation caches of six different Intel microarchitectures, identifying new structures and replacement policies (Sect. 4).

2   Background

In this section, we provide background on address translation, cache-memory desynchronization, and cache replacement policies.

2.1   Address Translation

Modern processors use memory virtualization to simplify and optimize system memory management. This abstraction presents each process running on a machine with its own dedicated, contiguous blocks of virtual address space to operate on. Such functionality is made possible by dynamic translation of virtual addresses to physical addresses. The memory management unit (MMU) is the dedicated hardware component responsible for managing these translations.

The high level translation process is generally well documented by CPU manufacturers. Translation information is stored in main memory in the form of *page tables*, which are structured as a multilevel directed tree. Each page table is indexed by a specific portion of the virtual address. A page table entry (PTE) points to either the next (lower) level page table or the requested physical page.

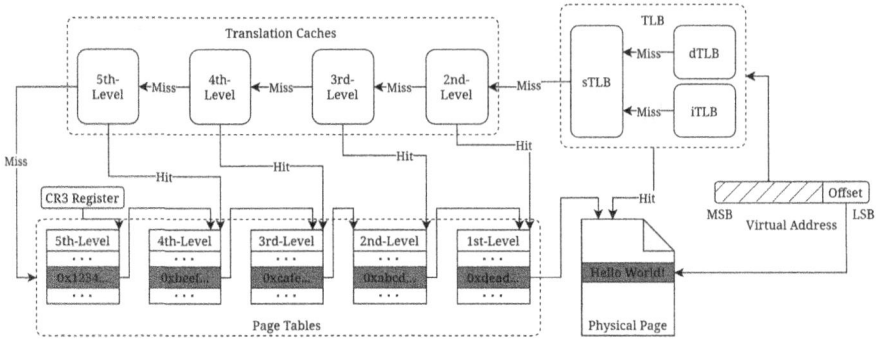

Fig. 2. Schema of the address translation process, showing the interaction of translation caches and the page table hierarchy.

To translate an address, the MMU traverses these page table structures in what is known as a *page table walk*. Figure 1 depicts a page table walk for a system that uses five-level paging, as implemented in the x86_64 architecture. First, the MMU finds the physical address of the root page table by reading the CR3 register. The prefix of the virtual address is split into five 9-bit slices, which index the page tables of successive lower levels. The MMU begins the page table walk by using the most significant 9-bit slice to index the root (or fifth-level) page table, selecting a PTE that points to the next lower-level page table. The process continues iteratively until the page with the requested data is found. The remaining address bits are used as an offset within that page.

2.2 MMU Caches

Regularly performing multi-level page table walks that retrieve each entry from physical memory can severely degrade system performance. To avoid this where possible, CPUs cache recently resolved addresses in a dedicated structure called TLB. That is, the TLB stores the PTEs of recently translated addresses, allowing the MMU to avoid the page table walk for these addresses. While this avoids most walks, TLB misses can still add significant overhead [6, 23].

To reduce the overhead of TLB misses, processors cache partial address translation information. One microarchitectural design for implementing this, and the focus of this paper, is with translation caches [5], which store higher level PTEs that match prefixes of recently resolved addresses. The MMU uses a prefix of the virtual address to index these caches [18], allowing it to avoid the initial part of the page table walk and resume the walk from the cached location. As depicted in Fig. 2, the MMU first looks up the virtual address in the TLB. In the case of a hit, it uses the address to bypass the whole page table walk. Otherwise, the MMU proceeds to progressively higher levels of translation caches, until a match is found from where it then starts the page table walk. In practice, the MMU typically queries the translation caches for all levels in parallel, and uses the result of the most specific (lowest level) hit found.

Alternative designs for caching partial address translation information also exist, in AMD Opteron processors for example, the page walk cache is a fully associative cache which stores the PTE [6], indexed by its physical address.

2.3 Desynchronization

To reverse-engineer the TLB, Tatar et al. [32] use desynchronization, which is an approach adapted from cache storage channels [16] for address translation. The core idea of cache storage channels is to break cache coherency, creating a scenario wherein a cached value differs from its corresponding memory contents. Reading the desynchronized data then signals a cache hit or miss, depending on whether the cached value or stored memory content is read. This provides a mechanism to observe whether certain entries have been evicted from the TLB.

Persistent desynchronization is possible because the TLB does not enforce memory coherency. Therefore, when Tatar et al. overwrite a PTE in one of the first-level page tables stored in memory, the corresponding TLB entries are not subsequently invalidated or changed. With this, they directly change the physical address to which the virtual address translates to. If the desynchronized entry is present in the TLB, translating an address using the corresponding PTE results in its original physical address. However, if the entry has been evicted from the TLB, the same virtual address translates to a different physical address. By selecting physical addresses with different contents, TLB hits and TLB misses can be distinguished.

2.4 Replacement Policies

When placing new data into a full cache, the CPU selects which entry to replace based on a given *replacement policy*. Ideally, the replacement policy chooses the entry for eviction that will be used most distantly in the future [9]. Replacement policies use past access behavior to take a best guess at this.

Permutation policies [1] are a popular class of replacement policies that maintain an eviction order for all of the blocks stored in a cache set. This order can be represented using *permutation vectors*, where the right-most elements are chosen for eviction first. A permutation vector π_i indicates how each position in the original order is updated upon an access to the entry at position i.

Pseudo least-recently used (PLRU) is a commonly used example of such a permutation policy. PLRU aims to approximate the *least-recently used* (LRU) policy, while reducing the amount of resources needed to track the least recently used entry. In the frequently used tree-based PLRU policy, the entries are organized in a binary tree, where each node keeps track of which of its subtrees were used least recently. Figure 3 depicts a tree-based PLRU and the corresponding permutation vectors. As illustrated in the figure, the permutation vector $\pi_3 [1]$ is 0. This means that upon a hit to the entry at position 3 (d), the entry at position 0 (b), is updated to position 1 (according to its index). Along with the other positional updates and the other permutation vectors, this permutation behavior defines the tree-based PLRU replacement policy.

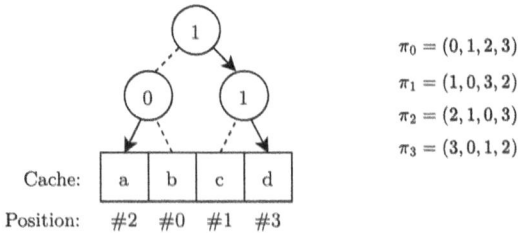

$$\pi_0 = (0, 1, 2, 3)$$
$$\pi_1 = (1, 0, 3, 2)$$
$$\pi_2 = (2, 1, 0, 3)$$
$$\pi_3 = (3, 0, 1, 2)$$

Fig. 3. A sample state of the PLRU replacement policy and the identifying permutation vectors [32].

3 Strategy

We present our approach for reverse-engineering translation caches. We observe that TLB desynchronization [32] operates on fundamental properties of caching and can therefore reliably distinguish between cache hits and misses. We build on the technique, applying it to translation caches implemented within Intel CPUs, which we choose for their prevalence and dominance in the CPU market [2, 26, 30].

We first detail how to apply TLB desynchronization to translation caches, and achieve translation cache desynchronization. We discuss optimizations that we extend on previous approaches with to further reduce the noise of measuring cache hits and misses in translation caches.

3.1 Translation Cache Desynchronization

The TLB desynchronization strategy of Tatar et al. [32] provides a means of reverse-engineering the TLB. Unlike previous approaches based on timing, TLB desynchronization operates on fundamental cache properties making it significantly more precise and robust.

To apply desynchronization to translation caches we overwrite an intermediate-level PTE in memory instead of a first-level, as would be done for TLBs. Address translations that use the newly-desynchronized PTE consequently use a new page table for the next-lower level which has been desynchronized. We prepare this page table such that it provides mapping for the same address ranges as the original one, but these mappings point to different physical addresses. Following Tatar et al., we ensure that the content of the physical pages of the new mapping differs from that of the original mapping. This allows us to differentiate between a translation cache hit and a translation cache miss with a single read from the address range:

- If the affected PTE is in the translation cache after desynchonization, translating an address that uses it accesses the original physical address.
- If this entry is evicted, translating the same virtual address results in a different physical address (the one we overwrite with in desynchronization).

In order for these follow-up observations to reliably distinguish translation cache hits and misses, we must ensure that the translation process (page table walk) indeed reaches the target page table level. This will not occur in scenarios where address translation is served from the TLB or from lower-level translation caches, which we do not desynchronize. In such scenarios the page walk process skips the target page translation level which we target, ignoring the desynchronized translation cache.

3.2 Prefix Alignment

We now discuss how we can use a property that we call *prefix alignment* to ensure that the address translation process always touches the PTE of interest.

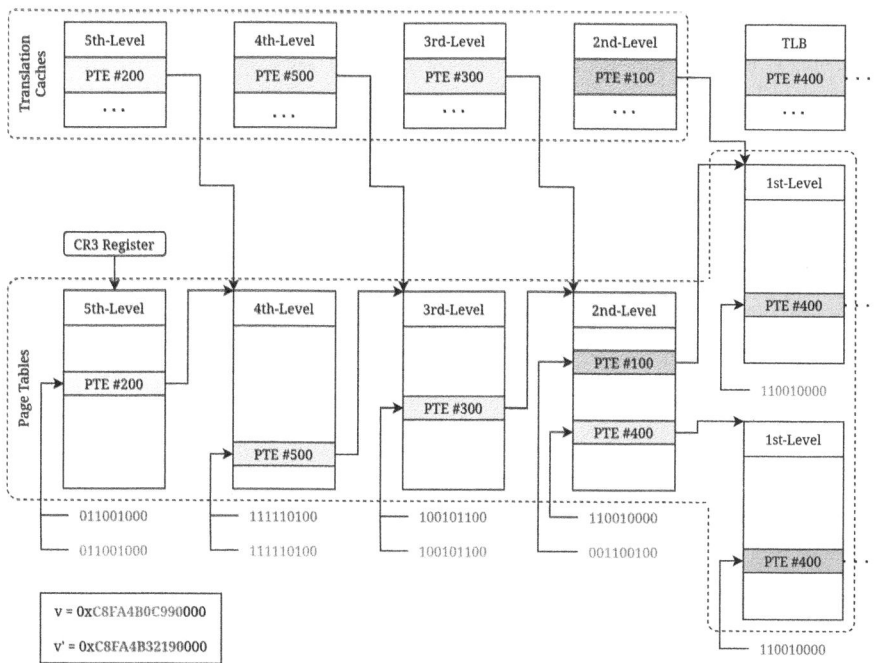

Fig. 4. Page table layout for two virtual addresses, v and v', that are prefix aligned for down to the third-level page table, with separate second-level translation cache and TLB entries.

Suppose we want to measure whether the third-level PTE of a virtual address v is cached. Assume that the lower-level PTEs of v are present in either the TLB or the second-level translation cache. After having desynchronized the corresponding translation cache entry, we need to trigger a translation process that touches the third-level PTE of v to measure a MMU cache hit or miss. Simply re-translating the same address v is not an option, because the translation process will take the shortcut offered by the TLB or otherwise by the second-level translation cache.

A naïve way to avoid this problem would be to evict the first- and second-level PTEs from the TLB and second-level translation cache, respectively. While this approach can work, it introduces system noise and increases the complexity of the technique.

We opt for an alternative approach, which instead circumvents re-hitting the TLB and PTEs at lower levels than our target. Instead of re-translating v, we translate a *prefix-aligned* address v' that shares all upper-level PTEs with v, down to and including the third-level PTE. Figure 4 depicts the page tables corresponding to two such example addresses. In order for the lower-level translation cache and TLB entries for v and v' to not interfere with one another, the second-level PTEs of v and v' must differ. Hence, in a system with five-level paging, the two virtual addresses v and v' share the three top-most significant 9-bit slices, but are different in the following 9-bit slice, which is used to index the second-level page table. Because the MMU selects a translation cache or TLB entry based on the longest matching prefix of the translated address, the MMU cannot use the lower-level translation cache and TLB entries of v when translating v'. With this approach we force the subsequent page table walk to include the target-level PTE and are able to check whether it is cached or not.

This method is applicable for any target-level PTE by ensuring the target- and all upper-level PTEs are shared, and the next lower-level PTEs of two or more addresses are different. Since page tables contain 512 PTEs, for each target level we can construct up to 512 prefix-aligned addresses that have unique PTEs on the next lower-level.

3.3   Limitations

As is the case in past work [35], desynchronization cannot differentiate between translation cache hits and page walk cache hits. To the best of our knowledge, Intel does not implement page walk caches. Earlier AMD processors did implement it [6], but AMD's documentation is not clear about the current implementation [4].

4   Reverse Engineering

We reverse-engineer translation caches by using translation cache desynchronization. We design and evaluate experiments to determine their behavior and implementation details. We extend the approach of Tatar et al. [32] with additional properties relevant to translation caches. With this we discern the existence of certain cache structures, their hierarchy, and relationships.

All experiments focus on reverse engineering the named properties on a single translation cache level at a time, but they are applicable to all levels.

Table 1. System configurations of our test environment.

Processor	Alder Lake i7-1260P	Rocket Lake i7-11700KF	Kaby Lake i5-7500	Skylake i7-6700	Haswell 4415U
Hardware					
Cores	4/8[a]	8	4	4	2
Hyperthreading	✓/✗[b]	✓	✗	✓	✓
Memory size	32 GB	64 GB	16 GB	32 GB	4 GB
Memory type	DDR4	DDR4	DDR4	DDR4	DDR4
Software					
OS	Debian 12	Ubuntu 22.04.3 LTS	Arch Linux	Debian 12	Ubuntu 22.04.4 LTS
Kernel version	6.1.0-22-amd64	5.15.0-116-generic	6.9.10-arch1-1	6.1.0-18-amd64	6.5.0-41-generic
Architecture	x86-64	x86-64	x86-64	x86-64	x86-64

[a] Alder Lake implements four performance and eight efficiency cores.
[b] Only performance cores support hyperthreading.

We experiment with five processor models, summarized in Table 1. On the Alder Lake system we also differentiate between *performance* and *efficiency* cores, often referred to as *P-Cores* and *E-Cores*. They differ in cycle frequency and hyperthreading support and implement different microarchitectures. Thus, our experiments cover six microarchitectures. We note that none of these systems support five-level paging. In Table 2 we summarize all of our results.

4.1 Cache Existence

We first verify the existence of translation caches at various target levels. According to the Intel developer manual, the MMU can implement any or all level translation caches [18].

To confirm the existence of a translation cache at a given target level, we trigger translation for an address. We then desynchronize the target level PTE and finally access a prefix aligned address to check whether the address is cached. If the target-level translation cache exists, reading from a prefix aligned address still results in its original physical address. However, when the translation cache does not exist, reading from the prefix aligned address results in a different physical address, as the MMU resorts to accessing the in-memory PTE to translate the address.

Our experiments show that the MMU typically implements all levels of translation caches. If translation caches are omitted, this concerns the uppermost levels.

4.2 Cache Hierarchy

Our next aim is to find out whether the translation cache consists of multiple layers, similar to the TLB or the CPU instruction and data caches. These typically implement multiple layers where the first layer is split between instructions and data, whereas lower layers are shared between both.

Shared Layer. First, we determine whether the target translation cache implements a shared layer. Using an instruction fetch, we trigger translation for a

158 P. Ertmer et al.

target address and desynchronize its target-level PTE. Then, we use a data read to trigger translation of a prefix aligned address. If read loads from the original physical address associated to the prefix aligned address, the translation cache implements a shared layer. Otherwise, we conclude that the translation cache is split between data and instructions. We find that all translation caches implement a shared layer, except for the second-level translation caches of Alder Lake's efficiency cores.

Split Layer. Additionally, we implement an experiment to determine the existence of a split layer. First, we bring a target PTE to the translation cache using one access type and desynchronize it. Afterwards, we evict the related translation cache entry from a potential shared layer by triggering many translations for different addresses using the other access type. Finally, we use the original access type again, to trigger translation of a prefix aligned address. If accessing this prefix aligned address still results in the originally associated data, the translation cache must implement a split layer. If not, we conclude that the translation cache does not implement a split layer. The experiment results indicate that none of the microarchitectures we examine implement a split layer.

Our observations suggest that Alder Lake's E-cores do not implement a second-level translation cache, because it implements neither a shared nor a split layer. However, this is a contradiction to the translation cache existing. Analyzing the case, we conclude that the second level translation cache must be implemented as a semi-split cache. That is, entries that are introduced by one access type can be *evicted* using the other access type, but a cache entry introduced by one access type cannot be *used* by the other access type.

4.3 Sets, Ways, and Hash Functions

To reverse-engineer the number of sets and ways, as well as the hash function implemented by translation caches, we first assume, without loss of generality, that the target translation cache is split into S sets of W ways. We further assume that the cache implements a hash function H : address $\rightarrow \{0, 1, \ldots, S-1\}$ that maps a given address to a set. We guess multiple combinations of S, W, and H, which are similar to results from past research.

We then use the guessed hash function H to generate multiple random groups of $W+1$ addresses whose target-level PTEs map to the same translation cache set and use these to validate the guess. Specifically, we trigger an address translation for each of the addresses in the group, loading them to the respective translation cache. We then desynchronize all of the translation cache entries of these addresses. Finally, we attempt address translation again to detect if any address prefix was evicted from the translation cache, indicating contention on a cache set. We then look for the combination of S, W, and H that minimizes S and W, with priority to minimize W.

Table 2 summarizes the results. Based on the suspicion that the hash functions are similar to those used for TLBs, we test the *LIN* and *XOR* hash functions [12,32]. However, we find that these are not implemented in translation

Table 2. Summary of the reverse-engineered properties of translation caches on different Intel processors.

Property	Alder Lake i7-1260P (P-Cores)	Alder Lake i7-1260P (E-Cores)	Rocket Lake i7-11700KF	Kaby Lake i5-7500	Skylake i7-6700	Haswell 4415U
Second-level translation cache						
Exists	✓	✓	✓	✓	✓	✓
Split Layer	✗	✗	✗	✗	✗	✗
Shared Layer	✓	✗	✓	✓	✓	✓
Sets	8	1	8	8	8	8
Ways	4	28–30	4	4	4	4
Hash function	LIN» 1	n/a[a]	LIN» 1	LIN»1	LIN» 1	LIN» 1
Replacement policy	PLRU	LRU type	HUPLRU	HUPLRU	HUPLRU	HUPLRU
Nested	✗	✗	✗	✗	✗	✗
Unified huge TLB	✗	✗	✗	✗	✗	✗
Supported PCIDs	0[c]	0[c]	0	0	0	0
Third-level translation cache						
Exists	✓	✓	✓	✓	✓	✓
Split Layer	✗	✗	✗	✗	✗	✗
Shared Layer	✓	✓	✓	✓	✓	✓
Sets	1	1	1	1	1	1
Ways	3–4	12–14	3–4	2–4	4	1–3
Hash function	n/a[a]	n/a[a]	n/a[a]	n/a[a]	n/a[a]	n/a[a]
Replacement policy	PLRU	LRU type	PLRU	(P)LRU	PLRU	(P)LRU
Nested	✗	✗	✗	✗	✗	✗
Unified huge TLB	✗[d]	✗	✗[d]	✗[d]	✗	n/a[e]
Supported PCIDs	0[c]	0[c]	0	0	0	0
Fourth-level translation cache						
Exists	✓	✗	✓	✗	✓	✗
Split Layer	✗	–	✗	–	✗	–
Shared Layer	✓	–	✓	–	✓	–
Sets	1	–	1	–	1	–
Ways	2	–	2	–	1	–
Hash function	n/a[a]	–	n/a[a]	–	n/a[a]	–
Replacement policy	MRH	–	MRH	–	n/a[b]	–
Supported PCIDs	0[c]	–	0	–	0	–

[a]Defining a hash function is not useful when the cache is directly mapped.
[b]Cannot explicitly define a replacement policy.
[c]Cannot test PCID support with NOFLUSH bit set to one.
[d]The huge TLB is not wiped by our eviction set.
[e]The machine does not have enough free memory to execute the experiment.

caches. Instead, we find that translation caches implement an undocumented hash function, which we call *LIN» 1* hash.

To calculate the target set t of a virtual address *VA*, *LIN» 1* computes $t = (tag_{VA} >> 1) \bmod S$. Thus, the translation cache index tag_{VA} of *VA* is shifted one bit to the right, essentially ignoring the least significant bit of the cache index on set selection. Hence, pairs of adjacent PTEs in a page table map to the same translation cache set.

4.4   Replacement Policy

To reverse-engineer the replacement policy of a target translation cache set, we follow the approach of Abel and Reineke [1]. We use knowledge of the number of ways W and the hash function H attained in our prior experiments to establish a known state in a target set s. We can achieve this by triggering translation for W addresses that, according to H, are known to map to s. Then, we need to determine the replacement order of the known state. We repeat the following steps for all $i \in \{1, \ldots, W - 1\}$:

1. desynchronize all entries contained in the target translation cache set
2. trigger translation for i independent addresses which also map to s
3. iteratively observe which entry is evicted by the i-th access

The position of the evicted entry in the replacement order corresponds to $W - i$. Afterwards, we can determine the permutation vectors with the following steps:

1. establish a known state with a known order
2. touch one of the PTEs in the target set to trigger a permutation
3. repeat the process presented before to determine the replacement order of the entries

Repeating this process for all possible W permutations, we obtain the permutation vectors that define the target set's replacement policy.

The results of this experiment are displayed in Table 2. On most examined platforms the algorithm cannot determine the permutation vectors. It only works for the second-level translation cache of Alder Lake's performance cores and for most of the third-level translation caches, except those of Alder Lake's efficiency cores. From manual analysis of more complex access patterns we find that these translation caches do not implement permutation policies. We identify two novel replacement policies, which we present in the following two sections.

HUPLRU. *Hit-updated PLRU* (HUPLRU) is the first replacement policy we discover and it is implemented on most second-level translation caches. Just like tree-based PLRU, it implements a binary tree to keep track of the recency of the entries in the cache. Each node in the tree serves as a decision point, pointing in the direction of the least recently used entry. In PLRU, each node in the tree is updated upon hitting an existing or inserting a new entry to point in the opposite direction of that entry. For HUPLRU this is different. The tree is only updated upon hitting an existing entry, but not upon inserting a new one.

We hypothesize that Intel has chosen to implement this variation of PLRU as an optimization strategy to prevent cache pollution. HUPLRU ensures that one-hit wonders do not pollute the cache, as a second-level PTE must be accessed at least twice to be retained. Such PTEs have high reuse probability because it indicates that memory is being iterated and more hits are to follow. Accessing a second-level PTE only once suggests a 2 MiB jump in the virtual addresses utilized. These are rare due to the principal of spatial locality and should not be retained in the cache.

MRH. On the fourth level translation caches of Rocket Lake and Alder Lake's P-cores, we identify another novel replacement policy, which we call *most-recently hit* (MRH). This is a variant of the more common *most-recently used* (MRU) replacement policy. As the name suggests, MRU chooses the most-recently used entry for eviction. The most-recently used entry is either the one that was most-recently inserted, or the one that was most-recently hit. For MRH, this is different: only the most-recently hit entry can be chosen for eviction. Inserting a new entry does not change the order of replacement.

In general, it seems counterintuitive to replace the entries that were just recently used. However, in our specific case this strategy aligns with observed translation cache behavior and access likelihoods. The third-level PTEs map memory regions of 1 GiB. In Sect. 4.5, we discover that regardless of fourth-level translation cache evictions, the corresponding third-level translation cache entries are retained. Hence, even after a fourth-level eviction, the third-level cache ensures that an access to the same region incurs minimal additional cost. Furthermore, we suspect that Intel chooses MRH over MRU because it is less prone to pollution through unused entries, which otherwise would never leave the cache.

4.5 Nesting

Next, we examine the relationships between the different levels. To this end, we introduce the notion of *nesting*. We say that translation caches are nested when it is guaranteed that all indices of entries present in a lower-level translation cache are prefixed by an index of an entry present in the next upper-level.

To investigate nesting, we trigger translation for a target address and desynchronize its target level PTE. Then we try to evict a higher-level translation cache entry associated with the target address. Finally, we read from a prefix aligned address, sharing the target level PTE of our target address. Measuring a cache miss indicates nesting, whereas a cache hit indicates that there is no such relationship between the different level translation caches. Our evaluation demonstrates that the translation caches we investigate are not nested.

4.6 Huge TLBs

In addition to the *normal* TLB that stores translations of 4 KiB pages, the MMU implements additional TLBs for 2 MiB and 1 GiB huge pages, which we call *huge* TLBs. These TLBs cache PTEs of the second- and third-level page tables mapping the corresponding huge pages. We now check whether these TLBs are unified with their corresponding translation caches [35].

To this end, we introduce a target PTE to the target level translation cache and desynchronize it. We then evict all entries from the 2 MiB or 1 GiB huge TLB, depending on whether we are investigating the second- or third-level translation cache respectively. Finally, we trigger translation for a prefix aligned address, sharing the target PTE. If this translation indicates a translation cache miss, it confirms that huge TLBs and translation caches are unified.

We find that translation caches and huge TLBs are not unified. Considering our refined findings for the structure of translation caches, this is not too surprising. While the 2 MiB and 1 GiB dTLB's structure aligns to some degree with the structure of the second- and third-level translation caches, we already found that translation caches are shared between instructions and data. Hence, the only conceivable relation is to have them unified with the huge sTLBs. However, as we can see in Fig. 5, their structure does not really align with the structure we have observed for translation caches.

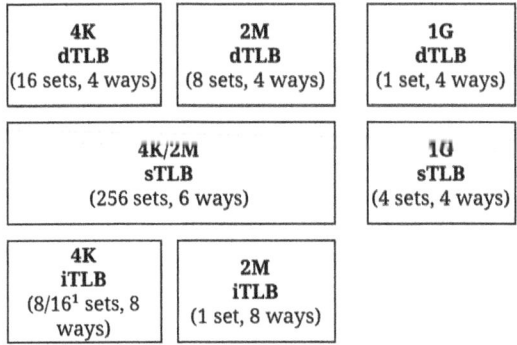

¹ 8 sets on Haswell and Skylake; 16 sets on Kaby Lake

Fig. 5. TLB sizes and layout reported by `cpuid` for all test systems, except Alder Lake and Rocket Lake.

4.7   PCID Support

Finally, we investigate how translation caches enforce process isolation. Intel CPUs manage this using *process context identifiers* (PCIDs). Tatar et al. [32] observe that the TLB supports only four of the 4096 possible PCIDs simultaneously. As soon as an additional PCID is used, the MMU invalidates all TLB entries related to the *least-recently used* PCID. They conclude that the MMU implements a PCID cache with four entries, and that the TLB entries also store an identifier relating them to one of the entries in that cache.

We investigate whether translation caches operate according to a similar principle. To switch PCIDs, we write to the CR3 control register. According to the Intel developer manual [18], setting bit 63 of the CR3 register hints the processor to not flush the translation caches when switching PCIDs. However, we find that whether or not we set bit 63, translation caches do not support PCIDs and are always flushed upon context switches.

If this were not the case, the number of supported PCIDs could be discerned with the following experimental procedure. First, trigger translation for a target address from a fixed PCID. Next, desynchronize its target level PTE and then switch to $n + 1$ different PCIDs. Finally, switch back to the fixed PCID and trigger translation for a prefix aligned address. The lowest value of n, for which this translation still indicates a cache miss, is exactly equal to the maximum amount of supported PCIDs.

4.8 Cross-Hyperthread PCID Support

One remaining question that subsequently arises is how process isolation is enforced between different hyperthreads running on the same core. As translation caches are implemented per core, and they do not make use of PCIDs, process isolation needs to be enforced in a different way.

To investigate this, we first introduce a new entry to the target translation cache and desynchronize it. Afterwards, we switch to the co-resident hyperthread, and switch to a different PCID from there. Then, in the original hyperthread, we trigger translation of a prefix aligned address, in order to determine whether our target entry is still present in the translation cache. A translation cache hit indicates that process isolation is properly enforced, because the PCID switch only invalidates the translation cache entries related to one hyperthread.

Our experiments show, that all translation caches do implement proper process isolation between hyperthreads. We assume that they make use of the hyperthread ID to relate entries to their corresponding hyperthreads [32], though a different design is conceivable. We leave it to future work to further investigate upon this. The behavior is quite interesting, because the processor needs to find and invalidate the specific entries related to the current hyperthread upon every context-switch.

4.9 Discussion

In this section, we discuss our reverse engineering results. In particular, we compare our results to the previously observed properties of translation caches by van Schaik et al. [35].

Previous Results. Two of the microarchitectures we examine, Haswell and Skylake, were also examined by van Schaik et al. [35]. The Skylake models examined are identical, but the Haswell models differ. We use the *Intel Pentium 4415U*, whereas they examine the *Intel Core i7-4500U*.

On both microarchitectures, we find different sizes for the second-level translation caches. Van Schaik et al. observe that each of these consists of 24 entries. However, with our experiment to determine the hash function, we instead observe that they consist of 32 entries in total. Additionally, contrary to van Schaik et al., we find that Skylake implements a fourth-level translation cache, and Haswell's third-level translation cache consists of only 1–3 entries instead of 3–4.

There are several explanations for these different findings. First, van Schaik et al. [35] rely on timing measurements to differentiate between translation cache hits and misses. Desynchronization differentiates translation cache hits and misses based on fundamental properties of the address translation process [32]. Because of this, desynchronization offers far more robust classification than timing measurements, which are susceptible to noise [32].

Amplifying this, van Schaik et al. have to consistently evict the PTEs from the CPU data caches, the TLB, and lower-level translation caches. Desynchronization does not require such techniques, as the CPU data caches do not interfere with desynchronization, and must run in kernel mode, allowing us to easily flush the MMU caches. Additionally, in Sect. 3.2, we propose prefix alignment, which allows us to touch a target-level PTE without flushing the lower-level MMU caches at all.

Furthermore, as we discover in Sect. 4.4, translation caches may implement complex replacement policies. As we cannot find any hints that van Schaik et al. [35] consider that, this might be another explanation for the differing results.

5   Talbot

We present Talbot, a tool that automates the processes for reverse engineering translation caches in Intel processors. Talbot implements all of the experiments we describe in Sect. 4. In this section, we discuss implementation challenges and limitations, and we evaluate the tool.

5.1   Challenges

Several non-trivial technical challenges arise in the effort of automating various reverse engineering processes.

Out-of-order execution, along with scheduled and asynchronous events, complicate running our experiments as they introduce interference and system noise. Where possible we mitigate these effects, for example by implementing strict pointer chasing for the relevant memory accesses, and disabling preemption and interrupts.

Additionally, due to the large memory areas mapped by upper-level page tables, we also have to optimize the required memory. We make use of overlapping memory areas and page table entries to implement memory management. Finally, we encounter some trouble with flushing the translation caches in between experiments, making the experiment results very unstable. We suspect that this is related to the replacement policy state that the translation cache is left in after flushing. This results in certain entries being evicted early, before the corresponding cache set is completely filled. Consulting with the Intel developer manual [18], we find that using mov instructions targeting the CR3 register is the most stable approach across our different test platforms.

5.2   Limitations

Talbot only supports four-level paging because none of the machines we tested feature five-level paging. We leave extending the tool for five-level paging to future work.

The set of hypothesized hash functions that Talbot supports is limited to *LIN*, *XOR*, and *LIN» 1*. When Talbot cannot attribute any of these hash functions as the tested system's correct candidate, it produces minimal eviction sets, using the single holdout method [22,28]. Similarly, Talbot only provides the permutation vectors when reverse engineering the replacement policy. Manual work is required to draw conclusions about the implemented policy. Analysts running the tool must have some understanding of the experiments it performs. All experiments are run, regardless of whether the translation cache exists or not. Furthermore, Talbot even prints out a hash function, if the translation cache is directly mapped.

All of the experiments our tool implements do not account for multi-layer caches. We did not encounter any test platforms that implement a multi-layer hierarchy. Regarding replacement policies, in its current state Talbot is limited to reverse-engineering permutation-based policies.

5.3   Evaluation

For our experiments, Talbot is particularly effective in identifying cache existence, hierarchies, and hash functions. Across multiple processor architectures, it reliably detects different architectures with high accuracy, though sometimes there is a need to run the tool multiple times and average the results. Talbot also consistently identifies features such as PCID support, nesting, and the relationship to huge TLBs. However, our test environment does not challenge these experiments, as these properties are consistent across the different microarchitectures tested.

Despite this, Talbot also has shortcomings, mostly related to the limitations we mentioned. Most prominently, the automated reverse engineering of replacement policies is not completely reliable. This is because the assumption that translation caches implement permutation policies rarely holds. Future work may try to adapt the approach of Vila et al. [36] for reverse-engineering replacement policies to the case of translation caches. As a more limited solution, we provide scripts that identify the HUPLRU and MRH replacement policies on the affected systems.

In summary, Talbot offers a robust means of quickly reverse engineering translation caches, with careful and knowledgeable interpretation of its results. The tool is a basis that can be further developed for reverse engineering diverse replacement policies and its inter-experiment relationships.

6   Conclusion

In this work, we adopt the TLB desynchronization strategy [32] to obtain translation cache desynchronization and reverse-engineer translation caches. We obtain a refined description of the implementation details and largely extend the understanding of translation caches, opposed to previous reverse-engineering efforts.

We go beyond the sizes of translation caches, and investigate more specific implementation details, including cache structure, replacement policy, process isolation capability, and relationships.

Our results show that the potential of attacks utilizing the address translation process, and especially translation caches, is not fully utilized yet. Future work can build upon the properties we expose and construct novel attacks or improve the performance of existing attacks.

Additionally, we implement Talbot, a tool to automatically reverse-engineer specific properties of translation caches on arbitrary Intel processors. Future work can use Talbot to evaluate the implementation of translation caches on other existing and future microarchitectures. Enhancing the tool's reverse engineering capabilities would be especially useful, in particular concerning more complex replacement policies.

Acknowledgments. This work was supported by an ARC Discovery Project number DP210102670; and the Deutsche Forschungsgemeinschaft (DFG, German Research Foundation) under Germany's Excellence Strategy - EXC 2092 CASA - 390781972.

References

1. Abel, A., Reineke, J.: Measurement-based modeling of the cache replacement policy. In: RTAS, pp. 65–74 (2013)
2. Alcorn, P.: AMD and Intel CPU Market Share Report: Recovery on the Horizon (2023). https://www.tomshardware.com/news/amd-and-intel-cpu-market-share-report-recovery-looms-on-the-horizon
3. Aldaya, A.C., Brumley, B.B., ul Hassan, S., García, C.P., Tuveri, N.: Port contention for fun and profit. In: IEEE SP, pp. 870–887 (2019)
4. AMD: AMD64 Architecture Programmer's Manual Volume 2: System Programming (2024)
5. Barr, T.W., Cox, A.L., Rixner, S.: Translation caching: skip, don't walk (the page table). In: ISCA, pp. 48–59 (2010)
6. Bhargava, R., Serebrin, B., Spadini, F., Manne, S.: accelerating two-dimensional page walks for virtualized systems. In: ASPLOS, pp. 26–35 (2008)
7. Bosman, E., Razavi, K., Bos, H., Giuffrida, C.: Dedup Est Machina: memory deduplication as an advanced exploitation vector. In: IEEE SP, pp. 987–1004 (2016)
8. Braun, B.A., Jana, S., Boneh, D.: Robust and efficient elimination of cache and timing side channels, arXiv:1506.00189 (2015)
9. Cormen, T.H., Leiserson, C.E., Rivest, R.L., Stein, C.: Introduction to Algorithms, 4th edn. MIT Press (2022)
10. Evtyushkin, D., Ponomarev, D., Abu-Ghazaleh, N.: Jump over ASLR: attacking branch predictors to bypass ASLR. In: MICRO, pp. 1–13 (2016)
11. Ge, Q., Yarom, Y., Cock, D., Heiser, G.: A survey of microarchitectural timing attacks and countermeasures on contemporary hardware. J. Cryptogr. Eng. $8(1)$, 1–27 (2018)
12. Gras, B., Razavi, K., Bos, H., Giuffrida, C.: Translation leakaside buffer: defeating cache side-channel protections with TLB attacks. In: USENIX Security, pp. 955–972 (2018)

13. Gras, B., Razavi, K., Bosman, E., Bos, H., Giuffrida, C.: ASLR on the line: practical cache attacks on the MMU. In: NDSS (2017)
14. Gruss, D., Lettner, J., Schuster, F., Ohrimenko, O., Haller, I., Costa, M.: Strong and efficient cache side-channel protection using hardware transactional memory. In: USENIX Security, pp. 217–233 (2017)
15. Gruss, D., Maurice, C., Fogh, A., Lipp, M., Mangard, S.: Prefetch side-channel attacks: bypassing SMAP and kernel ASLR. In: CCS, pp. 368–379 (2016)
16. Guanciale, R., Nemati, H., Baumann, C., Dam, M.: Cache storage channels: alias-driven attacks and verified countermeasures. In: IEEE SP, pp. 38–55 (2016)
17. Hund, R., Willems, C., Holz, T.: Practical timing side channel attacks against kernel space ASLR. In: IEEE SP, pp. 191–205 (2013)
18. Intel Inc.: Intel 64 and IA-32 Architectures Software Developer Manuals
19. Kocher, P., et al.: Spectre attacks: exploiting speculative execution. In: IEEE SP, pp. 1–19 (2019)
20. Lipp, M., et al.: Meltdown: reading kernel memory from user space. In: USENIX Security, pp. 973–990 (2018)
21. Liu, F., et al.: CATalyst: defeating last-level cache side channel attacks in cloud computing. In: HPCA, pp. 406–418 (2016)
22. Liu, F., Yarom, Y., Ge, Q., Heiser, G., Lee, R.B.: Last-level cache side-channel attacks are practical. In: IEEE SP, pp. 605–622 (2015)
23. McCurdy, C., Cox, A.L., Vetter, J.S.: Investigating the TLB behavior of high-end scientific applications on commodity microprocessors. In: ISPASS, pp. 95–104 (2008)
24. Osvik, D.A., Shamir, A., Tromer, E.: Cache attacks and countermeasures: the case of AES. In: CT-RSA, pp. 1–20 (2006)
25. Paccagnella, R., Luo, L., Fletcher, C.W.: Lord of the ring(s): side channel attacks on the CPU on-chip ring interconnect are practical. In: USENIX Security, pp. 645–662 (2021)
26. PassMark Software: PassMark CPU Benchmarks - AMD vs Intel Market Share (2024). https://www.cpubenchmark.net/market_share.html
27. Pessl, P., Gruss, D., Maurice, C., Schwarz, M., Mangard, S.: DRAMA: exploiting DRAM addressing for cross-CPU attacks. In: USENIX Security, pp. 565–581 (2016)
28. Qureshi, M.K.: New attacks and defense for encrypted-address cache. In: ISCA, pp. 360–371 (2019)
29. Razavi, K., Gras, B., Bosman, E., Preneel, B., Giuffrida, C., Bos, H.: Flip Feng Shui: hammering a needle in the software stack. In: USENIX Security, pp. 1–18 (2016)
30. Shilov, A.: Arm-Based CPUs Could Double Notebook PC Market Share by 2027 (2023). https://www.tomshardware.com/news/arm-based-cpus-set-to-double-notebook-pc-market-share-by-2027
31. Sprabery, R., Evchenko, K., Raj, A., Bobba, R.B., Mohan, S., Campbell, R.H.: A novel scheduling framework leveraging hardware cache partitioning for cache-side-channel elimination in clouds, arXiv:1708.09538 (2017)
32. Tatar, A., Trujillo, D., Giuffrida, C., Bos, H.: TLB;DR: enhancing TLB-based attacks with TLB desynchronized reverse engineering. In: USENIX Security, pp. 989–1007 (2022)
33. van der Veen, V., et al.: Drammer: deterministic Rowhammer attacks on mobile platforms. In: CCS, pp. 1675–1689 (2016)
34. van Schaik, S., Giuffrida, C., Bos, H., Razavi, K.: Malicious management unit: why stopping cache attacks in software is harder than you think. In: USENIX Security, pp. 937–954 (2018)

35. van Schaik, S., Razavi, K., Gras, B., Bos, H., Giuffrida, C.: RevAnC: a framework for reverse engineering hardware page table caches. In: EuroSec, pp. 1–6 (2017)
36. Vila, P., Ganty, P., Guarnieri, M., Köpf, B.: CacheQuery: learning replacement policies from hardware caches. In: PLDI, pp. 519–532 (2020)
37. Wang, Y., Paccagnella, R., He, E.T., Shacham, H., Fletcher, C.W., Kohlbrenner, D.: Hertzbleed: turning power side-channel attacks into remote timing attacks on x86. In: USENIX Security, pp. 679–697 (2022)
38. Yarom, Y., Falkner, K.: Flush+Reload: a high resolution, low noise, L3 cache side-channelattack. In: USENIX Security, pp. 719–732 (2014)
39. Zhang, Z., Cheng, Y., Liu, D., Nepal, S., Wang, Z., Yarom, Y.: PThammer: cross-user-kernel-boundary Rowhammer through implicit accesses. In: MICRO, pp. 28–41 (2020)
40. Zhang, Z., Tao, M., O'Connell, S., Chuengsatiansup, C., Genkin, D., Yarom, Y.: BunnyHop: exploiting the instruction prefetcher. In: USENIX Security, pp. 7321–7337 (2023)
41. Zhao, Z.N., Morrison, A., Fletcher, C.W., Torrellas, J.: Binoculars: contention-based side-channel attacks exploiting the page walker. In: USENIX Security, pp. 699–716 (2022)
42. Zhou, Z., Reiter, M.K., Zhang, Y.: A software approach to defeating side channels in last-level caches. In: CCS, pp. 871–882 (2016)

The HMB Timing Side Channel: Exploiting the SSD's Host Memory Buffer

Jonas Juffinger$^{(\boxtimes)}$ (ID), Hannes Weissteiner (ID), Thomas Steinbauer (ID), and Daniel Gruss (ID)

Graz University of Technology, Graz, Austria
jonas.juffinger@tugraz.at

Abstract. Over the past three decades, cache side channels evolved from specialized attacks on cryptographic implementations to generic techniques (e.g., Flush | Reload and Page Cache Attacks) on general-purpose operations. During the last decade, SSDs became the de facto standard persistent storage, where capacity is not the highest priority.

In this paper, we present a novel cache side channel, targeting the host-memory buffer (HMB) used by mid-range SSDs to cache translations from logical page addresses to physical page addresses.

We demonstrate that, compared to page cache attacks, our attacks are significantly faster as we can evict reliably in only 22 ms. Consequently, we propose a hybrid attack, using the slow page cache eviction as little as possible and using the HMB side channel for our main attack. We evaluate the HMB side channel in four practical attacks: First, we evaluate the capacity of the HMB side channel in a covert channel scenario, achieving up to 8.3 kbit/s channel capacity. Second, we demonstrate a UI redress attack using the HMB side channel, where the fake UI element covers the real one within 100 ms. Third, given that multiple pages from different security contexts are translated through the same HMB entry, we demonstrate blind templating attacks, that allow to spy on accesses to arbitrary other files whose translation is co-located in the same HMB entry. We use this to demonstrate a cross-VM covert channel and a remote side channel where an unprivileged process without network access exfiltrates data to a remote system over the network, through the HMB side channel by using an nginx web server as a confused deputy.

1 Introduction

The performance of modern systems is fundamentally limited by memory latency. Caches play a central role in modern systems to alleviate this bottleneck. However, while they reduce the latency of accesses to recently or frequently accessed memory, this latency difference can also be exploited in so-called side-channel attacks, at the beginning, mainly on cryptographic implementations [1,10]. Later, generic techniques like Prime+Probe [18] or Flush+Reload [25] enabled cache side-channel attacks on general-purpose computations [5,6].

© The Author(s), under exclusive license to Springer Nature Switzerland AG 2025
M. Egele et al. (Eds.): DIMVA 2025, LNCS 15747, pp. 169–190, 2025.
https://doi.org/10.1007/978-3-031-97620-9_10

Despite the growing popularity and the increasing performance of SSDs, they have received relatively little attention in terms of side-channel research. Liu et al. [14] studied the now-discontinued Intel Optane, which has several caches introducing distinct timing differences. Two works focused on SSD covert channels with attacker-controllable FPGAs reaching <1 bit/s [4,24]. Recently, Juffinger et al. [9] demonstrated that contention on commodity SSD can be exploited to build covert channels with over 1 kbit/s transmission rate and website fingerprinting with over 90 % accuracy. While page cache attacks also leak spatial information and can be faster [6], they were partially mitigated recently [22], reducing the temporal resolution to one measurement within several seconds due to the high cost of eviction, e.g., of a multi-gigabyte page cache. In concurrent work, we discussed the negative result of performing a Rowhammer attack through the HMB and the effect on bit flips on the HMB [8].

In this paper, we present a novel cache side channel, targeting the host-memory buffer (HMB) [17] used by mid-range SSDs to cache translations from logical to physical page addresses. We investigate four different SSD models supporting the HMB feature and reverse-engineer how these SSDs use it. Our investigations show that the HMB operates either full- or set-associatively, translating contiguous blocks of multiple megabytes of memory. We demonstrate how an attacker can observe the HMB state and how they can evict the HMB through accesses to arbitrary files. As a result, we obtain a side channel with a spatial resolution of a few megabytes, low compared to the 4 kB of page cache attacks [6].

However, our HMB attack is significantly faster as we can evict the HMB in only 22 ms or less. Additionally, we demonstrate how we can use the unprivileged FIEMAP ioctl system call to effectively increase the spatial granularity of a few megabytes to the 4 kB of page cache side channels by using the page cache to our advantage. Consequently, we propose a hybrid attack, using the slow page cache eviction as little as possible and the HMB side channel for our main attack.

We evaluate the HMB side channel in four practical attacks: First, we evaluate the capacity of the HMB side channel with an inter-process covert channel, achieving up to 8.3 kbit/s. Second, we demonstrate that we can monitor disk accesses to files on the SSD. We exploit this side channel in a UI redress attack using the HMB side channel within 100 ms. The goal of a UI redress attack is to detect when a program starts and then cover it with a fake UI element to, e.g., steal an entered password. Even without any sharing of files or memory, we can exploit that multiple SSD pages from different security contexts are translated through the same HMB entry in blind templating attacks. Third, we show a covert channel between virtual machines with up to 7.1 bit/s. Finally, we demonstrate a remote side channel where an unprivileged process without network access exfiltrates data to a remote system over the network with up to 8.9 bit/s, by exploiting nginx as a confused deputy.

In summary, the **contributions** are as follows:

- We are the first to analyze the host memory buffer for potential side channels and reverse engineer the behavior of different implementations.

– We show that differences in the access latency leak information about recently accessed pages that can be measured with very high accuracy.
– We present a novel blind templating attack that also works if the attacker has no read rights to the target files.

In Sect. 2, we provide background information on cache timing side channels and SSDs. In Sect. 3, we describe the HMB side channel and reverse-engineer how the SSDs use the main memory and how the HMB can be evicted. In Sect. 4, we evaluate the HMB side channel in attacks on accessible files, showing that yields a significantly higher performance than current page cache attacks. In Sect. 5, we demonstrate that even without accessible files an attacker can mount a blind templating attack using the HMB side channel. We discuss countermeasures in Sect. 6. We conclude in Sect. 7.

2   Background

In this section, we provide background on cache timing side channels, SSDs and flash memory, the flash translation layer and host memory buffer.

Cache Timing Side Channels. Caches are used in computers to store data that is likely to be accessed soon, to decrease the delay of these accesses. They are both, faster and smaller, than the main memory they cache data from. Due to this smaller size, they cannot store all the data from the main memory but only a small subset at any given time.

Caches introduce execution timing differences based on the current cache state. This effect is exploited in cache timing side-channel attacks like Prime+Probe [13,18] or Flush+Reload [25]. While these techniques have mainly been shown on CPU caches, they also can be applied to other caches [6,23]. Gruss et al. [6] demonstrated the that flush- and eviction-based attacks on the OS page cache are possible. The page cache buffers data from the disk, in blocks of pages. However, as also reported by Schwarzl et al. [22], the interfaces exploited by Gruss et al. [6] are now more restricted.

Solid State Drives and NAND Flash Memory. Solid state drives (SSDs) are persistent storage devices that use NAND flash memory to store data. They are faster than hard disk drives (HDDs), the number of random input output operations per second (IOPS) is magnitudes higher.

NAND flash memory restricts how data can be written. While data can always be read, settings bits to 1 (erasing) uses a different process than setting bits to 0 (programming). Erased memory stores all 1's, data is then written by setting bits to 0. Typically, the read and programming size is 512 B, 2048 B or 4096 B, called a page. Erasures happen in blocks of 32, 64 or 128 pages and take much longer. Flash memory has only a limited number of program-erase cycles. To wear out all blocks equally, the SSD controller performs wear-leveling.

Table 1. The SSDs used in our experiments.

SSD	Model	PCIe Revision	Size	HMB Size
$\mathcal{A}$	Samsung 990 EVO	4.0	2 TB	64 MB
$\mathcal{B}$	Lexar NM790	4.0	2 TB	40 MB
$\mathcal{C}$	Samsung 980	3.0	1 TB	64 MB
$\mathcal{D}$	Western Digital Blue SN550	3.0	1 TB	32 MB

Flash Translation Layer. Wear-leveling and other performance optimizations cause the data on an SSD to be scattered. Therefore, a persistent translation layer is required that translates the logical page addresses, the ones seen by the operating system, to the physical page addresses of actual storage locations.

If the FTL was only stored in the flash memory, every read would incur an additional read. Therefore, SSDs employ different caches for their FTL. Budget SSDs only have a small on-chip cache in the SSD controller. "Pro"-level SSDs contain a DRAM chip that is large enough to hold the entire FTL. Mid-range SSDs can use a feature called Host Memory Buffer as a trade-off.

Host Memory Buffer (HMB). HMB is an NVMe feature that allows SSDs to request main memory from the operating system [17]. The SSD can then access the HMB through direct memory accesses (DMA) over PCIe and cache parts of the FTL there. DMA accesses to the main memory over PCIe are slower than accesses to an integrated DRAM but faster than accessing the flash memory for each FTL entry. The HMB is not large enough to store the whole FTL.

HMB Prevalence. Overall, the HMB feature is not uncommon. We used techpowerup's SSD database [3] containing 869 SSDs from 109 manufacturers to understand the prevalence of the HMB feature in SSDs. Of these 869 SSDs, 694 use the PCIe interface. 37 % or 255 of all PCIe SSDs do not have a DRAM. Of these DRAM-less PCIe SSDs, 97.6 % or 249 support the HMB feature. This shows that the feature is very prevalent in DRAM-less SSDs, as it enables a performance gain at almost no cost.

FIEMAP Ioctl. The file extent map (FIEMAP) system call allows unprivileged processes to retrieve the "physical" block addresses of all fragments of a file [12]. In this case, "physical" correspond to the logical block addresses of the SSD as seen by the operating system that are then translated to the real physical addresses using the FTL.

3   The HMB Side Channel

The HMB caches recently used logical to physical translations of the SSD to reduce the access latency to the corresponding pages. We show that access

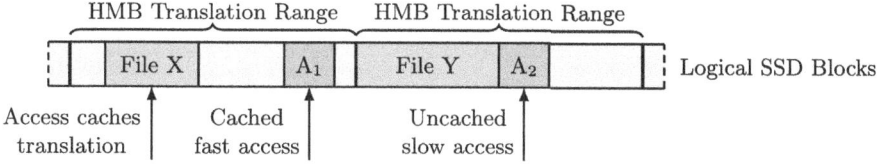

Fig. 1. Working principle of the HMB. An access to file X caches all logical-to-physical translations within a specific storage range in the HMB, including the translation for the page containing file fragment A_1. Therefore, a subsequent access to X or A_1 is faster than an access to Y or A_2.

latency differences are measurable and give an attacker valuable information about recently accessed pages on the SSD. In this section, we reverse-engineer the basic functionality of the HMB of four different SSDs and show that the HMB introduces a timing side channel that is observable from user space. Table 1 shows the SSDs we used for our experiments throughout the paper. Figure 1 shows the basic working principle of the HMB side channel. It shows an SSD's logical storage space as seen by the operating system and its file system driver. After accessing file X, the logical to physical translation to the SSD pages that contain file X but also to the file fragment A_1 are cached in the HMB. This decreases the subsequent access time to X and A_1 while the access to the uncached pages containing file Y or fragment A_2 are slower.

3.1 Reverse Engineering HMB Usage

In this section, we reverse engineer how the different SSDs use the HMB to cache their translations as shown in Fig. 2. We use the IOMMU of a AMD Zen3 Ryzen 7 5800X to observe the accesses from the SSD to the HMB with 4 kB page granularity, similar to Juffinger et al. [8].

Samsung 990 EVO. The Samsung 990 EVO splits the 64 MB HMB into four independent 16 MB cache sets, as shown in Fig. 2a. After accessing 66 GB of data, or 17 million pages, sequentially, the HMB is filled with translations. Afterwards, the SSD controller starts replacing the least recently used (LRU) translations. Bits 14 and 15 of the logical block address select a set.

A lot less *random* accesses are required to fill the HMB, as shown in as Fig. 2b. This is because a single access also prefetches many translations of surrounding pages, filling the HMB. There are also seemingly random HMB accesses in-between the linear patterns. These HMB accesses come from accesses to addresses within an already cached HMB translation range. To get size of these prefetches, we build an experiment where we access an addresses and then measure the cache state of a neighboring addresses with increasing distance. This shows that up to 15 360 pages, *i.e.*, translations to 60 MB of data, are cached at once. After 12 000 accesses in Fig. 2b, we only access addresses of set 2 by fixing bits 14 and 15 of the logical block address to 10.

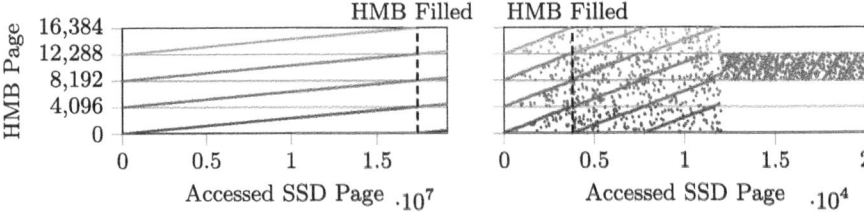

a. $\mathcal{A}$: Sequential SSD accesses causes a linear pattern of HMB accesses. The HMB is split into four sets.

b. $\mathcal{A}$: Random SSD accesses show the linear pattern plus "random" accesses to the HMB.

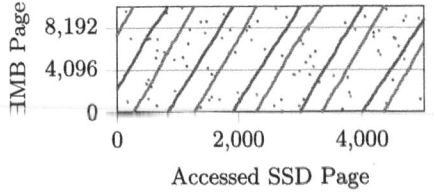

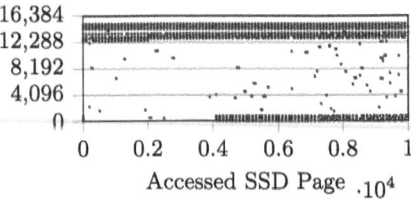

c. $\mathcal{B}$: Sequential and random (shown) SSD accesses show two LRU patterns.

d. $\mathcal{C}$: Sequential and random (shown) SSD accesses show a pattern to specific offsets at the start and end of the HMB range.

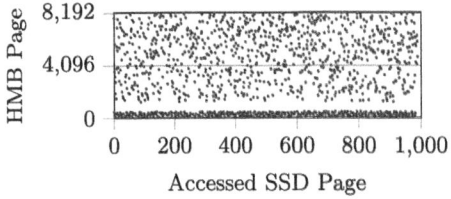

e. $\mathcal{D}$: Sequential and random (shown) SSD accesses show no particular pattern.

Fig. 2. The accesses from the SSD to the HMB caused by accesses to the SSD. The four SSDs behave very differently.

When the SSD is not accessed for 100 ms, it resets the HMB state. Afterwards, each cache set is again filled from the first page and all previously cached translations are not used anymore. The SSD seems to enter a low power state after 100 ms but we cannot think of a reason why this resets the HMB state.

Lexar NM790. The Lexar NM790 also uses LRU cache eviction, as shown in Fig. 2c. It also uses two cache-sets, but differently than the Samsung 990 EVO. Both cache sets can occupy every page of the HMB, but they seem to use an individual LRU counter. Bit 25 of the logical SSD address defines the set. We find in Sect. 3.3 that accessing only addresses of a single set greatly increases variance in the access timings. So it seems like these two sets enable some sort of load balancing in the SSD.

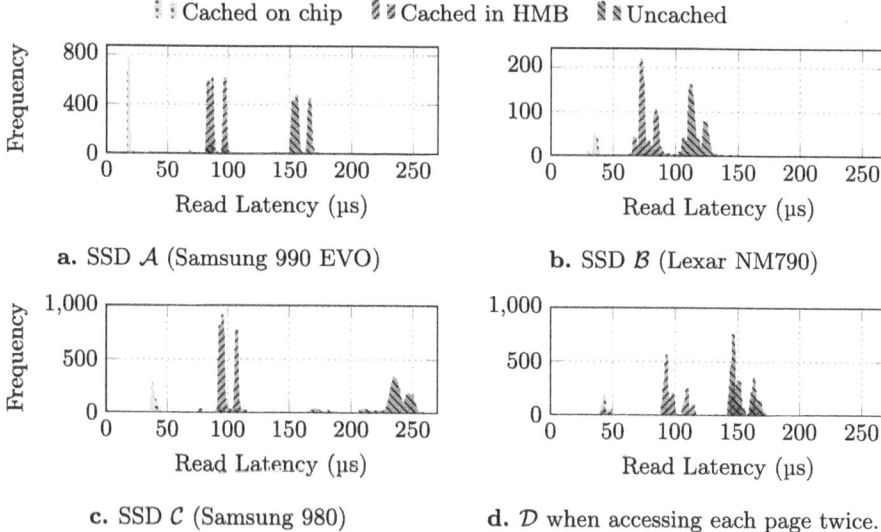

Fig. 3. The timing differences between a translation cached on-chip, cached in the HMB and an uncached translation are clearly distinguishable on our SSDs.

Samsung 980. The Samsung 980 does not show a clear usage pattern like the Samsung 990 EVO or Lexar NM790, neither ith sequential nor random accesses, as shown in Fig. 2d. Most parts of the HMB were only used very sparsely. The Samsung 980 also resets the HMB state after 100 ms.

WD Blue SN550. This SSD splits the HMB into two distinct parts, as shown in Fig. 2e. Every translation consists of two HMB accesses. One to the lower part, between HMB pages 48 and 488 and the other to the upper part above HMB page 1255. We think that the lower part contains a cache directory to store which translations are stored in the upper part of the HMB. The other SSDs seems to store this directory in the SSD controller, requiring more memory.

3.2 Cache Timing Differences

In this section, we show that the timing differences caused by the HMB cache are measurable and clearly distinguishable by user space applications, see Fig. 3. We start by measuring with direct access to the block device and verify that we can also distinguish the cache levels when accessing files as an unprivileged user.

For this experiment, we first "reset" the HMB state by accessing 50 000 random pages of the SSD. Then we access 500 additional random pages and store their addresses. As the translations to these addresses are now cached in the HMB, we put them in our "HMB cached" set. The last N_C of those, we put in the "on-chip" set. With the sets initialized, we randomly choose between accessing a random, very likely uncached page, a random page from the HMB set or a random page from the "on-chip" set. We repeat these access 200 times and continuously update the sets of HMB cached translations by adding all new performed uncached accesses and the "on-chip"-set by always keeping the last N_C accesses in this set. The size N_C is only an estimate for the on-chip cache size because knowledge of the exact size is not required for our attacks.

Samsung 990 EVO. Figure 3a shows the timing histogram. The timings are separated by $50\,\mu s$ with a threshold between uncached and HMB cached accesses of $125\,\mu s$. The on-chip cache has an approximate size of $N_C = 150$ entries.

Lexar NM790. Figure 3b shows the timing histogram. The cached accesses are still clearly separated from the uncached accesses with a threshold of $100\,\mu s$. The on-chip cache is smaller with an approximate size of only $N_C = 40$ entries.

Samsung 980. Figure 3c shows the timing histogram. The timings are again very clearly separated with a threshold of $130\,\mu s$. The Samsung 980 has a very small on-chip cache $N_C \approx 5$.

WD Blue SN550. Figure 3d show the timing histograms when accessing every page twice. Almost no translations are cached, when performing the experiment like on the other SSDs by only accessing each page once. With two accesses, translations get cached but there are still many uncached timings for recently accessed pages.

Unprivileged File Access. While we measured the histograms shown in Fig. 3 by directly accessing the block device, we verified that we measure the same timings when accessing files as an unprivileged user. As already shown by Juffinger et al. [9], the overhead from the file system is negligible.

3.3   HMB Eviction

Measuring the HMB state of a target file is destructive. After the read, the translation is cached and has to be evicted again quickly. In this experiment, we find the minimum number of accesses required to evict the HMB. To do this, we read a random page from the SSD, caching the translation in the HMB. Then we access an increasing number of pages, either randomly or sequentially. For the sequential accesses, we do not access directly neighboring pages but pages with a stride length of one HMB translation range. We use io_uring to submit the reads asynchronously to maximize IOPS. Finally, we access the randomly selected page again and measure the access time. Based on the thresholds from Sect. 3.2, we know if the page was evicted or not.

Samsung 990 EVO. Figure 4a shows eviction with random accesses to all or one cache set. When accessing random addresses, 4 000 accesses are required to evict a translation from the HMB. This takes on average 45 ms. However, due to the four sets, a more targeted eviction is possible. By only accessing addresses that

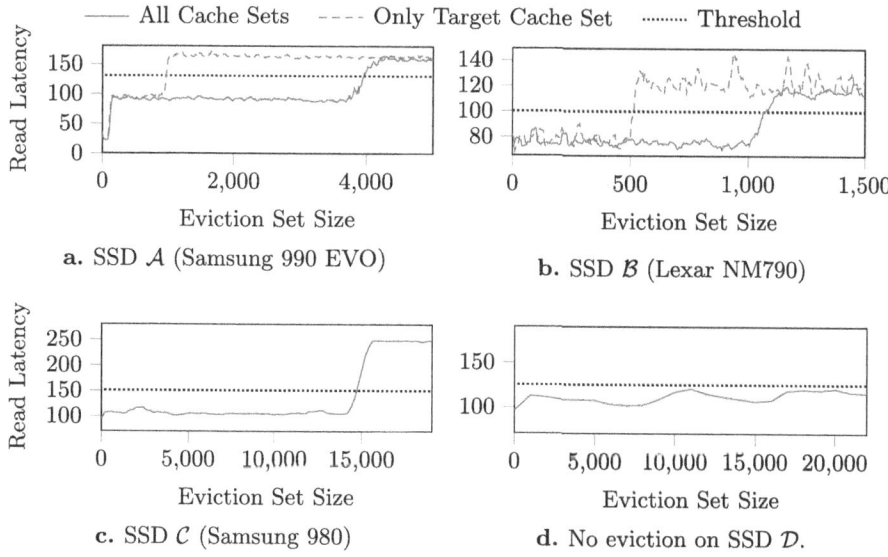

Fig. 4. Eviction with accesses to random pages of the SSD.

fall in the same set as the target, with bits 14 and 15 matching, the eviction only requires 1 000 accesses and takes on average 22 ms.

When accessing the SSD sequentially, 17 million accesses are required to evict the HMB. Our hypothesis is, that sequential accesses are optimized to not pollute the HMB with translations only used once. Translations evicted from the on-chip cache could, e.g., be dropped instead of being moved to the HMB.

Lexar NM790. Figure 4b shows eviction with random accesses to all or one cache set. It takes at least 1 150 random disk accesses, which takes on average 11 ms. When only accessing the same cache set as the target is in, by fixing bit 25, the eviction time is halved to 530 accesses (6 ms). However, when only accessing addresses from the same cache set, there is a significantly higher variance in the access timings: $\sigma = 29.5$ vs $\sigma = 15.7$ when accessing both sets, $n = 1000$.

Samsung 980. On the Samsung 980 it takes around 15 000 random accesses (340 ms) to evict a translation from the HMB, as shown in Fig. 4c. Sequential accesses, even for over 20 s, only very infrequently evict the target from the HMB.

WD Blue SN550. We did not manage to reliably and quickly evict a translation from the HMB on the WD Blue SN550 with neither sequential nor random accesses, as shown in Fig. 4d. Even when accessing more than a 100 000 pages, multiples times each, for over a second, our target page translations was still fast. Our hypothesis is, that the SN550 can store more detailed translations due to a bigger directory. We show in Sect. 3.1 that the SN550 accesses the lower

2 MB of the HMB for every translation, hinting its usage as a directory. While other SSDs store their directory on the controller, they only store translations of large blocks of addresses in the HMB. Storing translations of only small blocks, maybe even in page granularity, makes eviction very tedious.

4   Attacks on Accessible Files

In this section, we show a covert channel and a UI redress attack. In this threat model, the attacker has read permissions for the file they want to observe. This means that the attacker can measure the translation cache state of the file they want to observe directly by reading it. We show a covert channel between processes that share access to files in /usr and other shared directories in Sect. 4.1. The covert channel reaches up to 8.3 kbit/s. In Sect. 4.2, we show a UI redress attack that can detect when pkexec is started with a high accuracy.

As we could not reliably evict the HMB of the WD Blue SN550, we are not further evaluating our attacks with it.

Cache Eviction. The big advantage of the HMB side channel over the page cache side-channel is that evicting the HMB is a magnitude faster while causing no memory pressure. Gruss et al. [6] managed to evict the page cache in 149 ms, while it took Juffinger et al. [9] 347 ms with lower memory pressure. Evicting the HMB takes only 22 ms or less on the Samsung 990 EVO and Lexar NM790. This enables a higher sampling rate for more accurate attacks and smaller blind spots. For even faster eviction on the Samsung 990 EVO or Lexar, the attacker needs to know the cache set of the victim file. For this, we use the unprivileged FIEMAP ioctl to get the physical mappings of the file on the disk.

If the target we want to observe is in the page cache, we have to evict it. This is once at the beginning and then only after the event we monitor, like the execution of a binary, actually happens. We perform our HMB timing measurements with O_DIRECT to not load the target into the page cache.

Preventing False Positives. Due to the design of the HMB, multiple translations are cached together in what we call the HMB translation range. While we exploit this in blind attacks in Sect. 5, they can cause false positive measurements. A false positive happens if not our target file is accessed and its translation therefore cached in the HMB, but any other file that is in the same HMB translation range. We can reduce the chance of this happening with the *help* of the page cache.

Again, with the unprivileged FIEMAP ioctl, we can get all files we have read permissions for that are in the same HMB translation range. On a not strongly fragmented file system, the chance that we have the permissions to read neighboring files is high. If our target is, for example, a binary or library file, they probably reside next to other binary or library files which are typically readable on Linux. After learning which files could cause fault-positive measurements, we read them to load them into the page cache. Therefore, every other program reading any of these files does not access the SSD but gets the file served from the page cache. We then add these files to our page cache eviction set, so they are not evicted when we have to evict our target file from the page cache.

Table 2. All covert-channel results.

Threat Model	SSD	# Co-Locations	Transmission Rate	Bit Error Rate	Channel Capacity
Process (Sect. 4.1)	$\mathcal{A}$	256	4.4 kbit/s	3.9 %	3.3 kbit/s
	$\mathcal{B}$	512	8.8 kbit/s	0.7 %	8.3 kbit/s
	$\mathcal{C}$	256	594 bit/s	1.4 %	532 bit/s
Virtual Machines (Sect. 5.1)	$\mathcal{A}$	1	7.1 bit/s	0 %	7.1 bit/s
	$\mathcal{B}$	1	1.9 bit/s	0 %	1.9 bit/s
	$\mathcal{C}$	1	1.7 bit/s	0 %	1.7 bit/s
Remote (Sect. 5.2)	$\mathcal{A}$	1	10 bit/s	1.5 %	8.9 bit/s
	$\mathcal{B}$	1	4 bit/s	3.0 %	3.2 bit/s
	$\mathcal{C}$	1	2.7 bit/s	0.9 %	2.5 bit/s

4.1 Covert Channel Across Processes

In this section, we show a covert channel across processes that can transmit data with up to 8.3 kbit/s, as shown in Table 2.

Threat Model. Two processes are on the same machine without any means of communication. They have access to a shared set of files, e.g., in /bin, /lib or /usr and a precise clock source like rdtsc or clock_gettime. If no process has access to enough files to evict the HMB, at least one of the processes must be able to create a file for HMB eviction. No other privileges are required.

Implementation. The basic idea is to find a set of files both processes have access to and use each file to transmit a bit of the message. This enables for the transmission of multiple bits per time slice and HMB eviction. For this to work, two requirements for these files must be met: First, each file used for the transmission must be in its own HMB translation range, so they are all independent of each other. Second, the two processes must agree on the files used for the transmission and the order of these files.

File Independence. If two files used to transmit individual bits were in the same HMB translation range, setting one bit would always also cache and, therefore, set the other bit. We use the unprivileged FIEMAP ioctl to get the physical block location of each file on the disk. The sender and receiver can then select files that are spaced out enough on the disk. Large or fragmented files can also be used to transmit multiple bits if they span multiple HMB translation ranges. We do not perform *false positive prevention* for our covert channel.

File Selection and Order. Both, the sender and receiver must use the same files in the same order to transmit the message. To achieve this, we only use files that are world readable, if a file is readable due to owner or group permissions, it is not used. To use the files in the correct order, we lexicographically order the full file paths and then randomize them with a seed known to the sender and receiver. The randomization ensures that we do not perform too many sequential accesses during transmission which could trigger HMB optimizations of the SSD.

Transmission. The sender and receiver open the files with the O_DIRECT flag to evade the page cache. With both processes sharing a precise clock source, we can use a time-sliced approach like prior work [4,9,13,19,24]. This means that the transmission is divided into time slices known to the sender and receiver.

In the first time slice, the sender accesses all files corresponding to 1-bits, to cache their translation. We use *two* files per bit. In the next time slice, the receiver accesses the files and records slow timings from *both* files as a 0 and one or two fast timings as a 1. The third time slice is used by the sender to evict the HMB. Sender and receiver can swap their roles for bi-directional communication.

We use io_uring for asynchronous accesses for sending, receiving, and eviction. While asynchronous accesses are considerably faster, issuing them too fast can cause contention in the SSD [9], slowing down the accesses and interfering with our cache timing measurements. Therefore, we find the optimal delay between the measurement (receive) accesses for all three SSDs. We also had to add a shorter delay between the sending accesses.

We record the raw transmission rate and bit-error rate to compute the true channel capacity, the maximum transmittable information over a noisy channel. A bit-error ratio of 0 or 1 corresponds to a perfect transmission; 0.5 means no information was transmitted. We use the binary symmetric channel model to compute the true channel capacity T as $T = C \cdot (1 + (1-p) \cdot \log_2(1-p) + p \cdot \log_2(p))$ where C is the raw transmission rate and p the bit-error rate.

Results. All results are shown in Table 2. On the Samsung 990 EVO we transmit 128 bits using 256 files, reaching a raw transmission rate of 4.4 kbit/s with a 3.9 % bit error rate, resulting in a channel capacity of 3.3 kbit/s. We use a 10 μs delay between the send accesses and 50 μs between the measurement accesses. We only access SSD pages in a single set to make the eviction four times faster.

On the Lexar NM790 we got an even lower error rate while doubling the number of files used for the transmission to 512. With them, we transmit 256 bits per time slice for a raw transmission rate of 8.8 kbit/s with a bit error rate of 0.7 %. This results in a channel capacity of 8.3 kbit/s. The raw transmission rate is not only increased by the larger number of files used, but also by the smaller required delays between the send and receive accesses, 3 μs and 20 μs respectively. This enables us to access all 512 files for sending and receiving, in the same time, we can send and receive 256 files on the Samsung 990 EVO. We use SSD pages from both cache sets. Using a single set increased the variance of the timing measurements and the error rate, decreasing the channel capacity.

The very slow eviction that takes over 350 ms on the Samsung 980, makes its covert channel considerably slower. Accessing the 256 files for sending and receiving takes up only 15 % of each time slice. However, the bit error rate is also very low with only 1.4 %. With a raw transmission rate of 594 bit/s, this results in a channel capacity of 532 bit/s.

4.2 Authentication UI Redress Attack

In this section, we show that we can detect when a file is accessed on the disk with high accuracy. We use this for a user-interface redress attack [2,6,20]. In an UI redress attack, the attacker waits for a program to be started to then draw their own fake window over the genuine one. In our example, we detect the start of pkexec to steal the password of the victim user. pkexec displays a password promt to execute GUI applications with super user privileges. For this to work well, the latency between the genuine drawn and the fake window drawn over it must be small. The short eviction time of our HMB side channel enables this low latency detection. It takes at most 100 ms to detect the start of pkexec.

False Positive Prevention. To prevent false positives, which would be strange for the victim and could make them suspicious, we load the other files in the same HMB translation range in the page cache. To test the viability of this app-roach, we use a privileged user to find all co-located files for both a 2 MB and a 60 MB HMB translation range. In the 2 MB surrounding /usr/bin/pkexec find that all 15 co-located files are in /usr/bin. All of these files are read-able by a user without elevated privileges. In the 60 MB range, we find 1 969 files. Constantly keeping all of these files in the page cache is time consuming. We therefore exclude files that are unlikely to be read, like man files for other locales (e.g., /usr/share/man/id/man1/), firmware images in /lib/firmware/, or /lib/x86_64-linux-gnu/fwupd-plugins-5/. This leaves us with 793 files, overall 46.5 MB large, to keep in the page cache.

Some of these files are log files. Keeping them in the page cache does not prevent disk accesses if the file is written. To prevent false positives from writes, we check the last modification date of all of our 793 files when we measure a HMB cache hit. If one or more of the files where written in the last few seconds, we count the cache hit as a false positive.

Implementation. We implement a page cache eviction algorithm similar to Gruss et al. [6]. However, we only have to perform page cache eviction once at the beginning of the monitoring to evict pkexec. We, additionally, perform page cache eviction every minute because our monitoring could miss an execution of pkexec due to the blind spot during HMB eviction.

In the monitoring loop, we start by sleeping for 100 ms, during this time pkexec could be executed. This means that it takes us at most 100 ms to detect the execution of pkexec. During this sleep interval the Samsung SSDs require periodic reads to keep the HMB active and preserve the HMB state, we perform one 4 kB read each millisecond. Then, we perform our measurement by reading the first page of `pkexec`. Our attacks also works by reading a co-located file like `ping`. It could be less suspicious if a program opens `ping` instead of `pkexec`.

Performing the measurement caches the translation in the HMB. This creates a blind spot until the HMB is evicted. For the Samsung 990 EVO and Lexar NM790 we know the correct cache set, because we know the physical block address from `FIEMAP`. This enables quick HMB eviction.

Results. On the Samsung 990 EVO SSD the eviction takes on average 23.2 ms ($n = 1000$, $\sigma = 0.18$ ms). This means that one whole measurement window takes $100\,\text{ms} + 23.2\,\text{ms} = 123.2\,\text{ms}$. In this window, we are blind for 23.2 ms or 18.8 %. This is a trade-off between the relative size of the blind spot and the time it takes at most to detect the execution of `pkexec`. On the Lexar NM790, the eviction takes only 6 ms resulting in a blind spot of only 6 %. We chose a sleep duration of 100 ms, because this means that with a 30 fps screen we are at most 3 frames too late to draw our window over the `pkexec` window.

Because our attack process sleeps a lot and also block device accesses are not CPU intensive it, does not cause easily noticeable CPU usage. The attack causes on average 3.3 % CPU usage and 32 MB/s disk usage.

The slow eviction on the Samsung 980 takes approximately as long as page cache eviction. This defeats the main advantage of the HMB side channel over the page cache side channel, making the page cache an easier and more portable attack target.

5 Blind Attacks

In this section, we show how attacks can be performed if the attacker does not have read permissions for the files it wants to observe. Because the attacker cannot read the file it cannot use the `FIEMAP` ioctl to get the exact mappings of the file on the disk. Therefore, we fall back to a templating approach, where the attacker learns which parts of the attacker file are in the same HMB translation range as the victim files. We first show a covert channel between two virtual machines in Sect. 5.1, reaching up to 7.1 bit/s. In Sect. 5.2, we show how an attacker can exploit nginx as a confused deputy to build a remote covert channel over the network. This covert channel reaches up to 8.9 bit/s. For both covert channels, we only use a single file for transmission, for simplicity, instead of the 256 and 512 for the process covert channel.

5.1 Covert Channel Across Virtual Machines

In this section, we show a covert channel between two virtual machines. We achieve up to 7.1 bit/s on the Samsung 990 EVO.

Threat Model. Two unprivileged processes are inside two virtual machines. They want to communicate because, e.g., one process wants to transmit a secret to the other one. The processes have no means of communication. Both processes have the permission to create files on their system. They also have a access to a shared clock. The VM disk is not cached on the host, as recommended, for example, by Red Hat [7]. To enable repeated tries to co-locate, the VM disk must also be configured with TRIM support or even better, with automatic trim when zeros are written [11]. We run our experiments on KVM with libvirt.

Implementation. For this attack to work, we need HMB co-location between the VM disks for one VM to be able to influence the cache state of the other.

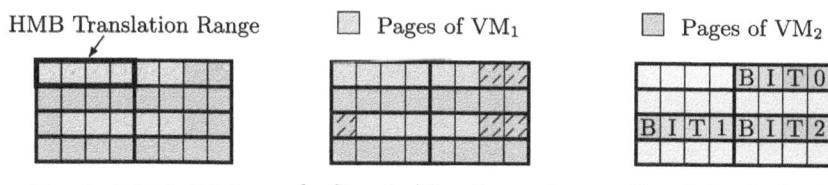

a. Step 1: Initial disk layout of sender ☐ and receiver ☐ files.

b. Step 2: After the sender accesses the whole file, the receiver gets all pages with cached timings: ▨.

c. Step 3: In the final step, the sender gets its pages with cached timings.

Fig. 5. The algorithm used to find all pages within the sender and receiver files that are in one HMB translation range. With five overlapping pages, five bits can be transmitted in one time slice.

Co-location. The communicating parties previously agree on a time when they start their covert communication. At the respective time, sender and receiver create a file containing random data. We assume that the operating system in the virtual machine, periodically uses TRIM to return unused disk space to the host, this is the default behavior, e.g., in Ubuntu. Therefore, the disk files of both virtual machines are sparse and grow with the created files, causing fragmentation and interleaving of the disk files.

We find to achieve maximum fragmentation and most overlaps between the two VM disk by writing in a three step process. In each VM, we write a 16 kB block, flush it to the disk using the 'sync' syscall and finally, punch a hole into this block, keeping only the first 4 kB. We repeat these three steps until we wrote between 50 MB and 200 MB. The steps are not synchronized between the VM.

Co-location Detection. The virtual machine guest cannot get FIEMAP data of the host, therefore, it has to resort to another algorithm to detect which pages (4 kB blocks) of the files are within one HMB translation range, as shown in Fig. 5. The sender repeatedly access a number of pages of their file to cache the translations. The receiver accesses their whole file and measures the access times. If the receiver measures a cached timing they store the offset. Then the receiver accesses the stored offsets that are potentially overlapping and the sender reads and measures the timings. After doing this for the whole file, the sender and receiver both know which offsets of their files are in the same HMB translation ranges. In a last step, the bit order must be measured. For this, the sender sends 1 bit every second and the receiver records the order in its file.

Transmission. We use the same time-sliced transmission as in the covert channel across processes, see Sect. 4.1. However, we only implement the transmission with a single bit per time slice for simplicity.

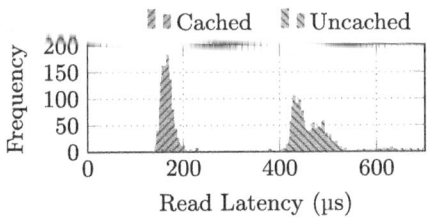

Fig. 6. Timing histogram measured from within a VM on SSD $\mathcal{C}$ (Samsung 990 EVO). It seems like libvirt multiplies the timing difference between the cache states, cf. Fig. 3b.

Results. To measure the overhead introduced by the disk virtualization of libvirt, we transmit a toggling bit between two virtual machines and measure and plot the timings on receiver side. Figure 6 shows the resulting histogram. The timings are clearly more separated than in the native scenario, cf. Fig. 3b. On the Lexar NM790, the timings are even more separated, 569 µs ($n = 1000$, $\sigma = 267$ µs) vs 4881 µs ($n = 1000$, $\sigma = 2298$ µs). We ruled out other reasons for this timing like the page cache. Our hypothesis is, that our highly fragmented disk files span many different HMB translation ranges and the read-ahead of the VM operating system, therefore, multiplies the uncached timing.

Co-locations. For the covert channel to work, at least one fragment of the VM disk files each must be co-located in the same HMB translation range. The technique described above maximizes the chance of this happening. We run the file creation five times, always starting from a trimmed VM disk. On average we get 14091 ($n = 5$, $\sigma = 11002$) co-locations on the Lexar NM790. In one case no co-location was found. The larger HMB translation ranges of the two Samsung SSDs increase the number of co-locations respectively.

Transmission. We transmit random data to measure the transmission and error rate of the covert channel. Because we only transmit a single bit per time slice for simplicity, the transmission is considerably slower than with the process covert channel. However, the high average number of co-locations we found between the VMs would allow for multi bit permission with higher transmission rates.

On the Lexar, the large delay for an uncached accessed and the resulting slow eviction time of 500 ms results in a slow transmission of 1.9 bit/s. However, due to the large separation between cached and uncached timings, we did not measure a single bit error over 400 transmitted bits. On the Samsung 990 EVO, we achieved a raw transmission rate of 7.1 bit/s, again without any errors over 400 transmitted bits. The HMB eviction takes 120 ms. On the Samsung 980 we, we achieved a raw transmission rate of 1.7 bit/s, again without any errors

5.2   Remote Covert Channel Attack

We show that an attacker can communicate through a benign website by inducing timing differences into HTTP responses. Schwarzl et al. [21] showed that timing differences of only 60 μs are measurable over 14 hops on the internet. This matches the timing differences on the Samsung SSDs.

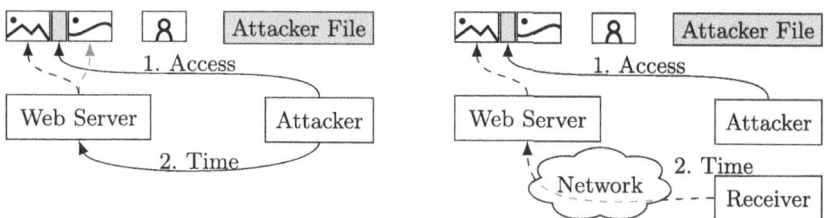

a. Templating: The attacker accesses all offsets of the attacker file and learns about co-located assets by timing the web server.

b. While the attacker transmits a known sequence the receiver probes all assets and finds the one used for the transmission.

Fig. 7. The remote covert channel exploiting a benign web server as a confused deputy. The translation to one of the assets is either cached in the HMB or not.

Threat Model. We assume a threat model with two processes running on a system: First, a networked application, e.g., a web server like nginx, that is reachable by a remote attacker. Second, an unprivileged application that is isolated from the network but can access the web server through localhost, *i.e.*, the process has access to secret information but no outside network access. We assume that both application have access to a (non-overlapping) set of files on the disk, *i.e.*, they can perform disk operations but not directly communicate through any shared writeable or read-only files. For example, the attacker process has *no* access to the website assets served by the web server. This is typical, with

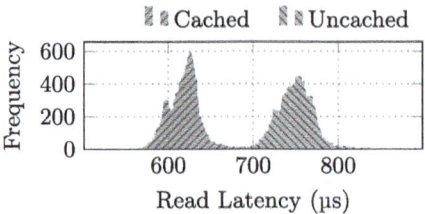

Fig. 8. Timing histogram of the Samsung 990 EVO measured over the network, exploiting nginx as a confused deputy. The distributions are spread out but overlap only barely, cf. Fig. 3a.

the web server running with its own user. The web server uses direct disk I/O for large files, see Sect. 5.2. If the attacker had access to website assets, the templating phase could be skipped.

Communication Protocol. Figure 7 shows the attack overview. The basic idea is that the sender process either caches or not caches the translations to assets of the website. For this, the sender process must co-locate a file with a websites asset. If the receiver then accesses this asset they can infer if the translation was cached or not based on the response time of the HTTP request.

Implementation. For this attack to work we need HMB translation range co-location, direct I/O and low measurement overhead on the receiver.

Co-location. To gain reliable co-location with many website assets, we create a large file, filling the whole disk for a brief moment. After creation of the file, we instantly free the first 90 % using FALLOC_FL_PUNCH_HOLE. The high disk pressure only persists for less than a second. In a final step, we punch more holes into the file to keep only one page per HMB translation range. Using the FIEMAP ioctl, we know which file offsets to free. This further reduces the size of our file, while we keep at least one page in many HMB translation ranges. If we do not yet have co-location with a usable website asset, there is a high chance that new assets will be co-located with fragments of our file.

Direct I/O. For this attack to work, we again have to evade the page cache. The nginx web server can be configured to use direct disk I/O with O_DIRECT on Linux when serving large files [16]. This has the advantage, that these big files then do not pollute the page cache. The nginx documentation recommends to enable it for files with 4 MB or more [16].

    To minimize overhead and noise from the network transmission, we want to transmit as little data as possible. However, the asset served with O_DIRECT are multiple megabytes large. To circumvent this restriction, we use the range requests feature of HTTP [15]. It allows the client to request only a specified part of a file. We find that nginx always accesses the disk, when requesting at least 32 kB. Shorter ranges are still cached somewhere as we did not see nginx access the disk when requesting a shorter range more than once.

Templating. Before the transmission can start, the sender must find website assets co-located with its file fragments. First, the sender traverses the whole website to find files larger than 4 MB. Then it transmits a known pattern through up to 512 fragments of its file using the HMB and measures the access times to the assets though localhost. If the assets are large enough to span multiple HMB translation ranges, HTTP range requests are used to check multiple file offsets.

Sender. We again split one time slice in three parts. One for the sender to evict the HMB, the next to cache the required HMB translations, and the last for the receiver to measure the access time. We synchronize the sender and receiver using `clock_gettime` with the `CLOCK_REALTIME` clock source. After evicting the HMB, the sender accesses the co-located file fragment if the bit to transmit is a **1**, caching the translation in the HMB. While waiting, the sender periodically accesses the disk to prevent it from going into a low power state.

Receiver. The receiver measures the timing of the HTTP request, using the `Range` header to limit the response to 32 kB. While the receiver is waiting for the response to the HTTP request, Linux halts the core the receiver is running on, adding large timing variations to the timing measurement of the request. To mitigate this, we keep the core awake by running a busy loop in the sibling SMT thread while waiting for the response. At the beginning of the transmission, the receiver also has to traverse the whole website to get all large assets and then measure all of them to detect the one that is transmitting a known sequence.

5.3   Results

We measured the covert channel using nginx/1.18.0 between two machines in the same network connected by one switch. Figure 8 shows the timing histogram of the Samsung 990 EVO. The added timing overhead from nginx and the network is not large enough for the two distributions to overlap, except for some outliers. This allows us to transmit 10 bit/s with an error rate of 1.5 %, resulting in a true capacity of 8.9 bit/s. The faster eviction on the Lexar NM790 allows for a raw transmission rate of 20 bit/s. However, the cached and uncached timing overlap, therefore, we use 5 measurements per bit, decreasing the raw transmission rate to 4 bit/s. With an error rate of 3.0 % this results in a true capacity of 3.2 bit/s. The slower eviction of the Samsung 980 slows down the covert channel to a raw capacity of 2.7 bit/s with an error rate of 0.9 %, for a true capacity of 2.5 bit/s.

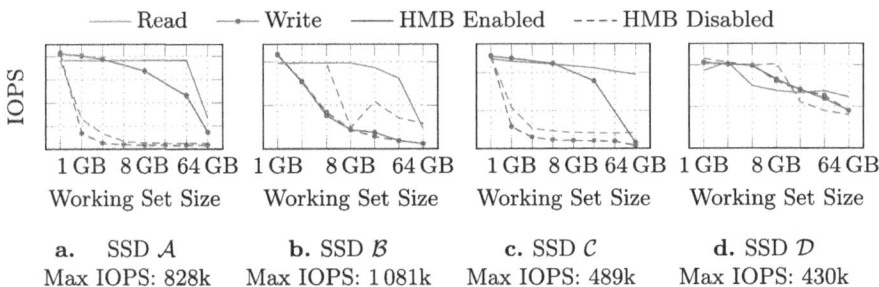

Fig. 9. The impact of disabling the HMB on random 4 kB IOPS at different working set sizes. The queue depth for the measurement is 32 with 16 parallel threads.

6   Countermeasures

There are ways to mitigate the HMB side channel or make exploitation harder.

Disabling the HMB. Every SSD with the HMB feature must also work without host memory resources [17], albeit with reduced performance. On Linux the HMB can be disabled for all SSDs in the system with the kernel command-line parameter nvme.max_host_mem_size_mb=0, on Windows it is possible by setting the registry key HKEY_LOCAL_MACHINE\SYSTEM\CurrentControlSet\Control\ -StorPort\HMBAllocationPolicy to 0.

Figure 9 shows the performance degradation when disabling the HMB. We measured the maximum achievable random IOPS when performing random 4 kB reads or writes with a queue depth of 32 and 16 threads on different working set, *i.e.*, file sizes. On SSDs $\mathcal{A}$ and $\mathcal{C}$, the IOPS of read and write accesses already decline by 80 % with a working set size of only 1 GB. They are practically unusable without the HMB. SSD $\mathcal{B}$ seemingly uses the HMB only for reads, they stay fast up until a working set size of 4 GB. We did not measure any difference in the write IOPS. On SSD $\mathcal{D}$ we measured a performance improvement when disabling the HMB over multiple measurements. However, it is also the SSDs least affected by the side channel. We did not see any differences in the sequential accesses on any of the SSDs. These results show, that depending on the workload and SSD, disabling the HMB can have a significant impact on the performance.

Modifying HMB Usage. We also see very different behavior of the HMB. The two newer PCIe 4.0 SSDs use LRU cache replacement which makes eviction very fast. On the Samsung 980 eviction takes more than 10 times longer and on the WD Blue SN550 we could not reliably evict the HMB. These older HMB implementations may have other disadvantages but they are superior security wise as they make attacks slower or impossible.

Making Software Interfaces Privileged. We use O_DIRECT to bypass the page cache for our timing measurements. Disabling it would make the attacks slower by multiple magnitudes, as the page cache would have to be evicted for every measurement. However, it is also used by benign applications like databases or as shown in Sect. 5.2 by nginx, for performance reasons. The usage of O_DIRECT could be controlled through a new capability that restricts its use to select programs. The same is true for the FIEMAP ioctl. Restricting untrusted applications from using it would greatly impede our attacks on accessible files.

7 Conclusion

In this work we show that the HMB feature of NVMe SSDs introduces exploitable timing differences measurable from use space. It is supported by almost all DRAM-less NVMe SSDs and we expect most of them to be exploitable. We evaluate four SSDs and reverse engineer how they use the HMB and analyze the timing differences and quick HMB eviction. Then we show four attacks exploiting the HMB. In the first two attacks we have access to the files we want to observe and build a covert channel with up to 8.3 kbit/s and a UI redress attack that detects the start of an application in less than 100 ms. In the blind attacks we use templating to find co-locations in the HMB and show a covert channel across VMs and network covert channel exploiting nginx as a confused deputy. Finally, we show that the HMB side channel can be mitigated by disabling the HMB, albeit with a significant performance impact on some workloads and SSDs.

Acknowledgements. This research is supported in part by the European Research Council (ERC project FSSec 101076409), and the Austrian Science Fund (FWF SFB project SPyCoDe 10.55776/F85). Additional funding was provided by generous gifts from Red Hat and Google. Any opinions, findings, and conclusions or recommendations expressed in this paper are those of the authors and do not necessarily reflect the views of the funding parties.

References

1. Bernstein, D.J.: Cache-Timing Attacks on AES (2005). http://cr.yp.to/antiforgery/cachetiming-20050414.pdf
2. Fratantonio, Y., Qian, C., Chung, S.P., Lee, W.: Cloak and dagger: from two permissions to complete control of the UI feedback loop. In: S&P (2017)
3. Gabriel, F.: SSD Database (2025). https://www.techpowerup.com/ssd-specs/
4. Giechaskiel, I., Tian, S., Szefer, J.: Cross-VM covert- and side-channel attacks in cloud FPGAs. ACM Trans. Reconfigurable Technol. Syst. (2022)
5. Gras, B., Razavi, K., Bosman, E., Bos, H., Giuffrida, C.: ASLR on the line: practical cache attacks on the MMU. In: NDSS (2017)
6. Gruss, D., et al.: Page cache attacks. In: CCS (2019)
7. Herrmann, J., Zimmerman, Y., Parker, D., Radvan, S.: Red hat enterprise Linux 7 - virtualization tuning and optimization guide (2019)

8. Juffinger, J.: An analysis of HMB-based SSD Rowhammer. In: UASC (2025)
9. Juffinger, J., Rauscher, F., La Manna, G., Gruss, D.: Secret spilling drive: leaking user behavior through SSD contention. In: NDSS (2025)
10. Kocher, P.: Timing attacks on implementations of Diffe-Hellman, RSA, DSS, and other systems. In: CRYPTO (1996)
11. libvirt: Domain XML format (2025)
12. Linux: Fiemap (2024). https://docs.kernel.org/filesystems/fiemap.html
13. Liu, F., Yarom, Y., Ge, Q., Heiser, G., Lee, R.B.: Last-level cache side-channel attacks are practical. In: S&P (2015)
14. Liu, S., Kanniwadi, S., Schwarzl, M., Kogler, A., Gruss, D., Khan, S.: Side-channel attacks on optane persistent memory. In: USENIX Security (2023)
15. Mozilla: HTTP range requests — MDN (2 2025). https://developer.mozilla.org/en-US/docs/Web/HTTP/Range_requests
16. nginx: directio — Docs ngx_http_core_module (2025). https://nginx.org/-en/-docs/-http/-ngx_http_core_module.html#directio
17. NVM Express, Inc: NVM Express, rev 1.2.1 (2016)
18. Osvik, D.A., Shamir, A., Tromer, E.: Cache attacks and countermeasures: the case of AES. In: CT-RSA (2006)
19. Pessl, P., Gruss, D., Maurice, C., Schwarz, M., Mangard, S.: DRAMA: exploiting DRAM addressing for cross-CPU attacks. In: USENIX Security (2016)
20. Rydstedt, G., Gourdin, B., Bursztein, E., Boneh, D.: Framing attacks on smart phones and dumb routers: tap-jacking and geo-localization attacks. In: 4th USENIX Conference on Offensive Technologies (2010)
21. Schwarzl, M., Borrello, P., Saileshwar, G., Müller, H., Schwarz, M., Gruss, D.: Practical timing side channel attacks on memory compression. In: S&P (2023)
22. Schwarzl, M., Kraft, E., Gruss, D.: Layered binary templating. In: ACNS (2023)
23. Song, L., et al.: The early bird catches the leak: unveiling timing side channels in LLM serving systems. arXiv (2024)
24. Trochatos, T., Etim, A., Szefer, J.: Covert-channels in FPGA-enabled SmartSSDs. ACM Trans. Reconfigurable Technol. Syst. (2023)
25. Yarom, Y., Falkner, K.: Flush+reload: a high resolution, low noise, L3 cache side-channel attack. In: USENIX Security (2014)

Cohere+Reload: Re-enabling High-Resolution Cache Attacks on AMD SEV-SNP

Lukas Giner$^{(\boxtimes)}$ (ID), Sudheendra Raghav Neela (ID), and Daniel Gruss (ID)

Graz University of Technology, Graz, Austria
{lukas.giner,sudheendra.neela,daniel.gruss}@tugraz.at

Abstract. Confidential computing platforms, e.g., AMD SEV-SNP, allow running mutually distrusting workloads on the same hardware with the protection of several isolation mechanisms: data is encrypted in RAM, and access to unencrypted data is architecturally prevented. Furthermore, access and cache line operations are restricted, mitigating attacks like Flush+Reload. The hypervisor can access the encrypted data of virtual machines, e.g., for migration purposes. This creates a coherency challenge around modifications between encrypted and decrypted cache lines. AMD enforces coherency between these two cache lines by removing one when the other is *accessed*.

In this paper, we present Cohere+Reload, a novel side-channel attack exploiting AMD's coherency for encrypted memory. We discover two types of leakage in the coherency mechanism: First, coherence conflicts leak victim operations on a spatial granularity of a 2 kB block. Second, the timing correlates with number and location of accesses the victim performed within the confidential virtual machine, allowing to infer how often or where within a coherence partition victim accesses were performed, with a maximum spatial resolution of 256 bytes. We evaluate Cohere+Reload in two synthetic and two real-world attacks: In synthetic attacks, we demonstrate that Cohere+Reload can observe the control flow and access locations in workloads within a confidential virtual machine. We present a real-world attack on mbedTLS RSA, leaking 4096 key bits in a single-trace attack, with 99.7 % of bits correct. We present another real-world attack on OpenSSL AES exploiting disalignments on a cache line granularity: In a first round T-table attack we achieve an accuracy of 100 % in only 1500 encryptions and with a novel correlation attack an accuracy of 92.81 % in 12000 encryptions. We conclude that the coherence approach for AMD SEV-SNP should be re-evaluated and discuss further potential mitigations.

1 Introduction

Modern processors have a multi-layered memory hierarchy for data, including code. Data can reside in registers, in cache lines in L1, L2, or L3 cache, or in the RAM. Some processors have even further cache layers, e.g., an L4 cache. While

M. Egele et al. (Eds.): DIMVA 2025, LNCS 15747, pp. 191–212, 2025.
https://doi.org/10.1007/978-3-031-97620-9_11

caches are crucial for the performance of modern computers, they also inherently introduce timing side channels that distinguish cached from non-cached data.

The most widely known attacks are Prime+Probe [37] and Flush+ Reload [50]. Flush+Reload [50] works by constantly flushing a cache line from the cache, using the processor's flush instruction, and measuring how long it takes to reload the cache line. Flushing a cache line requires read access to the memory, e.g., read-only shared memory with the victim, which is typically not available across virtual machines or in the context of confidential computing. Prime+Probe does not require shared memory. Prime+Probe [37] works by measuring how much time it takes to constantly re-fill a specific cache set. If a victim access falls into the same cache set, the timing increases.

Confidential computing is an emerging compute paradigm where a confidential workload, running isolated inside a virtual machine, is isolated from all other workloads and from the host. More specifically, it is part of the threat model that the hypervisor can be malicious or compromised but the confidential virtual machine remains secure. Confidential computing platforms, e.g., Intel TDX and AMD SEV-SNP, still share the underlying hardware across mutually distrusting workloads running in virtual machines. However, vendors introduced several isolation mechanisms to protect workloads: For instance, data is encrypted in RAM, decrypted on-the-fly when moved into the caches, and the processor prevents direct access to unencrypted data in caches or registers. The hypervisor cannot access unencrypted memory of the confidential virtual machine. Furthermore, cache line operations are restricted, mitigating attacks like Flush+Reload.

Despite the strong isolation, some functionality requires access from the hypervisor to the encrypted data, e.g., migration of virtual machines in the cloud. Consequently, the hypervisor can access the encrypted data of virtual machines. However, this implies that data can be in the caches twice: once encrypted for the host, and once unencrypted for the confidential virtual machine. Clearly, this creates a coherency challenge as a virtual machine may modify a cache line, i.e., the cache contains a modified unencrypted cache line and an outdated encrypted cache line. Google reported that there is a coherency mechanism on Intel TDX [1] for this purpose, where accesses with one key flush all other copies of the address with different keys from the cache. AMD pursued a similar approach by enforcing coherency between the unencrypted and encrypted cache lines by removing one when the other is accessed. Still, AMD does not operate on the granularity of a cache line, as we show in this work.

In this paper, we present Cohere+Reload, a novel attack exploiting that AMD's coherency approach introduces a surprisingly powerful side channel. We thoroughly analyze AMD's coherence mechanism for encrypted memory and discover two properties that form the basis of our Cohere+Reload attack: First, there is a significant timing difference between cache misses and coherence conflicts on a spatial granularity of a 2 kB coherence partition, i.e., half a page. This timing difference directly reveals whether a victim confidential virtual machine just accessed a specific memory location. Second, the amplitude of the timing coarsely correlates with the number of accesses the victim performed. It is also more finely correlated with the location of single victim accesses, allowing to infer which out of 8 alignments within a coherence partition the victim access had, *i.e.*, we have a maximum spatial resolution of 256 bytes, which is in the same order of magnitude as Flush+Reload with a spatial resolution of 64 bytes.

We evaluate Cohere+Reload in two synthetic and two real-world attacks: The two synthetic attacks demonstrate that Cohere+Reload can observe the control flow in workloads within a confidential virtual machine, *i.e.*, identify the target of a jump; and that Cohere+Reload can observe which data locations a workload within a confidential virtual machine accessed, naturally within the limits of the spatial resolution of Cohere+Reload. We present an attack on mbedTLS RSA-4096 and show that we can leak all 4096 key bits in a single-trace attack, with a Levenshtein distance of less than 11 bits on average. Finally, we present a novel attack on OpenSSL AES that exploits disalignments on a cache line granularity. Based on this insight, we mount a first round attack with and accuracy of 100 % in only 1500 encryptions. We recover all upper nibbles of AES with a novel correlation attack with an accuracy of 92.81 % in 12000 encryptions. We conclude that the coherence approach for AMD SEV-SNP should be re-evaluated and discuss further potential mitigations.

Disclosure. We disclosed our results to AMD, who addressed our findings in security bulletin AMD-SB-3010.

Contributions. In summary, our main contributions are:

- We introduce Cohere+Reload, a novel attack exploiting that AMD's coherency with a spatial granularity of a 2 kB per coherence partition, and 8 distinguishable alignments within a partition, yielding a maximum spatial resolution that is on par with Flush+Reload.
- We evaluate Cohere+Reload in two synthetic attacks demonstrating that we can leak control flow and data access from a confidential virtual machine on AMD SEV-SNP.
- We present an attack on mbedTLS RSA-4096 and show that we can leak all 4096 key bits in a single-trace attack, with a Levenshtein distance of less than 11 bits on average.
- We present a novel attack on OpenSSL AES that exploits disalignments on a cache line granularity, yielding an accuracy of 100 % in only 1500 encryptions in a first-round attack and an accuracy of 92.81 % in 12000 encryptions in a novel correlation attack.

Outline. We provide background in Sect. 2. We define our threat model in Sect. 3. We present our novel Cohere+Reload attack in Sect. 4. We template target pages in Sect. 5. We present an attack on mbedTLS RSA in Sect. 6 and an attack on OpenSSL AES T-Tables in Sect. 7. We present an attack on control flow and data accesses in Sect. 8. We discuss potential mitigations in Sect. 9 and conclude in Sect. 10.

2   Background

In this section, we provide background on trusted-execution environments, side-channel attacks, and coherence in the context of memory encryption.

2.1   Trusted-Execution Environments

The goal of trusted-execution environments (TEEs) is to provide confidentiality and integrity for code and data on a system even on a compromised system [2,5,23,24]. Older TEEs often focus on personal and mobile computers, e.g., Intel Software Guard Extensions (SGX) [23]. The TEE runs a small trusted workload in a signed enclave [23]. These enclaves run on the same CPU as regular applications. To prevent access from a compromised host system, SGX prevents access to the encrypted enclave memory and register state.

More recent TEEs focus on cloud use cases and virtual machines (VMs). Instead of protecting a small workload, the idea is to move entire VMs into the TEE, which are then called confidential virtual machines (CVMs) and protect them from a malicious or compromised host [10], e.g., AMD Secure Encrypted Virtualization (SEV) [3] and Intel Trust Domain Extensions (TDX) [22].

AMD SEV protects memory contents of CVMs by encrypting any data moved out of the CPU, e.g., to DRAM or disk [27]. Still, there are many attacks on SEV. In particular the basic SEV design was demonstrated to provide too little protection for the guest state [19,47] and memory [12,19,35,48]. AMD addressed this issue with the *Encrypted State (ES)* and *Secure Nested Paging (SNP)* SEV extensions, protecting guest state and memory integrity.

Like AMD SEV-SNP, Intel supports CVMs through their Trust Domain Extensions (TDX) [24]. Guest memory and state are encrypted and managed by the TEE. The host can only interact with the guest through well-defined secure interfaces. For fast inter-process communication, the memory has both private encrypted parts and shared parts that are equally accessible to the host.

2.2   Side-Channel Attacks

Side channels can be used to attack systems even if there are no software or hardware vulnerabilities or they are not known. Side channels instead exploit side effects of the implementation such as timing [28], power consumption [29], or radiation [39]. Older works focused on cryptographic primitives [7,9,28], leaking keys of vulnerable cryptographic implementations of e.g., AES [7,37], RSA [9,50], or ECDSA [49]. More recent works often focus on larger systems to leak information from one system component, e.g., kernel information [21], user input [18,34,41], and system activity [16].

Many side-channel attacks target caches as they can be probed without privileges at a high temporal (*i.e.*, nanosecond to microsecond range) and spatial resolution (*i.e.*, 64 B) while being comparably robust against noise. Most importantly, they allow for use generic attacks that are not tailored to specific applications and victim programs. Consequently, the community developed a set of generic attack techniques that follow a uniform naming pattern based on the attack components, e.g., Prime+Probe and Flush+Reload. One of the first generic attack techniques was Evict+Time [37], in which an attacker runs and times a victim process twice, once with evicting a target cache line from the cache by performing a larger number of memory accesses that collide in the cache, e.g., due to set associativity, and once without. A statistically higher execution time means the cache line was used by the victim. Instead of timing the victim, Prime+Probe [37] times the (evicting) memory accesses, *i.e.*, they time how long it takes to (re-)prime a cache set. If it takes more time, more cache lines were replaced by the victim execution. Prime+Probe is one of the most widely used attack techniques besides Flush+Reload [50]. In the Flush+Reload attack, an attacker flushes a cache line using a dedicated flush instruction, and measures the time it takes to reload the memory location in order to decide whether or not the victim used it. Variations of Flush+Reload include Evict+Reload [15], which substitutes eviction for flushing, and Flush+Flush [17], which measures the timing of the flush instruction instead of the reload, thereby revealing a similar timing difference. Similarly, for Prime+Probe, several attack techniques and variations have been presented more recently, such as Prime+Abort [11], Prime+Scope [38], and Spec-o-Scope [20].

2.3   Coherence Between Ciphertext and Plaintext

Modern CPUs feature memory encryption technology, such as Intel's Total Memory Encryption (TME) [25] and AMD's Secure Memory Encryption (SME) [27]. Memory encryption is a crucial aspect of trusted execution environments, including Intel TDX and AMD SEV, designed to safeguard sensitive data. These built-in memory encryption systems encrypt data before it is written to the main memory and decrypt it when loaded into the CPU caches.

Given the nature of trusted computing, the untrusted hypervisor must interact with ciphertexts to facilitate operations such as migrating the guest machine to another server. To enable direct access to encrypted data, AMD's encryption unit includes a short-circuit path that forwards the data without decrypting it. Each data access's physical address encodes a so-called encrypted bit (C-bit), which indicates whether this short-circuit path should be utilized. This leads to an important question: how is coherence maintained when the hypervisor actively requests ciphertext while the guest is processing plaintext?

AMD states that coherency between the ciphertext and plaintext depends on the hardware [4] as some hardware enforce coherency while others do not. In the systems where coherence is not enforced, the hypervisor must flush the encrypted data from all CPU caches. In other systems, hardware supports coherency across encryption domains and software does not have to flush encrypted data, and the presence of this feature can be determined by the CPUID bit "CoherencyEnforced" (see AMD Architecture Programmer's Manual [4], 7.10.6).

In our systems, all AMD EPYC CPUs support SEV-SNP and include this coherence feature. Consumer Ryzen CPUs only support SME [30] without automatic hardware coherence between ciphertext and plaintext. For SEV systems without SNP, not having hardware coherence poses significant risks, allowing potential fault-attack-like exploitation during the write-back of ciphertext.

With the introduction of AMD SEV-SNP, the hypervisor can no longer directly write to an encrypted page [2]. Each guest page undergoes a procedure to assign it in a reverse page map, indicating its ownership to a given guest VM. Once a guest accepts a page, the hypervisor retains only read access, which is ciphertext. Despite these advancements, maintaining coherence between ciphertext and plaintext remains essential. On Intel TDX, it is known that accesses to a ciphertext flushes all other copies of the address from the cache [1]. In Sect. 4, we present our initial analysis of how coherence is managed on AMD systems.

3   Threat Model

Exploiting the Cohere | Reload mechanism requires a ciphertext view and a plaintext view on the same memory region. Outside of SME, this can only happen when a hypervisor maps an SEV guest page (ciphertext view), as guests have no option to map pages outside their allocated memory. Our threat model is therefore a malicious hypervisor trying to extract information from an encrypted SEV, SEV-ES or SEV-SNP guest. In this scenario, the hypervisor has control over all parts of the CPU that are not part of the attestation. This includes control over CPU frequency, disabling hardware prefetching and selecting a suitable DRAM interleaving setting at boot (see Sect. 4.1). While Cohere+Reload attacks can be performed even without stabilizing the frequency or disabling prefetching, like many cache attacks [6,31,33,45,46], it is simplified by these settings and they will be used throughout the paper.

4   Cohere+Reload

In this section we examine the behaviour of coherence for AMD memory encryption. In all of our tests, hardware cache coherence works the same in SME as it does in SEV, SEV-ES or SEV-SNP. Therefore we will conduct all basic experiments in SME for simplicity, unless specifically mentioned.

As a first step, we configure our systems (cf. Table 1) for transparent secure memory encryption (TSME). This means all pages will be encrypted by default, denoted by a bit in the physical address, e.g., bit 51. To get a ciphertext view of a page, we create a second mapping where this bit is not set. When we now measure access times to a cache line in the ciphertext mapping, we can clearly distinguish three cases: hits, misses and coherence conflicts (see Fig. 1). We cause a normal miss by flushing the line with `clflush` before measuring it, and a coherence conflict by accessing the same line in the plaintext mapping. We attribute the latency increase to the fact that when there is a plaintext line to evict, this has to happen before the load is completed, to ensure coherency. We also observe, as expected, that this effect is entirely symmetrical; it does not matter which mapping is used as the observer. The coherence also holds for code pages that were cached through code execution.

Table 1. Test systems.

CPU	Architecture	SME	HW Coherence	SEV	VM page flush MSR
2x AMD EPYC 7443	Zen 3	✓	✓	SNP	✓
AMD EPYC 7313P	Zen 3	✓	✓	SNP	✓
AMD EPYC 8024P	Zen 4c	✓	✓	SNP	X

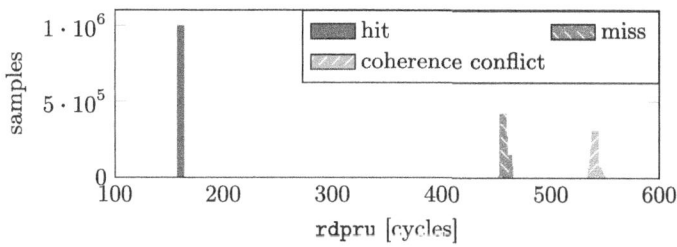

Fig. 1. Access timing histogram for accesses that are hits, misses (flushed) or conflicts caused by SME coherence.

This basic hit/conflict behaviour constitutes the first part of the Cohere+ Reload primitive (see Sect. 4.2 for the second).

4.1 Eviction Pattern

Contrary to Intel TDX, AMD memory encryption does not enforce its coherence with single line granularity. Instead, we find that any access *always* triggers the eviction of 32 out of 64 cache lines on a 4 kB aligned section of memory. We notice that in all of our machines' default configurations, the page is not simple split into two contiguous 2048 B halves, but instead shows an alternating pattern of 256 byte (4 cache lines) *coherence blocks* between the two *coherence partitions* (see Fig. 2a). Concretely, this means that an access to one or multiple plaintext (or ciphertext) addresses in the first (or second) coherence partition of a page will always trigger an eviction of *all* ciphertext (or plaintext) addresses in the first (or second) partition of a page. This limits the channel's spatial resolution compared to Flush+Reload, though it speeds up page profiling (see Sect. 5). We run this experiment on different physical and virtual pages, different page sizes (4 kB and 2 MB) and between different cores. We find that the pictured pattern is always the same. However, two of our machines' (EPYC 7443 and EPYC 7313P) mainboard menus expose a boot setting for "*DRAM interleaving size*". When we change it from its default of 256 B to 512 B, 1024 B or 2048 B, we can see the coherence eviction pattern changing to match (see Fig. 2), except for "off", 1024 B in the case of the EPYC 7443 system, and 4096 B (Table 2). While we do not know why the coherence mechanism is implemented as it is, we suspect some form of load balancing consideration w.r.t. DRAM.

198   L. Giner et al.

4.2   Access Delay Time

Since we have seen in Sect. 4 that the presence of a single plaintext cache line
increases the access time for a ciphertext line in the same coherence partition,
it stands to reason that more cache lines in the same partition might take even
longer to evict. Indeed, we find that the access delay on the evicting party's side
is related to the number of accessed lines in the coherence partition. When we
access from 0 to 32 lines of plaintext in the same coherence partition and measure
a ciphertext access, we see a monotonically increasing access time (Fig. 3a). But
we do not observe a strictly monotonic increase, instead we see plateaus every 4
cache lines. When we look at the individual distribution of access times for each
number of accesses (Fig. 3c), these groupings are visible. The first three access
groupings are somewhat separated (that is, measuring after 1,2 or 3 accessed
lines), but from then on there are quartets of consecutive numbers of accesses
that display very similar access times.

Investigating further, we can see that Fig. 3a and Fig. 3c are actually a special
case of timings when the accessed cache lines are contiguous, i.e., we do not skip
lines within a coherence partition up to our chosen number of accesses. Measuring
the ciphertext access times for single plaintext evictions of different plaintexts we
find the cause of this behaviour: different offsets within a coherence block have
distinct timings. As Fig. 3b shows, when there is only one plaintext access in
the coherence domain the access time of the ciphertext depends on the position
of the plaintext access within the coherence block and repeats from block to
block. This means that while *on average* a higher number of accessed cache lines
within a partition will increase the conflict eviction time, some combinations

Table 2. DRAM interleaving size and coherence pattern block size on our two systems.

DRAM interleaving setting	off	256	512	1024	2048	4096
block size EPYC 7443	2048	256	512	1024	2048	256
block size EPYC 7313P	2048	256	512	2048	2048	256

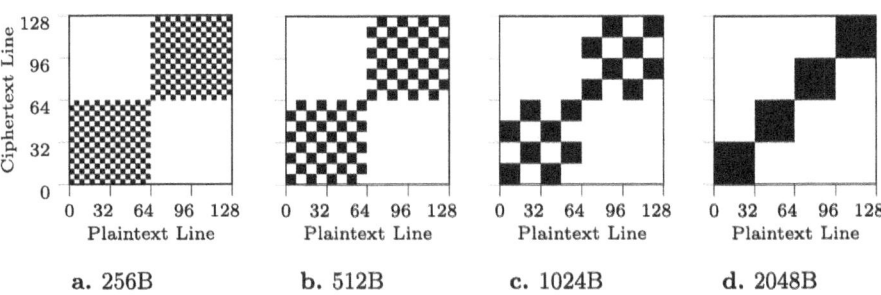

a. 256B   b. 512B   c. 1024B   d. 2048B

Fig. 2. Eviction Pattern for different *DRAM interleaving size* setting over 8 kB phys-
ically contiguous memory. A plaintext cache line is evicted by (and evicts) all corresponding ciphertext cache lines in black.

of lines will lead to far higher delays than others. This pattern depends on the core number *within a core complex* that loaded the plaintext address, but is the same between core complexes. We believe the pattern comes from the topology of the core complexes. When the block size is increased, the timing pattern also expands, though only up to 8 lines. For pattern sizes of 1024 B and larger, the pattern begins to repeat after 512 B, *i.e.*, 8 cache lines.

This timing behaviour is the second aspect of the Cohere+Reload primitive. We will further explore this effect in an attack in Sect. 7.2.

4.3   Cohere+Reload Compared to Other Cache Attacks

In this section, we compare Cohere+Reload with two cache attacks: Flush+ Reload and Flush+Flush. Our results, presented in Table 3, show that Cohere+ Reload is a fast attack, comparable to the two Flush-based cache attacks across three metrics: hit time, miss time (conflict time), and blind spots. Using the methodology presented in prior work [40], we measure each metric 100 000 times on three systems: 2x EPYC 7443 (Zen 3), EPYC 7313P (Zen 3), and EPYC 8024P (Zen 4c). We disabled the hardware prefetchers and fixed the frequency on all three machines. To measure the metrics, we spawn two threads on different

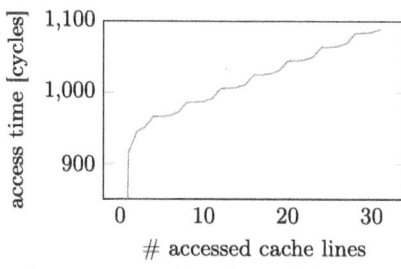

a. Average access time for n plaintext accesses in coherence partition.

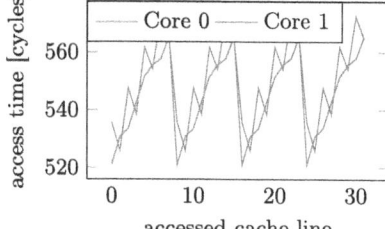

b. Average access time for one plaintext access at position n in partition.

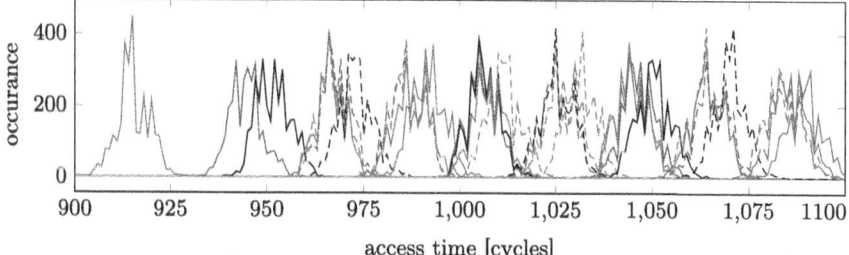

c. 32 histograms of access times for $n = 1$ to 32 plaintext accesses. Certain numbers of accesses have very similar timing distributions.

Fig. 3. Ciphertext conflict access times for different numbers of prior plaintext accesses on congruent addresses. Which addresses and how many where accessed changes the conflict eviction time.

Table 3. Comparison of Cohere+Reload with Flush+Reload and Flush+Flush.

System	Flush+Reload			Flush+Flush			Cohere+Reload		
	Hits [cycles]	Misses [cycles]	Blind Spots [%]	Hits [cycles]	Misses [cycles]	Blind Spots [%]	Conflict [cycles]	Hit [cycles]	Blind Spots [%]
EPYC 7443	122 $\sigma=13$	389 $\sigma=130$	65.1%	482 $\sigma=96$	367 $\sigma=90$	3.2%	428 $\sigma=103$	125 $\sigma=130$	0.06%
EPYC 7313P	150 $\sigma=0$	658 $\sigma=173$	73.2%	833 $\sigma=124$	632 $\sigma=31$	3.5%	800 $\sigma=167$	151 $\sigma=0$	0.53%
EPYC 8024P	145 $\sigma=0$	497 $\sigma=117$	74.8%	623 $\sigma=92$	484 $\sigma=59$	1.6%	623 $\sigma=131$	146 $\sigma=1$	0.02%

We compare Cohere+Reload with Flush+Reload and Flush+Flush across three metrics: hit time, miss/conflict time, and blind spots. The measurement for all three metrics is repeated 100 000 times on each system on different physical cores. Cohere+Reload has a much smaller blind-spot size and comparable hit and conflict times.

physical cores: a victim thread which randomly accesses a predetermined memory location, and an attacking thread that mounts the attack, measuring the metric.

On all three systems, we notice that Flush+Reload has a large blind spot—between 65–75% of victim accesses were missed by the attacker. Flush+Flush has a much smaller blind spot, with only 1.5–3.5% of victim accesses being missed by the attacker. Cohere+Reload has a minuscule blind spot, with less than 0.5% of victim accesses being missed by the attacker.

To measure the attack time, we consider both the hit and miss (conflict) timings. We see that the hit timings of Cohere+Reload are comparable to the hits of Flush+Reload, and the conflict timings of Cohere+Reload are comparable to Flush+Flush. This shows that Cohere+Reload is as fast as comparable attacks with a much smaller blind spot size, making it a very reliable side-channel attack.

5   Target Page Templating

In a standard SEV-SNP scenario, the host has no knowledge about where the guest maps which data in its virtual memory range. For an attack, the first step is therefore to locate pages of interest in the guest. The same result could be achieved with page access flags, though this is an alternative approach that does not require the flushing of TLB entries. The only requirement for this step is that a victim page access can be reliably triggered (e.g., establishing a connection that causes an RSA encryption).

We implement this for RSA by using a network call to the guest that triggers an encryption. In our test, we filter 4 GB of VM virtual memory with a sieve of sorts. Starting with all pages, we repeatedly cause the guest to access or not access the page of interest while measuring each page from different core with Cohere+Reload. We access each page with a single split load to detect

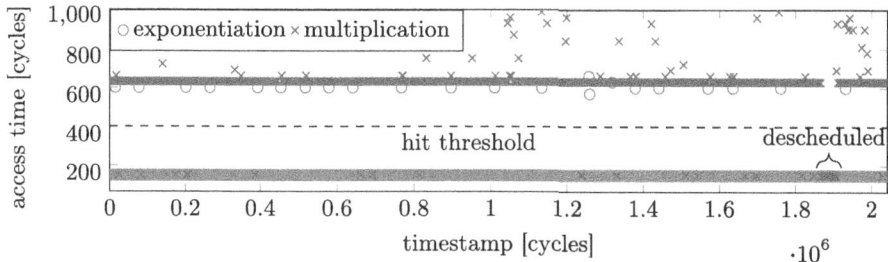

Fig. 4. Part of a raw Cohere+Reload trace of an RSA encryption. When *exponentiation* shows a conflict, a new exponentiation loop was started. When *multiplication* shows many in a row, the victim was most likely descheduled. (Color figure online)

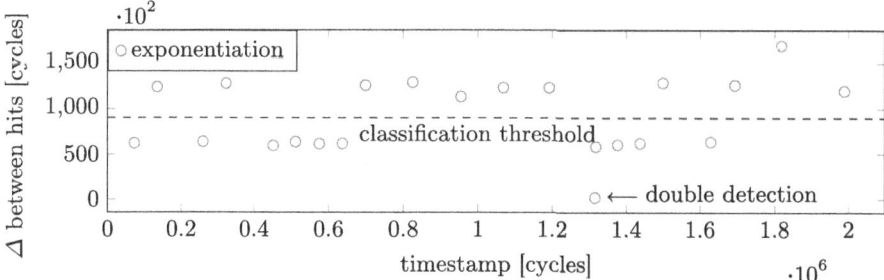

Fig. 5. Time difference between hits on *exponentiation*. The two bands show where a multiplication was executed (large difference) and the key bit is 1 or where it was not and the bit is 0 (small difference).

coherence eviction in both coherence partitions at once, as we find that the penalty for accessing both partitions is only ≈ 30 cycles. Pages that do not show the expected hits or conflicts are discarded from the list and the next step operates on the reduced list. After only 4 sieve steps, $1\,048\,510$ pages can be reduced to an average of 7.36 pages in 10.6 s in 100 experiments, with the two RSA pages of interest always being among them. Finding the correct page from there is trivial, as only those two pages show the expected access pattern during an encryption (see Sect. 6).

6 High Frequency Code Attack - RSA

We attack the square-and-multiply `mbedtls_mpi_exp_mod` implementation of RSA in Mbed-TLS v3.0.0. For the purposes of this demonstration, we configure it to use a maximum window size of 1. While the specific version is not crucial as long as the algorithm is the same, note that because of the coherence pattern, Cohere+Reload requires a suitable code layout. That is, the code that can be attacked needs to be aligned suitable within a coherence partition, while code that would hinder the attack needs to fall into the other partition. In this example, we find that a 256 B pattern is unsuitable, but switching to a 512 B pattern with the DRAM interleaving setting results in a working attack.

Our victim is an RSA encryption service in an SEV-SNP guest triggered by the attacker, which runs on the host. For the attack, the host program records traces of two code locations. First, the second partition in the `mpi_exp_mod` function. This contains the beginning of the loop that iterates over each key bit. Second, we trace the `mpi_montmul` function that does the squaring *and* multiplication operations. Our attack traces starts when the `mpi_exp_mod` is called for the first time and records long enough to capture the entire encryption (240000 samples). Figure 4 shows a section of such a trace. The `mpi_exp_mod` signal (blue) carries most of the key information, as it ideally detects an eviction precisely once per processed bit. This lets us infer whether or not `mpi_montmul` was called in addition to the square function (indicating that the key bit was 1) by the time delay to the next detection. The time difference between two conflicts in this signal is about twice as long when a '1' bit is processed. While this alone allows us to recover most of the key, we can correct some mistakes with the signal in `mpi_montmul` (pink). As the algorithm spends most of its time in this function, we detect almost all conflicts. However, when the algorithm is paused for any reason (e.g., scheduling), we see periods of hits on this address that we can then use to correct the primary signal. Figure 4 shows one occurrence of this near the end. In minor post-processing we also detect the precise start and end of the encryption and remove double detections for single bits that are too close together (see Fig. 5). Over 100 runs, our attack recovers randomly generated 4096 bit keys with a Levenshtein distance of 10.7 ± 3.94 (μ, σ) with a single trace (see Fig. 6). In terms of attack performance, this is on par with related works attacking RSA [13,14,36,42,44].

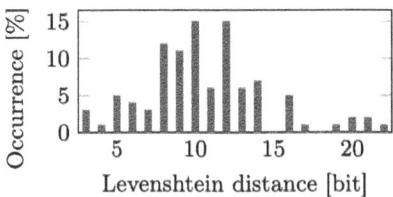

Fig. 6. The Levenshtein distances in bit for 100 single-trace RSA 4096 bit key recovery attacks.

7   AES T-Tables

In this section, we evaluate Cohere+Reload on the AES T-table implementation of OpenSSLv3.4 with 128 bit keys. Specifically, the `AES_encrypt` function which uses T-tables in lieu of hardware support (*i.e.*, AES-NI). Similar to the RSA attack (Sect. 6), we choose this implementation as it has been used extensively to evaluate prior side-channel attacks and is therefore a well-understood attack target [13, 14, 32, 36, 42, 44, 51].

The well-known first- and last-round attacks on AES T-tables [7, 8, 26] are both based on access probabilities. For a first-round cache attack, each of the four T-tables' cache lines (16 per table with 16 entries each) are measured as hits or conflicts after an entire encryption. As an encryption consists of 40 total accesses to each table (10 rounds with each 4 accesses), over a sufficient number of random plaintexts the probability for each line to be accessed at the end of an encryption is $1 - \frac{15}{16}^{40} = 92.43\%$. This can be distinguished from lines that are *always* accessed. In the first round of the encryption, the tables are accessed according to the result of $P_n \oplus K_n$, where K_n is byte n in the original key and P is the plaintext. Fixing one plaintext byte therefore allows an attacker to control which line is always accessed. After measuring enough encryptions, only this line will show no conflicts, hence we can infer the upper nibble of each key byte.

The resolution of Cohere+Reload, however, is not a single cache line. The partition size of half a page is simply not enough to perform this attack merely by distinguishing hits from conflicts, and even other tables on the same page influence each other. Even if one performed enough measurements to detect a non-accessed coherence partition (quite unlikely with $P_{accessed} = 1 - \frac{1}{2}^{160}$), this would only leak one bit per key byte.

However, we notice that while the default T-table placement in memory is aligned to cache lines with default compilation options, it is not necessarily aligned to a page. From this simple fact, we are able to conduct a limited standard first-round attack (Sect. 7.1) as well as a novel variation on the first-round attack based on a bias in the number of accessed cache lines instead of their location (Sect. 7.2).

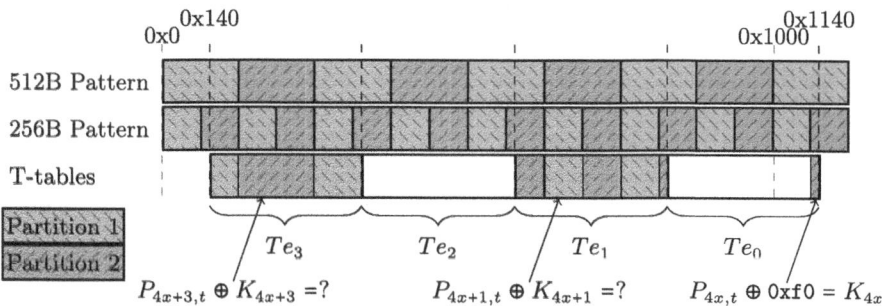

Fig. 7. AES T-table memory alignment in OpenSSL and memory coherence partition patterns for 256 B and 512 B. Annotations show where first round T-tables are accessed depending on the key- and plaintext bytes. Te_1 and Te_3 demonstrate the 256 B/512 B patterns respectively (only 1 pattern can be active for all memory for each system boot), and Te_0 shows the second partition on the next page.

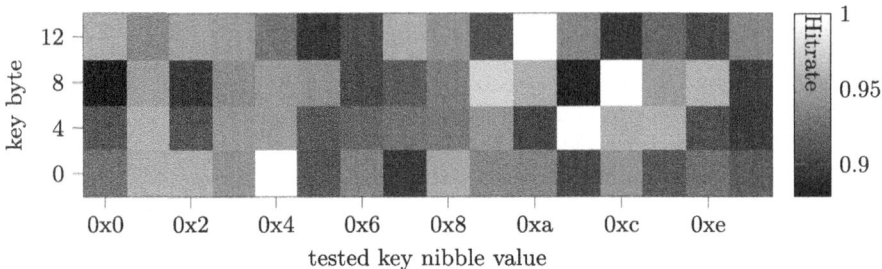

Fig. 8. Heatmap of conflicts in an AES disaligned T-table attack on key bytes 0, 4, 8, 12 with a total of 1500 encryptions. Correct key nibbles are 0x4, 0xb, 0xc and 0xa.

7.1 Disaligned T-Table First-Round Attack

For this first attack, we only look at table Te_0, which deals with key/plaintext bytes 0,4,8 and 12 in the first round. As we can see in Fig. 7 bottom right, Te_0 spans 5 cache lines into a new page. For a Flush+Reload attack, this would not make a difference, as the attack either works on all cache lines or it does not work at all (e.g., because the target cannot be accessed). For Cohere+Reload however, this enables an attack. With a coherence pattern size of 256 B, this means that the last line of Te_0 is the only one that accesses the second coherence partition on that page. With this disalignment we can convert the Cohere+Reload primitive into what is essentially a Flush+Reload primitive for this implementation, an improvement in the same vein as the attack described by Spreitzer et al. [43].

For random plaintexts, each cache line will be used by an encryption in 92 % of cases. In our case, the second coherence partition of the second page will measure this percentage for the last line of Te_0. The T-table access in the first round depends only on the XOR of the plaintext and key bytes, e.g., $Te_0 = P_{\{0,4,8,12\}} \oplus K_{\{0,4,8,12\}}$. As only an XOR product of 0xf in the upper nibble accesses the measured partition, the correct key nibble can thus be derived with $(P_{x,t} \oplus \text{0xf0}) \wedge \text{0xf0} = K_x$ for the test plaintext byte $P_{x,t}$ for which Cohere+Reload shows a 100 % hit ratio.

We can recover all upper nibbles for these 4 key bytes in 1500 encryptions with 100 % accuracy. Figure 8 shows a heatmap of one such attack with a total of 1000 encryptions, clearly displaying the key correct key nibbles 0x4, 0xb, 0xc and 0xa.

7.2 First-Round Correlation Attack

For the standard T-table attack, we use the fact that there is a 92 % chance that any given cache line in a table will be accessed with a random plaintext. We have established above that the spacial resolution of Cohere+Reload is not enough to us this in a standard first-round cache attack. However, we know each key- and plaintext byte combination accesses a specific cache line with 100 % certainty. This in turn means either the first or second coherence partition will be accessed for each plaintext byte in the first round. Looking at a single key byte at a time, we can easily calculate a pattern of partition 1 and partition 2 accesses for each plaintext and each key. Concretely, we can create a template vector for each possibly key byte value of the form $\{0, 0, 0, 1, 1, 1, 1, 0, 0, 0, 0, 1, 1, 1, 1, 0\}$. Each number denotes the partition that $P_n \oplus K_n$ will access, in this case for $K_n = 0$ and a pattern size of 256 B. We can now pick a fixed plaintext and change only one plaintext byte at a time and record the number of total accessed cache lines in the T-tables that fall within the one of the partitions. With this, we can now calculate the Pearson correlation coefficients between the templates and the access count vector. The highest correlation will show the template corresponding with the correct upper nibble of the tested key byte.

When we generate the template vectors, we find that they contain a different number of unique templates for the 256 B and 512 B coherence patterns. Depending on the cache line offset within a page, there are at most 8 unique templates for a 256 B pattern and 16 for the 512 B pattern. This means for all odd cache line offsets of the T-tables, we can recover 3 or 4 bits per key byte, depending on the chosen pattern size. The pattern size 256 B yields one bit less, since the disaligned coherence pattern within each table repeats once (see Fig. 7) and we can therefore not distinguish the most significant bit.

While this attack benefits from chosen plaintexts (see above), a sufficient number of (mostly) random known plaintexts will work just the same. By adding the number of partition accesses to a bucket for each plaintext byte value and for every byte of a random plaintext, each encryption can contribute to the recovery of all 16 key bytes instead of just one. With many encryptions, this creates a 16×16 matrix with one correlation vector for each key byte. After enough encryptions, the bias in the average number of accessed cache lines in a coherence partition outweighs the initial noisiness of random plaintexts and the key bytes can be recovered. Therefore we consider this a known-plaintext attack.

Cohere+Reload provides for two methods of measuring the number of accesses. Firstly, the access time for a single read, as described in Sect. 4.2. In theory, over enough measurements with randomized plaintexts, the average access time should provide a proxy measure for the number of cache lines that were accessed within a partion. Unfortunately, we could not make this method work with AES. Fortunately, we can make use of the minimal blind spot and high frequency of Cohere+Reload and mount a trace attack.

Unlike in the case of RSA, there is very little time between accesses to the T-tables in the OpenSSL implementation. When we try to trace an encryption normally, we only observe 10–15 accesses per partition, for a total of $\approx$20–30 accesses out of the $\approx$135 expected accesses (less than 160, as some accesses fall on the second page). Since in our threat model (cf. Sect. 3) we are the hypervisor, we can however employ a little trick to slow down our victim. Even in SEV-SNP, the hypervisor has control over the bits in the page table entries, including the uncacheable bit. We find that by making both the function code and the stash page uncacheable, we can slow down our victim considerably. This allows us to record around 100 total memory accesses for a single encryption, as we can

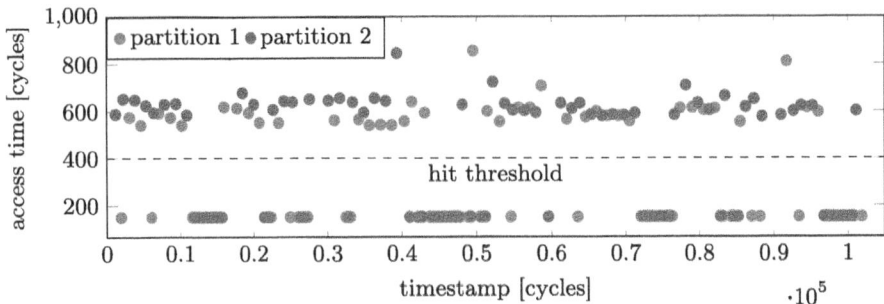

Fig. 9. Cohere+Reload access trace for a single AES encryption.

see in Fig. 9. Though still shy of 160, we cannot reliably see all 16 individual accesses in the first round and infer the table accesses directly. We can, however use the number of hits to the partitions as a proxy. Even though some hits will contain two or more accesses, on average the number of accesses in a partition will be biased by the first round. To reduce the noise, we only look at the first 16 accesses to both partitions, and extract the number of accesses to one of them as our signal. We choose 16 as it is the upper limit to how many hits we can see within the first round, and even if accesses from the second round are included, they only add noise. We can see this signal plotted together with the correct template in Fig. 10. In this case, we achieve a very high correlation coefficient of $\rho = 0.99$.

Figure 11 shows our results for both pattern sizes. We can see that recovering 3 bits is more robust, as the templates are more different to each other. With 4000 traces, we correctly recover 3 bit for an average of 14.5 key bytes in our first guess. Adding second guesses, this rises to 15.3. Using the 512 B pattern, we can recover an average of 14.85 nibbles per byte with 12000 encryptions with first guesses and 15.4 when we also include second guesses.

Since this is a correlation based attack, we can identify weakly correlating key nibbles, or those where several candidates are close, and record more traces only in these cases to minimize overall traces.

Tracing AES with this time resolution (without repeating encryptions) is something we do not believe can be easily achieved *across cores or even sockets* with other cache attacks like Flush+Reload or Flush+Flush, as Cohere+Reload can monitor an entire page with only two addresses (cf. Sect. 4.3). Though compared to the standard our attack takes longer (e.g., Flush+Reload can work with only a few hundred to low thousands of encryptions as shown in Sect. 7.1), it also has slight advantages. Firstly, as a trace-based attack it is not hindered by software prefetching, as each read resets the cache line. Secondly, this attack functions the same for 128 bit keys as it does for 192 bit or 256 bit keys, as the additional two or four rounds do not affect the beginning of the attack, whereas for other attacks the probabilities become less favorable.

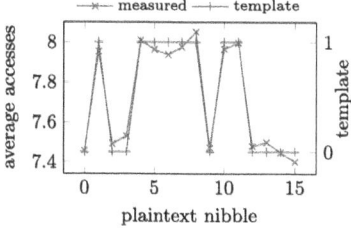

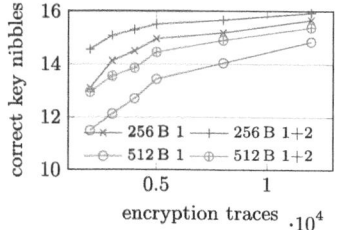

Fig. 10. Correlation attack template vector for key nibble 0xa vs. average access counts for correct key nibble guess with 8000 encryptions. $\rho = 0.99$.

Fig. 11. Average correct key nibbles for a given number of AES traces for the top guess (1) and the top 2 guesses combined $(1 + 2)$. $n = 100$ per point.

8   Load-Time Based Attacks

In Sect. 4.2, we observed that Cohere+Reload-timings to different cache lines in the same coherence block have discernible timing differences, *i.e.*, an attacker can conclude which cache line was accessed by measuring the time taken to access a cache block. In this section, we use this observation to mount two synthetic attacks: the first reveals which part of an array is accessed while the second detects which case of a `switch` statement is taken. We test these attacks on an AMD EPYC 8024P (Zen 4c) with hardware-prefetching disabled.

Our experimental setup consists of two processes, an attacker and a victim, which can access the same physical page as a ciphertext or plaintext mapping respectively. This page is either a dynamically allocated array storing values (the first attack), or it the victim's code (the second attack). We assume that the region of interest is one cache-block large (256 B) and assume that the physical address is aligned to this value. For the attack on code execution, we ensure that each case of the switch is one cache-line long and all four cases are within the same cache block. By measuring the access time with Cohere+Reload, we can now infer which line in a coherence block was accessed by the victim, which in turn can let us infer control- or data flow. With the attack on code, a source of noise is the (speculative) fetching of instructions from the next case by the instruction prefetcher. To overcome this, we use the `ret` instruction at the end of every case to indicate that the next set of instructions will not be executed.

For the purposes of the experiment, the attacker and victim alternately access the physical address. The attacker records the access times and the victim accesses only one random cache line. Each cache line is accessed 100 000 times, resulting in 400 000 measurements. Figure 12a shows the access time distribution to the coherence block when the monitored address corresponds to a dynamically allocated array, *i.e.*, *data*. In Fig. 12b, we see the access times to the coherence block when the physical address corresponds to a switch and each case is on a different cache line, *i.e.*, *code*.

With these measurements we can make several observations. First, access times to code (Fig. 12b) is noisier than data (Fig. 12a), even in our example with a `ret` at the end of every case. In further tests we learn that the distributions become visually separable when the attacker can choose to repeat the same input. Since the `ret` instruction also improves the result but does not make it perfect, we believe this is a combination of misspeculation and a race condition between how fast the front end can fetch data vs. how soon the `ret` is decoded.

In a real attack, exploitability depends heavily on the control an attacker has over the target branch. If a single secret bit can be reliably repeated, it only takes a few samples to train predictors and receive a clean signal (cf. Sect. 4.2). If a sequence of bits can reliably be repeated in the same order (e.g., a key that is used bit by bit), Fig. 12b shows that by combining repeated measurements we can produce distinct distributions, even when attacking instructions.

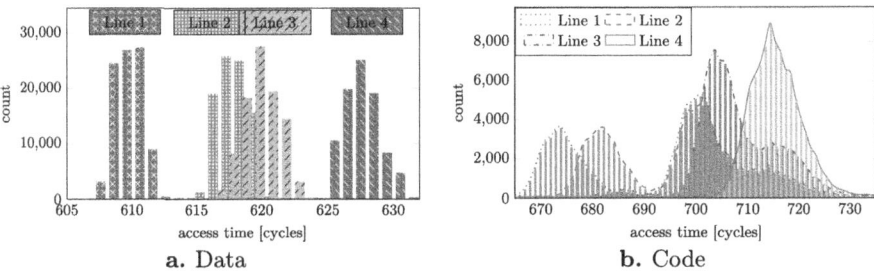

a. Data   **b.** Code

Fig. 12. Cohere+Reload access times to a coherence block when it contains victim data (Fig. 12a) and victim code (Fig. 12b).

9   Mitigation

AMD disabling the coherence feature will undoubtedly mitigate Cohere+Reload, though `VMPAGE_FLUSH`, if present, may lead to similar leakage, as it may well have the same time dependence. Here, the fundamental question is if the host's ability to read the ciphertext at all is even necessary for SEV-SNP. The SEV-SNP attestation flag `CiphertextHidingDRAM` suggests it may not always be necessary, as it disallows reading of the ciphertext by the hypervisor when active. As we have no hardware that supports this feature, we can only speculate that the coherence mechanism could be disabled when this setting is enabled, depending on how it is implemented.

On the developer side, hardened implementations could change the memory type of critical sections (e.g., T-tables) to *uncacheable*. While this slows down execution, our experiments show that *loads* to uncacheable memory do not trigger the coherence mechanism.

10   Conclusion

In this paper, we introduced Cohere+Reload, a novel side-channel attack exploiting AMD's coherency for encrypted memory. We exploit two types of leakage in the coherency mechanism: First, we exploit coherence conflicts, leaking victim operations on a spatial granularity of a 2 kB block. Second, we exploit timing correlations with the number and location of accesses, reaching a maximum spatial resolution of 256 bytes. In our synthetic attacks, we showed that Cohere+Reload can observe control flow and access locations in workloads within a confidential virtual machine. As a benchmark we mounted an attack on mbedTLS RSA, leaking 99.7 % of the 4096 key bits in a single-trace attack. We also mounted an attack on OpenSSL AES exploiting disalignments on a cache line granularity, achieving an accuracy of 100 % in only 1500 encryptions in a first round T-table attack and an accuracy of 92.81 % in 12000 encryptions with a novel correlation attack. Our work shows that coherence mechanisms can undermine the confidentiality of confidential virtual machines. Consequently, we believe that vendors need to weigh the necessity of coherence, as discussed above, against the opening of this side channel for future implementations.

Acknowledgments. We want to thank Andreas Kogler. After tinkering with cache coherency with SEV for some time at TU Graz, Andreas found an effect and suggested that we investigate it. This research is supported in part by the European Research Council (ERC project FSSec 101076409), and the Austrian Science Fund (FWF SFB project SPyCoDe 10.55776/F85 and FWF project NeRAM 10.55776/I6054). Additional funding was provided by a generous gift from Intel. Any opinions, findings, and conclusions or recommendations expressed in this paper are those of the authors and do not necessarily reflect the views of the funding parties.

References

1. Aktas, E., Cohen, C., Eads, J., Forshaw, J., Wilhelm, F.: Intel trust domain extensions (TDX) security review (2023). https://services.google.com/fh/files/misc/intel_tdx_-_full_report_041423.pdf
2. AMD: AMD SEV-SNP: strengthening VM isolation with integrity protection and more. (2020). https://www.amd.com/content/dam/amd/en/documents/epyc-business-docs/white-papers/SEV-SNP-strengthening-vm-isolation-with-integrity-protection-and-more.pdf
3. AMD: AMD Secure Encrypted Virtualization (SEV) (2024). https://developer.amd.com/sev/
4. AMD: AMD64 Architecture Programmer's Manual (2024)
5. ARM: Arm confidential compute architecture (2024). https://www.arm.com/architecture/security-features/arm-confidential-compute-architecture
6. Ashokkumar, C., Venkatesh, M.B.S., Giri, R.P., Roy, B., Menezes, B.: An error-tolerant approach for efficient AES key retrieval in the presence of cache prefetching–experiments, results, analysis. Sādhanā **44** (2019). https://doi.org/10.1007/s12046-019-1070-8

7. Bernstein, D.J.: Cache-Timing Attacks on AES (2005). http://cr.yp.to/antiforgery/cachetiming-20050414.pdf
8. Bonneau, J., Mironov, I.: Cache-collision timing attacks against AES. In: CHES (2006)
9. Carmon, E., Seifert, J.-P., Wool, A.: Photonic side channel attacks against RSA. In: HOST (2017)
10. Confidential Computing Consortium: A Technical Analysis of Confidential Computing (2022)
11. Disselkoen, C., Kohlbrenner, D., Porter, L., Tullsen, D.: Prime+abort: a timer-free high-precision L3 cache attack using intel TSX. In: USENIX Security (2017)
12. Du, Z.-H., et al.: Secure encrypted virtualization is unsecure (2017). arXiv:1712.05090
13. Gast, S., et al.: SQUIP: exploiting the scheduler queue contention side channel. In: S&P (2023)
14. Gast, S., Weissteiner, H., Schröder, R.L., Gruss, D.: CounterSEVeillance: performance-counter attacks on AMD SEV SNP. In: NDSS (2025)
15. Gruss, D., Bidner, D., Mangard, S.: Practical memory deduplication attacks in sandboxed JavaScript. In: ESORICS (2015)
16. Gruss, D., et al.: Page cache attacks. In: CCS (2019)
17. Gruss, D., Maurice, C., Wagner, K., Mangard, S.: Flush+Flush: a fast and stealthy cache attack. In: DIMVA (2016)
18. Gruss, D., Spreitzer, R., Mangard, S.: Automating attacks on inclusive last-level caches. In: USENIX Security, Cache Template Attacks (2015)
19. Hetzelt, F., Buhren, R.: Security analysis of encrypted virtual machines. ACM SIGPLAN Not. **52**(7), 129–142 (2017)
20. Horowitz, G., Ronen, E., Yarom, Y.: Spec-o-Scope: cache probing at cache speed. In: CCS (2024)
21. Hund, R., Willems, C., Holz, T.: Practical timing side channel attacks against kernel space ASLR. In: S&P (2013)
22. Intel: Intel Trust Domain Extensions (2021). https://software.intel.com/content/dam/develop/external/us/en/documents/tdx-whitepaper-v4.pdf
23. Intel: Intel Software Guard Extensions (Intel SGX) (2024). https://www.intel.com/content/www/us/en/products/docs/accelerator-engines/software-guard-extensions.html
24. Intel: Intel Trust Domain Extensions Module Base Architecture Specification (2024). https://www.intel.com/content/www/us/en/developer/tools/trust-domain-extensions/documentation.html
25. Intel: Intel Total Memory Encryption White Paper (2025). https://www.intel.com/content/www/us/en/architecture-and-technology/vpro/hardware-shield/total-memory-encrpytion.html
26. Irazoqui, G., Inci, M.S., Eisenbarth, T., Sunar, B.: Wait a minute! A fast, cross-VM attack on AES. In: RAID (2014)
27. Kaplan, D., Powell, J., Woller, T.: AMD memory encryption (2016)

28. Kocher, P.: Timing attacks on implementations of Diffe-Hellman, RSA, DSS, and other systems. In: CRYPTO (1996)
29. Kocher, P., Jaffe, J., Jun, B.: Differential power analysis. In: CRYPTO (1999)
30. Lendacky, T.: What processors support SEV? #1 (2019). https://github.com/AMDESE/AMDSEV/issues/1#issuecomment-581426096
31. Li, L., Huang, J., Feng, L., Wang, Z.: PREFENDER: a prefetching defender against cache side channel attacks as a pretender. IEEE Trans. Comput. (2024)
32. Lipp, M., et al.: PLATYPUS: software-based power side-channel attacks on x86. In: S&P (2021)
33. Maurice, C., et al.: Side: SSH over robust cache covert channels in the cloud. In: NDSS (2017)
34. Monaco, J.: SoK: keylogging side channels. In: S&P (2018)
35. Morbitzer, M., Huber, M., Horsch, J., Wessel, S.: Severed: subverting AMD's virtual machine encryption. In: EuroSec (2018)
36. Mushtaq, M., Mukhtar, M.A., Lapotre, V., Bhatti, M.K., Gogniat, G.: Winter is here! A decade of cache-based side-channel attacks, detection & mitigation for RSA. Inf. Syst. **92**, 101524 (2020)
37. Osvik, D.A., Shamir, A., Tromer, E.: Cache attacks and countermeasures: the case of AES. In: CT-RSA (2006)
38. Purnal, A., Turan, F., Verbauwhede, I.: Prime+scope: overcoming the observer effect for high-precision cache contention attacks. In: CCS (2021)
39. Quisquater, J.-J., Samyde, D.: ElectroMagnetic analysis (EMA): measures and counter-measures for smart cards. In: E-smart (2001)
40. Rauscher, F., Fiedler, C., Kogler, A., Gruss, D.: A systematic evaluation of novel and existing cache side channels. In: NDSS (2025)
41. Ristenpart, T., Tromer, E., Shacham, H., Savage, S.: Hey, You, get off of My cloud: exploring information leakage in third-party compute clouds. In: CCS (2009)
42. Schwarz, M., Gruss, D., Weiser, S., Maurice, C., Mangard, S.: Malware guard extension: using SGX to conceal cache attacks. In: DIMVA (2017)
43. Spreitzer, R., Plos, T.: Cache-access pattern attack on disaligned AES T-tables. In: COSADE (2013)
44. Wan, J., Bi, Y., Zhou, Z., Li, Z.: MeshUp: stateless cache side-channel attack on CPU mesh. In: S&P (2022)
45. Wang, D., Qian, Z., Abu-Ghazaleh, N., Krishnamurthy, S.V.: PAPP: prefetcher-aware prime and probe side-channel attack. In: DAC (2019)
46. Wang, Z., Peng, S., Jiang, W., Guo, X.: Defeating hardware prefetchers in flush+reload side-channel attack. IEEE Access **9**, 21251–21257 (2021)
47. Werner, J., Mason, J., Antonakakis, M., Polychronakis, M., Monrose, F.: The severest of them all: inference attacks against secure virtual enclaves. In: AsiaCCS (2019)
48. Wilke, L., Wichelmann, J., Morbitzer, M., Eisenbarth, T.: SEVurity: no security without integrity–breaking integrity-free memory encryption with minimal assumptions. In: S&P (2020)
49. Yarom, Y., Benger, N.: Recovering OpenSSL ECDSA Nonces using the FLUSH+RELOAD cache side-channel attack. Cryptology ePrint Archive, Report 2014/140 (2014)

212 L. Giner et al.

50. Yarom, Y., Falkner, K.: Flush+reload: a high resolution, low noise, L3 cache side-channel attack. In: USENIX Security (2014)
51. Zhao, M., Suh, G.E.: FPGA-based remote power side-channel attacks. In: S&P (2018)

Poster: Extracting Cryptographic Keys from Windows Live Processes

León Abascal and Ricardo J. Rodríguez(✉) (ID)

Universidad de Zaragoza, Zaragoza, Spain
{labascal,rjrodriguez}@unizar.es

Abstract. Cryptographic keys are a fundamental aspect of modern system security, but when compromised, they become a critical vulnerability, especially in ransomware attacks. Paradoxically, these keys must be available in memory at runtime to function, creating a unique opportunity for defensive tools. We introduce `KeyReaper`, an open-source tool designed to locate cryptographic keys in active Windows processes using advanced memory analysis. Unlike traditional approaches that rely on static memory dumps, `KeyReaper` performs dynamic analysis in real time, restricting the search to process heap memory to improve efficiency and accuracy. It employs robust key identification heuristics to minimize false positives and is designed for seamless integration with Endpoint Detection and Response systems. `KeyReaper` also encourages extensibility: its open-source nature allows researchers and practitioners to enhance its capabilities with custom key detection algorithms. We validated our approach through extensive experiments involving both proof-of-concept ransomware and real-world samples, demonstrating the effectiveness of key extraction and decryption success. Our tool provides a practical path to strengthening ransomware mitigation strategies.

Keywords: malware · Windows · digital forensics · cryptography

1 Motivation and Context

Ransomware has become one of the most significant threats in the cybersecurity landscape, as highlighted by multiple industry reports [5–7]. This type of malware primarily targets data availability [4], encrypting files or entire systems, and demanding a ransom in exchange for decryption keys.

The success of ransomware attacks depends largely on the use of strong cryptographic algorithms. However, operational errors in key management have led to the failure of some ransomware campaigns [1], including the use of flawed custom algorithms or poor key management practices [15]. In contrast, ransomware families such as `WannaCry` have employed robust and properly implemented encryption schemes that make recovery virtually impossible without the corresponding decryption key [2]. In this sense, the confidentiality and integrity of cryptographic keys are critical to the attack's effectiveness.

© The Author(s), under exclusive license to Springer Nature Switzerland AG 2025
M. Egele et al. (Eds.): DIMVA 2025, LNCS 15747, pp. 213–219, 2025.
https://doi.org/10.1007/978-3-031-97620-9_12

Cryptographic keys are typically generated using secure libraries (e.g., Crypto++) or system APIs (e.g., CryptoAPI for Windows) [2,8]. Due to the nature of the encryption process, these keys must reside in the ransomware's memory during its execution. This creates a limited, but interesting, window of opportunity to extract them, especially if their structure follows recognizable patterns. According to [3], key generation typically takes place on the victim's device, ensuring that the keys will be present at some point during the attack.

In response to this opportunity, we present KeyReaper, a novel tool designed to extract cryptographic keys from active Windows processes using advanced memory analysis techniques. Unlike existing approaches that rely on static memory dumps [10,11,15], our solution enables dynamic key extraction in real time by focusing on the process's heap memory, where such keys are most likely to reside during execution.

2   KeyReaper: System Design

In this section, we present the design of KeyReaper, our memory analysis tool developed to extract cryptographic keys from active Windows processes. It consists of three main modules and is designed to be used in an Endpoint Detection and Response workflow where a process has already been flagged as ransomware.

2.1   Workflow Description

Figure 1 illustrates the tool's workflow. The process begins with the ransomware actively running on a system, performing encryption operations and storing keys in memory. Let us clarify that KeyReaper does not include any ransomware detection mechanisms; rather, it is designed to operate after a process has already been flagged as malicious by an external security solution, such as an EDR system or antivirus engine. The tool consists of three main components that work sequentially to recover these keys.

The first module, *Process Capture*, pauses the execution of the ransomware process. This pause is essential to prevent future malicious behavior, including the potential destruction of encryption keys once the attack is complete. By stopping the process mid-execution, we ensure memory stability and a consistent state for extraction.

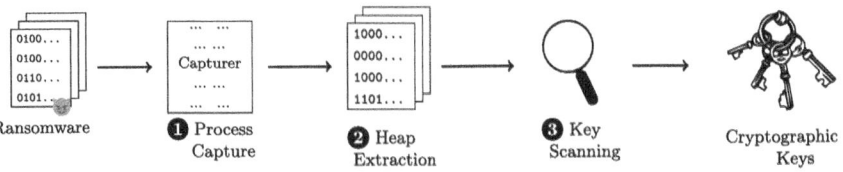

Fig. 1. System architecture and workflow of KeyReaper.

The second module, *Heap Extraction*, recovers the memory contents of the paused process. The tool specifically focuses on heap regions, as these are commonly used for dynamic allocation of cryptographic keys and related metadata. Focusing on the heap improves performance and reduces noise.

The final stage, *Key Scanning*, analyzes the extracted memory using a series of interchangeable pattern recognition-based detection algorithms. These algorithms are designed to identify the structural signatures of cryptographic keys and enable the detection of symmetric and asymmetric keys. Upon successful identification, the keys can be extracted, allowing the victim's files to be decrypted without paying a ransom.

2.2 Design Highlights

This live memory-based approach offers several advantages over traditional post-incident forensics. By operating on active processes, the tool avoids issues such as memory paging, process termination, or data loss, which can occur when acquiring and analyzing a memory dump [10]. Its modular architecture also allows for the easy integration of additional analysis techniques, increasing adaptability to future ransomware variants. Additionally, the tool provides a command-line interface that supports options to pause, resume, or terminate processes, making it ideal for automated use in real-time incident response systems.

To facilitate key detection, KeyReaper includes several analysis modules based on well-established forensic techniques. One module implements the AES round key derivation method of [11], which reconstructs AES round keys from memory. This allows the detection of AES keys even when they are not explicitly present, by identifying characteristic patterns of the derived keys. While computationally intensive, it is still highly effective at detecting AES keys, which are widely used in modern ransomware families.

Another module is based on the technique originally presented in [15], which focuses on detecting keys generated using Microsoft's CryptoAPI. Although deprecated, CryptoAPI remains supported on all versions of Windows and is frequently used in ransomware. Our improved version of this method uses DLL injection and inter-process communication to extract key material from the target process with minimal disruption.

The tool's focus on heap memory is aligned with the goal of recovering dynamically generated keys, which only exist at runtime. Keys hardcoded in the binary or retrieved from a Command and Control server are beyond the scope of this tool and would require alternative recovery methods. By focusing on runtime memory, the tool achieves better performance and reduces false positives.

To accurately enumerate heap regions, KeyReaper uses the Process Environment Block (PEB). It also optionally employs pattern-based analysis of undocumented Windows heap structures, such as _HEAP and _HEAP_SEGMENT, available since Windows Vista. These structures provide reliable indicators of the boundaries and layout of heap segments, especially in fragmented memory.

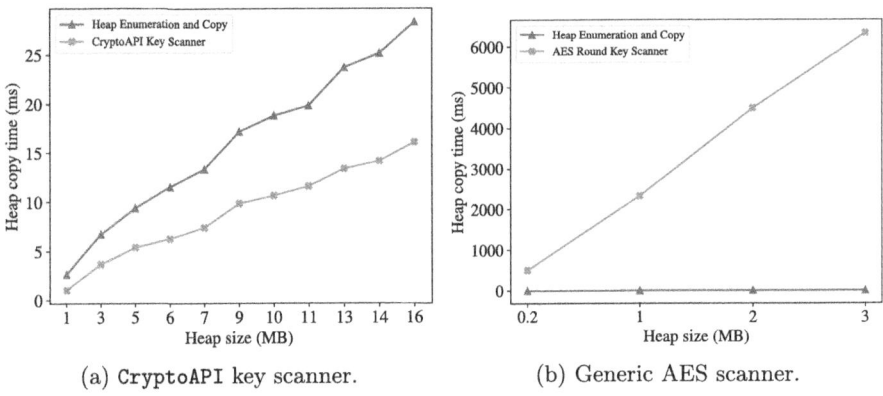

(a) `CryptoAPI` key scanner.   (b) Generic AES scanner.

Fig. 2. Comparison of performance for the two key scanning modules.

To ensure responsiveness during live memory analysis, `KeyReaper` is implemented in pure C++ for optimal performance. The tool uses native Windows API calls to interact with remote processes, such as `NtSuspendProcess` to pause execution, `NtQueryInformationProcess` for process introspection, and `ReadProcessMemory` for safe memory access. Additional functions such as `GetProcAddress` are employed for dynamic symbol resolution. Memory regions are extracted using the `TitanEngine` library, which provides a high-level interface to low-level memory operations. For pattern matching within memory dumps, our implements the Boyer-Moore-Horspool algorithm, a fast and efficient method for substring searching in binary data.

3   Experimentation

To evaluate the reliability and effectiveness of `KeyReaper`, we conducted four types of complementary experiments. These experiments aim to assess: (i) the tool's reaction time, (ii) its resistance to false positives, (iii) the validity of the extracted keys, and (iv) its performance in real-world ransomware scenarios.

To evaluate reaction time, we developed a validation program that generates 100 cryptographic keys per run. We measured heap extraction and scanning times using `QueryPerformanceCounter`, repeating each configuration 1000 times with varying heap sizes. As shown in Fig. 2a and Fig. 2b, the AES scanner, although implementation-agnostic, is significantly slower. In contrast, the CryptoAPI scanner offers faster and more versatile performance.

To evaluate robustness against false positives, we introduced high-entropy noise into the heap using pseudo-random data in our validation program. We also evaluated the scanners with benign open-source software (e.g., `Audacity`, `Notepad++`) during typical usage sessions. In tests on synthetic and structured data, the tool consistently recorded a zero false positive rate, indicating high selectivity and reliability.

Next, we evaluated the accuracy of the extracted keys. Using the CryptExportKey [14] function, we exported the CryptoAPI-generated keys from our validation program. By parsing the exported key blobs and stripping their headers [13], we isolated the raw key material and compared it to the result generated by KeyReaper. This comparison was repeated for over 1,000 iterations, and in all cases, the keys matched exactly with no false detections.

Finally, we evaluated the tool in real-world ransomware scenarios using two prevalent malware samples: Avaddon and WannaCry. Both rely on CryptoAPI for key management. Avaddon generates a persistent AES session key for encryption [15] and continues monitoring the file system for new files, remaining active until manually terminated. WannaCry, on the other hand, creates an RSA key pair as a session key and uses it to encrypt the AES keys of each file [9]. Both samples were run in a controlled, isolated virtual environment (Windows 10 Pro, 4 GB RAM, 50 GB HDD). After confirming normal ransomware behavior (file encryption and displaying the ransom note), we paused execution at strategic times to extract the keys.

To confirm successful decryption, we placed a known plaintext file in the victim's environment, calculated its SHA256 hash, and resumed ransomware execution to enable encryption. After pausing the malware, we used the extracted key to decrypt the file and verified the result by comparing the SHA256 hash with the original.

These results confirm that its scanning modules reliably identify valid keys without false positives, and it performs well in both synthetic and real ransomware scenarios. Together, these findings demonstrate the tool's viability as a practical solution for dynamic key extraction during ransomware incidents.

4   Future Work

We plan to extend KeyReaper with support for additional cryptographic algorithms, including Salsa20. We are also developing mechanisms for detecting keys generated by popular libraries such as OpenSSL and Crypto++.

Future releases will include memory dump analysis for offline forensic investigations and benchmarking against existing similar solutions. Additionally, we aim to expand support beyond the NT heap to cover the segment heap introduced in Windows 10 [12].

Ongoing research on CryptoAPI will improve the detection of asymmetric keys, and we plan to evaluate the tool with legitimate cryptographic software for auditing and compliance testing. To improve robustness, we also plan to study heap manipulation and evasion techniques used by ransomware.

Acknowledgments. This research was supported in part by grant PID2023-1514 67OA-I00 (CRAPER), funded by MICIU/AEI/10.13039/501100011033 and by ERD-F/EU, by grant TED2021-131115A-I00 (MIMFA), funded by MICIU/AEI/10.13039/501100011033 and by the European Union NextGenerationEU/PRTR, and the University of Zaragoza, by grant *Proyecto Estratégico Ciberseguridad EINA UNIZAR*, funded by the Spanish National Cybersecurity Institute (INCIBE) and the European Union

NextGenerationEU/PRTR, and by grant *Programa de Proyectos Estratégicos de Grupos de Investigación* (T21-23R), funded by the University, Industry and Innovation Department of the Aragonese Government. We used OpenAI's ChatGPT to improve the grammar, clarity, and coherence of the paper.

References

1. Aboud, M.A., Mariyappn, K.: Investigation of modern ransomware key generation methods: a review. In: 2021 International Conference on Computer Communication and Informatics (ICCCI), pp. 1–5, January 2021
2. Bajpai, P., Enbody, R.: An empirical study of key generation in cryptographic ransomware. In: 2020 International Conference on Cyber Security and Protection of Digital Services (Cyber Security), pp. 1–8, June 2020
3. Bajpai, P., Sood, A.K., Enbody, R.: A key-management-based taxonomy for Ransomware. In: 2018 APWG Symposium on Electronic Crime Research (eCrime), pp. 1–12, May 2018
4. Cawthra, J., Ekstrom, M., Lusty, L., Sexton, J., Sweetnam, J.: Data integrity: identifying and protecting assets against ransomware and other destructive events, December 2020. https://csrc.nist.gov/pubs/sp/1800/25/final. Accessed 30 Jan 2025
5. Cisco: Cisco cyber threat trends report: from trojan takeovers to ransomware roulette, June 2024. https://learn-cloudsecurity.cisco.com/umbrella-library/cyber-threat-trends-report. Accessed 9 April 2025
6. CrowdStrike: CrowdStrike 2025 global threat report, February 2025. https://go.crowdstrike.com/rs/281-OBQ-266/images/CrowdStrikeGlobalThreatReport2025.pdf. Accessed 9 Apr 2025
7. European Union Agency for Cybersecurity: ENISA Threat Landscape 2024, September 2024. https://www.enisa.europa.eu/sites/default/files/2024-11/ENISA%20Threat%20Landscape%202024_0.pdf. Accessed 9 Apr 2025
8. Genç, Z.A., Lenzini, G., Ryan, P.Y.: Security analysis of key acquiring strategies used by cryptographic ransomware. In: Proceedings of the Central European Cybersecurity Conference 2018. CECC 2018. Association for Computing Machinery, New York, NY, USA (2018). https://doi.org/10.1145/3277570.3277577
9. Hsiao, S.C., Kao, D.Y.: The static analysis of WannaCry ransomware. In: 2018 20th International Conference on Advanced Communication Technology (ICACT), pp. 153–158 (2018)
10. Ligh, M.H., Case, A., Levy, J., Walters, A.: The Art of Memory Forensics: Detecting Malware and Threats in Windows, Linux, and Mac Memory, 1st edn. Wiley Publishing (2014)
11. Maartmann-Moe, C., Thorkildsen, S.E., Årnes, A.: The persistence of memory: forensic identification and extraction of cryptographic keys. Digit. Invest. **6**, S132–S140 (2009). The Proceedings of the Ninth Annual DFRWS Conference
12. Mark Vincent Yason: Windows 10 Segment Heap Internals (2016). https://www.blackhat.com/docs/us-16/materials/us-16-Yason-Windows-10-Segment-Heap-Internals.pdf. Accessed 1 Apr 2025
13. Microsoft Learn: BLOBHEADER structure, February 2021. https://learn.microsoft.com/en-us/windows/win32/api/wincrypt/ns-wincrypt-publickeystruc. Accessed 30 Jan 2025

14. Microsoft Learn: CryptExportKey function, October 2021. https://learn.microsoft.com/en-us/windows/win32/api/wincrypt/nf-wincrypt-cryptexportkey. Accessed 30 Jan 2025

15. Yuste, J., Pastrana, S.: Avaddon ransomware: an in-depth analysis and decryption of infected systems. Comput. Secur. **109**, 102388 (2021)

Obfuscation

Experimental Study of Binary Diffing Resilience on Obfuscated Programs

Roxane Cohen[1,2]([✉]), Robin David[1], Riccardo Mori[1], Florian Yger[3], and Fabrice Rossi[4]

[1] Quarkslab, Paris, France
[2] LAMSADE, CNRS, Université Paris-Dauphine - PSL, Paris, France
roxane.cohen@dauphine.eu
[3] LITIS, INSA Rouen Normandy, Rouen, France
[4] CEREMADE, CNRS, Université Paris-Dauphine - PSL, Paris, France

Abstract. Obfuscation is commonly employed to protect sensitive program assets in legitimate use cases or to conceal malicious behavior in the context of malware. By altering the binary code of a compiled program, obfuscation disrupts binary analysis techniques, such as binary diffing or similarity. However, there is little comprehensive academic research addressing the effects of obfuscation on binary analysis tools and quantifying its impact. In this study, we examine how different types of obfuscation influence binary diffing algorithms. Specifically, we demonstrate a clear relationship between the type of obfuscation and the performance of the diffing algorithms used. Our benchmarks emphasize that, contrary to common assumptions, intra-procedural and data obfuscations have a limited impact on binary diffing when applied alone. In contrast, inter-procedural obfuscations significantly affect the diffing process, degrading performances by up to 40 f1-score points when comparing low and high obfuscation levels. These results highlight the need for modular diffing approaches, where parameters and features can be fine-tuned to handle adversarial scenarios, such as obfuscation. To support this research, we have released a comprehensive dataset comprising pairs of clear and obfuscated compiled programs, along with metadata specifying the type and exact location of each obfuscation. This dataset is intended to facilitate further research in this area.

Keywords: Obfuscation · Binary Diffing · Machine Learning · Binary Similarity · Malware

1 Introduction

Binary diffing is essential for reverse engineering. It is used for malware diffing [33], patch analysis [43], program similarity [10], backdoor detection, anti-plagiarism and to detect statically linked libraries. It consists in identifying similarities between two binaries, typically at the function level. Most binary differs use the disassembly to match the corresponding functions. Likewise, efficient

M. Egele et al. (Eds.): DIMVA 2025, LNCS 15747, pp. 223–243, 2025.
https://doi.org/10.1007/978-3-031-97620-9_13

solutions have been proposed to tackle the sub-problem of binary similarity on standard binaries in a cross-architecture, cross-compiler or cross-optimization setting [25].

However, diffing obfuscated binaries remains an open research question. Static or dynamic obfuscation consists in hiding the true behavior of a program and its syntactical representation, without modifying its semantics. It is widely used to protect application algorithms, data, and more generally, program assets. For example, MBA (Mixed Boolean Arithmetic) [44] replaces simple arithmetic operations by complex but strictly equivalent ones. As a consequence, obfuscation may alter binary code and modify features used by binary differs to match functions and more generally to compare programs. The assembly instructions, function data or execution flow, and the relationships between the functions can both be affected. Therefore, current binary differs and similarity tools may not be adapted to this adversarial context, leading to degraded performances. Such a case may occur if only one of the binaries is obfuscated, or if both are.

Contributions. This paper provides the first thorough experimental study of binary diffing in the presence of obfuscated binaries. Our study covers a large panel of state-of-the-art differs and similarity tools. We show how each of them is affected by different obfuscation types, pointing out that intra-procedural and data obfuscations have a limited impact on the diffing results. Contrarily, inter-procedural obfuscation significantly limits differ's efficiency, with a decrease up to 40 f1-score points between obfuscation levels, as most diffing algorithms rely on the CG (Call Graph) to perform matches. Results are valuable from both reverse engineering and software protection perspectives. A reverser may take advantage of existing modular differs and prior knowledge about the applied obfuscation in order to obtain a more resilient binary diffing result. Conversely, inter-procedural obfuscation should be privileged for software protection purposes, as it impedes most differ's work. Our experiments are based on two real-world datasets, the state-of-the-art BinKit dataset [19] and a new dataset dedicated to diffing and obfuscation[1], containing more than 6,700 binaries obfuscated with two obfuscators and more than 10 obfuscation passes. A real-world example of the X-Tunnel malware illustrates the potential of such a robust diffing approach and shows its practical usefulness.

This paper is structured as follows: Sect. 2 provides an overview of key concepts and related works. The details of our experiments, including the dataset, experimental setup, and the binary diffing and similarity tools used, are outlined in Sect. 3. Section 4 discusses how binary diffing can be leveraged to enhance resilience against obfuscation. The two binary diffing experiments, focusing on cases where either one or both binaries are obfuscated, are presented in Sects. 5 and 6, respectively. BinKit's results are discussed in Sect. 7. In Sect. 8, we showcase a real-world example with the XTunnel malware, followed by a discussion in Sect. 9 and concluding remarks in Sect. 10.

---

[1] https://github.com/quarkslab/diffing_obfuscation_dataset.

2   Background and Related Works

2.1   Program Representation

Programs are often represented by their disassembly, with functions modeled as CFG (Control Flow Graph), where nodes are basic-blocks (sequences of instructions without branching). A CFG encodes the intra-procedural execution flow, while the CG (Call Graph) represents inter-procedural call relationships. Both compilation or obfuscation can significantly alter these graphs. For instance, inlining affects the CG while loop unrolling impacts the CFG.

2.2   Binary Diffing and Similarity

Binary diffing usually operates at the function level and outputs a correspondence between the functions of two programs, denoted as primary and secondary. Binary diffing can be formalized as a one-to-one assignment $\phi : (\mathcal{P}, \mathcal{S}) \longmapsto \rho$ where $\mathcal{P}$ is the primary function set, $\mathcal{S}$ is the secondary function set. Then, $\rho : \mathcal{P} \longrightarrow \mathcal{S}$ is an assignment function both partial and injective, so that each function of $\mathcal{P}$ should be matched to at most one function of $\mathcal{S}$. This diffing is performed without having access to sources and symbols. Other diffing definitions consider different program granularities, such as basic block like DeepBinDiff [10], or one-to-many assignment [20]. Two programs do not need to be semantically equivalent to be efficiently diffed and small changes or light patches, induced by program versioning or compilation difference, can be used to analyze updates between binaries. BinDiff [11,12] and Diaphora [20] are the most widely used binary differs[2]. BinDiff starts by matching known imported functions and propagates these matches to neighboring functions using the CG, discriminating between neighbors based on function similarity. Diaphora establishes a set of heuristics, from confident to unreliable, used to iteratively match functions. The best heuristics are related to the function address and name, its assembly and pseudo-code and the function hash, with less attention dedicated to CG matching and other similarity measures. Both differs require disassembly as a starting point for diffing.

The **binary similarity** problem, closely related to binary diffing, is applied in order to find the most similar function to f inside a pool of candidate functions. It is an active research field relying heavily on ML (Machine Learning) and is particularly used for vulnerability search and malware analysis. Binary similarity tools either use precomputed assembly-based features or learn them with DL (Deep Learning), such as GNN (Graph Neural Networks). In the former category, TIKNIB [19] computes similarity scores using a specific distance combining various handcrafted features, BinShape [36] starts by extracting features and sorts them to obtain the top-ranked ones that are given to a decision tree. DL techniques tend to dominate the research field and are inspired from NLP (Natural Language Processing): Asm2Vec [9] is based on a refined and improved version of the word2vec model [31]. Trex and JTrans [35,41] are directly inspired by the

---

[2] According to popularity metrics (Github downloads, stars, Google Trend).

recent success of transformers for large language models. GNN are also gaining more and more popularity as the latest research articles mostly use increasingly complex GNN: current approaches use a pretrained language model on assembly instructions as initialization for further GNN embeddings [26]. GMN (Graph Matching Networks) [23] is the first work that jointly learns graph embeddings on similar graph pairs rather than independent embeddings. The idea is further developed with refined GNN architectures or language models [13,39]. These models are subject to adversarial attacks [3].

Binary similarity and diffing share common aspects but have distinct goals. Similarity algorithms output scores between function pairs, while diffing finds an assignment between functions in two binaries, often using similarity scores to establish these matches. Thus, similarity approaches when combined with a matching algorithm can yield diffing results.

2.3   Obfuscation

Obfuscation enhances code security against reverse engineering. Native code obfuscation (C, C++) can be source-to-source, like Tigress [6], or integrated into compiler toolchains, like OLLVM [18]. Research has explored deobfuscation [7,38], obfuscation detection [14], and diversification [16] which goal is to produce variants that cannot be linked with each other. Static obfuscation protects from static analysis by altering program layout while dynamic obfuscation protects from runtime analysis. Different obfuscation passes have specific effects on programs [34]:

- Data obfuscation alters the data-flow, e.g., hiding a XOR operation.
- Control flow obfuscation blurs the program execution flow logic. It can be divided into :
  - **Intra-procedural** obfuscation mutates the CFG structure. For example, CFF (CFG Flattening) puts every basic block at the same level and uses a dispatcher to maintain the function logic [40].
  - **Inter-procedural** obfuscation alters the CG by modifying the relationships between function callers and callees. It is damaging as CG been demonstrated essential for program analysis [21] (e.g., a Merge pass fusions two functions).

While binary diffing, similarity, and obfuscation have been studied separately, no research tackles the problem of diffing obfuscated code. Although some binary similarity techniques include small obfuscated experiments, they are mostly limited to intra-procedural or data obfuscations from OLLVM, lacking generalization across various obfuscations and obfuscators [9,17,19,35,36]. The limitations of intra-procedural obfuscation in a diffing context have been demonstrated [42], and inter-procedural obfuscation has only been studied using a symbolic execution approach [24].

Several reverse engineering use cases can benefit from efficient diffing of obfuscated binaries. Specifically, knowledge transfer refers to the ability to use insights gained from analyzing one binary to infer information about its subsequent, potentially obfuscated versions. This concept defines a threat model in which an attacker can leverage knowledge from one binary to circumvent the obfuscation applied to a new version. Such an approach has been applied to analyze obfuscated Android bytecode [8]. We denote this framework, where a plain binary (unobfuscated) is diffed with an obfuscated counterpart, as `plain-obfuscated` whereas the `obfuscated-obfuscated` setting indicates that two different obfuscated variants are compared. Conditions to operate such a diffing are commonly encountered, especially for programs that tend to be frequently updated as it is the case for Android applications. Similarly, malware are iteratively modified and improved accross attack campaigns. Finding plain and obfuscated variants is a common use-case [1].

3   Experimental Framework

This section outlines the creation of our reference dataset and the experimental settings used for evaluation. The dataset was designed to address the limitations of the existing BinKit dataset [19] (see also Sect. 7).

3.1   Dataset

Native code obfuscation is challenging to work with as there exist few obfuscators that are free or open-source. Those that are accessible only offer a limited number of obfuscations. For example, the latest official OLLVM [18], based on LLVM-4, exclusively provides intra-procedural obfuscations: `BogusControlflow`, `InstructionSubstitution` and `CFF`. Moreover, most binaries used in the literature are basic code snippets, such as sorting algorithms, and do not represent the complexity of true C projects. To fill the gap, and inspired by the state-of-the-art dataset for binary similarity research [25], we created a large dataset of realistic obfuscated binaries, with both plain binaries, obfuscated versions, and associated ground truths.

Five realistic projects written in C are used: `zlib`, `lz4`, `minilua`, `sqlite` and `freetype`. Obfuscation is applied using Tigress-3.1 [6] and OLLVM [18]. For each obfuscator, various obfuscation passes are selected: if possible, inter-procedural, intra-procedural and data. Because combining passes is essential to enhance security, several obfuscation sequences are created: `Mix` (`CFF` and `EncodeArithmetic` and `OpaquePredicates`)[3], and `Mix + Split` when possible. The latter scheme represents a real-world scenario where all aspects (assembly, CFG, and CG) are altered, making the task of an (adversary) differ challenging. For each project and obfuscation pass, we iteratively obfuscate from 10% to 100% of available functions using a 10% incremental step to observe how differs behave with varying levels of obfuscation. Given the inherent randomness in

---

[3] Obfuscation pass names are unified as they perform the same modifications but may have been implemented differently, leading to similar passes that can be more or less virulent. See Table 7 for details.

Table 1. Obfuscated dataset sum-up. ✔ compilation success, ✗ no binaries, ∼ some binaries are not available *(depends on obfuscation level or random seed used)*

	Passes	Pass type	zlib	lz4	minilua	sqlite	freetype
Tigress	Copy	Inter	✔	✔	✔	✔	✔
	Split	Inter	✔	✔	✔	✔	✔
	Merge	Inter	✔	✔	✗	✗	∼
	CFF	Intra	✔	✔	✔	✔	✔
	Virtualize	Intra	✔	✔	∼	∼	✗
	Opaque	Intra	✔	✔	✔	✗	∼
	EncodeArithmetic (Enc.A)	Data	✔	✔	✔	✔	✔
	EncodeLiterals (Enc.L)	Data	✔	✔	✔	✔	✔
	Mix	Intra & Data	✔	✔	✔	∼	∼
	Mix + Split	All	✔	✔	✔	∼	∼
OLLVM-14	CFF	Intra	✔	✔	✔	✔	✔
	Opaque	Intra	✔	✔	✔	✔	✔
	EncodeArithmetic (Enc.A)	Data	✔	✔	✔	✔	✔
	Mix	Intra & Data	✔	✔	✔	✔	✔

most obfuscation techniques, this process is repeated 10 times. The resulting dataset, summarized in Table 1, contains 6,718 binaries and 8,910,962 functions, making it the first freely available dataset of realistic obfuscated binaries at this scale.

The binaries are compiled with -O0 and -O2 optimizations for the x86-64 architecture. The -O0 optimization prevents most compiler effects that could attenuate obfuscation, such as inlining obfuscated functions or removing MBA. This allows us to study obfuscation passes in their unaltered form. Conversely, -O2 binaries are more representative of real-world scenarios, despite potential biases. The goal is to demonstrate that diffing obfuscated binaries works well both in a controlled, bias-free environment and on real-world binaries. Using OLLVM and Tigress-3.1 introduces several constraints. OLLVM, originally dependent on LLVM-4, was ported to LLVM-14 for this work and is referred to as OLLVM-14. It offers only three intra-procedural obfuscations. Tigress-3.1 requires amalgamated C files, as its cilly-merge functionality is unreliable to be used in practice. This constraint influenced project choices, limiting them to `zlib`, `lz4`, `minilua`, `sqlite`, and `freetype`, which are available in amalgamated form. Some Tigress binaries could not be produced due to internal errors or compilation issues with GCC 12. Default parameters are used for Tigress, except for the `Split` obfuscation where SplitCount is set to 2, and only top, block, and deep are allowed for SplitKinds. We limit our tests to these parameters as the resulting dataset already contains 6,700 binaries.

3.2 Experimental Settings

To support these experiments, we use IDAPro-8.1, assuming it produces a proper disassembly with a correct functions recovery. Most binary differs and binary similarity tools also make this assumption as a basis for further analysis and tackling disassembly issues is out of scope of this research. Various binary differs are selected: BinDiff [11,12] and Diaphora-3.0.0 [20]. Both are tested with default parameters. QBinDiff-1.2.0 [29,30], which is a modular differ that can be fine-tuned depending on the user goal, is also evaluated with two different configurations, explained in Sect. 4. Moreover, several binary similarity tools are studied: GMN [23] is considered as state-of-the-art [25] as it outperforms other similarity tools, such as SAFE [27]. Asm2Vec [9], proven to be robust against some obfuscations, and JTrans [41], which has shown promising results, are also considered. GMN is trained to output similarity scores on *Dataset-1A* [25], which contains standard binaries, with the original GMN hyperparameters and attributes for training. The same holds for JTrans and Asm2Vec, except for Asm2vec for which the number of random walks is set to 3 and that assembly instructions are normalized. All these binary similarity approaches are combined with the matching HA (Hungarian Algorithm) [22], with a cost matrix built over embeddings, in order to obtain a proper diffing output, from their similarity scores. We use the same Euclidean distance for all the binary similarity tools, like in the GMN framework [23]. Evaluating a differ efficiency requires to compare its results to the expected matching, called ground truth. Such a ground truth is built as follows:

– Data and intra-procedural obfuscations are applied within the function scope. These techniques will preserve the function name.
– Inter-procedural obfuscation alters the CG structure. Each type of obfuscation has a specific ground truth. For example:
  • A function f in primary split into f_1 and f_2 in the secondary should be matched with both f_1 and f_2.
  • Functions f_1 and f_2 merged into a single function f in the secondary should be matched with both f_1 and f_2 in the primary.
  • A function f cloned into f_1 in the secondary should be matched with both f and f_1 in the secondary.

This process can be extended to any number of split, merged, or cloned functions.

The ground-truth is defined at the function level since both Tigress and OLLVM can target specific functions, preserving the original mappings. Function-level granularity is necessary because basic-block-level ground truth cannot be computed for several obfuscations, such as `Virtualization`, which significantly alters a function's control flow. Maintaining correspondence at the basic-block level becomes impractical. As a result, we focus exclusively on function-level differs, excluding basic-block-level tools like DeepBinDiff [10] or BinSim [32], where accurate ground-truth cannot be established.

This inter-procedural ground-truth penalizes one-to-one diffing methods, as differs recover at most one correct match (either none or one), thus limiting the recall. All diffing and similarity tools evaluated in this paper are subjected to this penalty. After determining matches based on function names, the binaries are stripped to remove symbols, including function names. To compare the ground truth assignments with the diffing results, three standard metrics are considered: recall (R), precision (P), and f1-score, defined as:

$$P = \frac{TP}{TP + FP} \qquad R = \frac{TP}{TP + FN} \qquad \text{f1-score} = \frac{2 \times P \times R}{P + R}$$

with TP denoting True Positive, FP False Positive and FN False Negative.

Precision denotes how many retrieved items are relevant whereas the recall indicates how many relevant items are retrieved. Maximizing the f1-score requires both the recall and precision to be high.[4] Notice that even though similarity scores may decrease, due to small patches or differences, differs may still find the correct matches. Two f1-scores can be computed: a general f1-score applied on the whole binary, whereas an obfuscated f1-score is limited to obfuscated functions only. The higher these f1-scores, the better the differ is. If a significant difference exists between the general and obfuscated f1-scores, it implies that binary differs are correctly able to match regular functions but that they face trouble to do the same for obfuscated functions.

Finally, experiments were conducted on a Debian-6.1.27-1 with an Intel Xeon E3-12xx v2, 20 cores and 70 GB of RAM. Binary similarity experiments require a GPU and were launched with a Nvidia RTX A6000.

4   Resilient Diffing

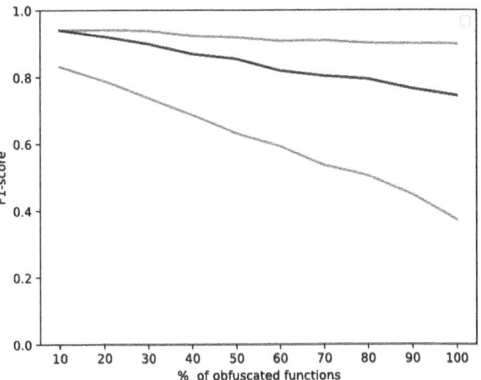

stable features (QBinDiff$_s$) ▦ all available features (QBinDiff) ▦ unstable features.

Fig. 1. Obfuscated `zlib` (Tigress CFF). QBinDiff f1-scores with respect to the % of obfuscated functions.

This Section illustrates how to obtain an obfuscation-resilient binary diffing on a concrete example. Extended results are available in Sect. 5 and 6. Existing differs

---

[4] The exact matching is known and using precision, recall and f1-score is consequently accurate. Binary similarity tools often rely on precision@k or recall@k as they search the most similar functions inside a pool of size k.

have not been designed to handle obfuscated binaries. Besides changing function, basic-block order strategies, BinDiff cannot be adjusted. Similarly, tuning Diaphora for obfuscation purpose is challenging due to the multitude of heuristics, some considered reliable and others not, for which there is no clear analysis of the order in which they should be applied and their influence on the final diffing process. QBinDiff was designed to be modular in order to leverage binary diffing for specific use cases. It enables the precise selection of features used to compute the similarity, which is then used to identify matches [4]. This modularity helps to obtain more accurate diffing results confronted with obfuscation, especially if its type is known. Such a knowledge can be obtained manually with reverse-engineering as some obfuscations, like CFF or MBA, exhibit very specific patterns, or with machine learning algorithms that directly infer if a function has been obfuscated and how [5]. Even though these machine learning algorithms do not exhibit perfect scores and suffer from several limitations, such as degraded performances for some optimization levels, they are still resilient enough to give an overview on the applied obfuscation within a binary. As a consequence, given an obfuscation found within a binary, it is possible to determine binary features that have likely been impacted by the obfuscation and choose to exclude them from the diffing process. Data obfuscations, such as MBA, significantly inflate the basic-block length without altering the CG and CFG structures, meaning that CFG and CG features such as the cyclomatic complexity or the number of children of a function will be unaffected. Similarly, intra-procedural obfuscations, such as Virtualization, strongly modify the CFG topology but do not impact the CG. Inter-procedural obfuscations, however, transform the CG topology and tend to also affect the corresponding CFG of functions impacted by the obfuscation. For instance, when two functions are merged, both the overall function relationships within the program and the CFG of the merged functions are restructured resulting in a unique CFG. As a consequence, only a limited set of features remain unaffected by this obfuscation type. We manually establish two obfuscation-dependent feature sets: a set consisting solely of unstable features directly impacted by an obfuscation, and a stable feature set, available in Table 8. These features sets have been determined by reasoning about each obfuscation and comparing unobfuscated functions to their obfuscated counterparts. If no knowledge of the obfuscation can be extracted from a binary, default features can be used, otherwise curated stable features should be selected. In this experiment, we aim to analyze the effect of these feature sets on the final diffing result. For each Tigress and OLLVM-14 obfuscations, QBinDiff is applied with these three configurations on the zlib project, compiled with -O0. The Fig. 1 highlights the difference between the standard QBinDiff differ (QBinDiff), where all the features are naively enabled, and a refined $QBinDiff_s$ on a Tigress CFF pass applied on zlib. Figures concerning other obfuscations are not shown as they depict the same trend. In particular, this experiment reveals how much a differ can be feature-dependent. The standard set of features should be used as a starting point and can be adapted to account for expert knowledge (for instance use the corresponding $QBinDiff_s$ in presence of intra-procedural obfuscation).

Moreover, it is fundamental to not use only features that are vulnerable to the same changes, as the unstable features provide degraded performances. For the rest of the paper, QBinDiff denotes the standard QBinDiff algorithm with the default feature set and illustrates the case where a reverse engineer does not have additional knowledge about the obfuscated binary. QBinDiff$_s$ indicates the adjusted QBinDiff with the stable feature set given a specific obfuscation and represents the best-case scenario, where preliminary knowledge about the obfuscation type was extracted with confidence.

5   Plain Against Obfuscated

This Section focuses on the `plain-obfuscated` experiment and examines the impact of obfuscation on binary diffing and similarity when only the second binary is obfuscated, while the first remains unmodified. This scenario is relevant for cases where newer binary versions are obfuscated, such as Android applications or malware.

Tables 2 and 3 present the OLLVM-14 and Tigress results, respectively, showing the outcomes for all obfuscation passes across five projects, averaged for 10%, 50%, and 100% obfuscation levels, with both -O0 and -O2 optimizations. The 10% obfuscation level simulates a situation where a defender has a limited obfuscation budget, either due to memory or computational constraints. This is common in embedded systems, where MCUs have limited storage and processing power, and only a small subset of critical functions are obfuscated for security. The 50% obfuscation level represents a guideline for achieving a reasonable tradeoff between security and efficiency. In practice, applying this amount of obfuscation often requires a deeper understanding of the criticality distribution over the functions, as well as the system constraints. Certain functions, that might precisely be the critical functions that one needs to obfuscate, could be highly sensitive to the obfuscation computational overhead, leading to prohibitive execution time for resource-constrained systems, such as MCUs. The 100% level depicts the worst-case scenario for an attacker.

To begin with, the most promising diffing tools are QBinDiff, QBinDiff$_s$ and JTrans, all of which achieve an f1-score of at least 0.80 for many obfuscation passes, particularly for data and intra-procedural obfuscations. Overall, real diffing tools like BinDiff, Diaphora, and QBinDiff tend to outperform some binary similarity methods combined with a matcher, such as GMN or Asm2vec. JTrans, however, does not face the same performance limitations as GMN or Asm2vec, thanks to its powerful embeddings. QBinDiff generally surpasses other diffing tools by balancing both CG structure and function similarity, whereas JTrans relies solely on function embedding similarity, and BinDiff uses CG structure as the foundation for match propagation.

Additionally, the differences between the general and obfuscated f1-scores depend on both the type of obfuscation and the diffing tool used. Similarity tools like GMN and Asm2vec struggle to maintain consistent f1-score range, often favoring matching unobfuscated functions over obfuscated ones. More effective

Table 2. Averaged f1-score comparison on OLLVM-14 obfuscations, `plain-obfuscated` setting. **First**, second and third best differs are displayed for each obfuscation and obfuscation level.

OLLVM-14 `plain-obfuscated`		General f1-score				Obfuscated f1-score			
		Mix	CFF	Opaque	Enc.A	Mix	CFF	Opaque	Enc.A
10%	Bindiff	0.81	0.78	0.78	0.79	**0.72**	0.71	0.69	0.75
	Diaphora3	0.79	0.78	0.78	0.80	0.45	0.59	0.61	0.78
	GMN	0.66	0.64	0.65	0.69	0.24	0.27	0.36	0.52
	Asm2vec	0.56	0.53	0.56	0.59	0.32	0.40	0.44	0.61
	JTrans	**0.85**	0.85	0.85	**0.86**	0.67	**0.81**	0.79	**0.82**
	QBinDiff	0.84	0.85	0.85	**0.86**	0.58	0.78	0.76	0.81
	QBinDiff$_s$	-	**0.86**	**0.86**	0.81	-	0.80	**0.81**	0.77
50%	Bindiff	0.72	0.76	0.74	0.79	**0.62**	0.67	0.65	0.73
	Diaphora3	0.64	0.70	0.71	0.79	0.40	0.57	0.59	0.78
	GMN	0.44	0.47	0.50	0.62	0.20	0.25	0.31	0.49
	Asm2vec	0.32	0.37	0.44	0.59	0.23	0.32	0.40	0.61
	JTrans	0.70	0.82	0.80	**0.86**	0.57	0.77	0.73	**0.81**
	QBinDiff	**0.74**	0.84	0.82	**0.86**	0.60	0.76	0.73	0.80
	QBinDiff$_s$	-	**0.85**	**0.85**	0.80	-	**0.79**	**0.79**	0.73
100%	Bindiff	0.53	0.65	0.65	0.78	0.47	0.56	0.57	0.72
	Diaphora3	0.40	0.50	0.61	0.79	0.35	0.49	0.56	0.77
	GMN	0.23	0.26	0.33	0.53	0.18	0.22	0.28	0.48
	Asm2vec	0.15	0.17	0.32	0.59	0.17	0.21	0.35	0.60
	JTrans	0.53	0.71	0.73	**0.86**	0.50	0.70	0.70	**0.81**
	QBinDiff	**0.65**	0.74	0.80	0.85	**0.59**	0.69	0.71	0.80
	QBinDiff$_s$	-	**0.77**	**0.84**	0.79	-	**0.73**	**0.78**	0.73

diffing tools, such as BinDiff and, to a lesser extent, Diaphora, reduce the gap between these scores to around 10 f1-score points for intra-procedural and data obfuscations. For JTrans and both versions of QBinDiff, the difference is even smaller, typically only about 5 f1-score points for lower levels of obfuscation. This is particularly true for QBinDiff$_s$ which incorporates features that make it more resilient to specific obfuscations. This result is surprising, as one might expect obfuscation to completely hinder binary diffing, but this is not the case. However, for inter-procedural obfuscations, this trend no longer holds, and all the diffing tools exhibit significant discrepancies between the f1-scores, sometimes exceeding 60 f1-score points. This is because either the tools fail to account for inter-procedural data, like similarity tools do, resulting in a major loss of information, or they rely on the CG for matching, which is altered during obfuscation and becomes unreliable, leading to poor performance, further exacerbated by the one-to-one mapping constraint.

Table 3. Averaged f1-score comparison on Tigress obfuscations, `plain-obfuscated` setting. **First**, second and third best differs are displayed for each obfuscation and obfuscation level.

Tigress plain-obfuscated		General f1-score										Obfuscated f1-score									
		Mix	Mix+Split	Copy	Merge	Split	CFF	Virtualize	Opaque	Enc.A	Enc.L	Mix	Mix+Split	Copy	Merge	Split	CFF	Virtualize	Opaque	Enc.A	Enc.L
10%	Bindiff	0.80	0.72	0.80	0.76	0.74	0.82	0.81	0.79	0.84	0.84	0.69	0.02	0.21	0.56	0.09	0.75	0.72	0.69	0.86	0.81
	Diaphora3	0.75	0.67	0.76	0.73	0.70	0.76	0.74	0.75	0.78	0.79	0.34	0.02	0.46	0.34	0.08	0.52	0.04	0.66	0.77	0.78
	GMN	0.51	0.46	0.57	0.52	0.47	0.53	0.48	0.53	0.54	0.59	0.04	0.01	0.34	0.15	0.02	0.08	0.01	0.25	0.16	0.49
	Asm2vec	0.49	0.42	0.51	0.55	0.46	0.49	0.45	0.51	0.56	0.57	0.15	0.01	0.28	0.35	0.03	0.20	0.03	0.45	0.52	0.58
	JTrans	0.80	0.74	0.82	0.78	0.76	0.84	0.80	0.81	0.85	0.85	0.55	0.02	0.54	0.35	0.04	0.72	0.36	0.74	0.87	0.88
	QBinDiff	0.84	0.77	0.86	0.82	0.81	0.87	0.83	0.84	0.89	0.89	0.55	0.05	0.25	0.59	0.19	0.73	0.35	0.69	0.92	0.91
	QBinDiff$_s$	-	-	0.85	0.84	0.82	0.89	0.89	0.79	0.83	0.84	-	-	0.25	0.60	0.22	0.85	0.79	0.65	0.90	0.86
50%	Bindiff	0.63	0.38	0.65	0.63	0.45	0.73	0.66	0.65	0.81	0.83	0.52	0.02	0.19	0.43	0.07	0.67	0.60	0.57	0.83	0.79
	Diaphora3	0.57	0.33	0.69	0.59	0.43	0.63	0.46	0.69	0.73	0.78	0.28	0.01	0.48	0.29	0.08	0.46	0.01	0.64	0.74	0.82
	GMN	0.30	0.19	0.49	0.35	0.26	0.32	0.27	0.38	0.38	0.58	0.02	0.01	0.36	0.13	0.02	0.06	0.00	0.20	0.13	0.50
	Asm2vec	0.27	0.16	0.41	0.40	0.26	0.30	0.18	0.40	0.49	0.59	0.10	0.00	0.29	0.28	0.04	0.14	0.01	0.39	0.50	0.64
	JTrans	0.64	0.41	0.71	0.56	0.47	0.76	0.52	0.74	0.63	0.85	0.46	0.01	0.54	0.16	0.03	0.69	0.23	0.71	0.67	0.90
	QBinDiff	0.68	0.45	0.75	0.70	0.56	0.79	0.68	0.74	0.87	0.89	0.49	0.03	0.26	0.51	0.17	0.72	0.53	0.66	0.90	0.90
	QBinDiff$_s$	-	-	0.74	0.73	0.58	0.87	0.81	0.68	0.82	0.83	-	-	0.26	0.58	0.20	0.84	0.77	0.59	0.88	0.87
100%	Bindiff	0.33	0.10	0.48	0.44	0.22	0.60	0.51	0.47	0.77	0.80	0.23	0.01	0.20	0.28	0.06	0.56	0.48	0.41	0.80	0.68
	Diaphora3	0.27	0.09	0.64	0.38	0.25	0.46	0.10	0.61	0.66	0.76	0.19	0.01	0.50	0.28	0.07	0.43	0.01	0.61	0.71	0.78
	GMN	0.10	0.05	0.42	0.23	0.13	0.12	0.11	0.24	0.25	0.57	0.01	0.01	0.35	0.12	0.02	0.05	0.00	0.19	0.12	0.49
	Asm2vec	0.08	0.04	0.29	0.29	0.13	0.11	0.02	0.32	0.43	0.56	0.08	0.00	0.24	0.32	0.03	0.11	0.00	0.37	0.48	0.65
	JTrans	0.46	0.21	0.62	0.32	0.28	0.68	0.20	0.66	0.60	0.83	0.43	0.01	0.54	0.14	0.03	0.68	0.16	0.68	0.66	0.89
	QBinDiff	0.40	0.19	0.65	0.57	0.36	0.71	0.49	0.63	0.85	0.87	0.33	0.02	0.26	0.48	0.15	0.70	0.46	0.61	0.89	0.87
	QBinDiff$_s$	-	-	0.64	0.61	0.39	0.84	0.72	0.56	0.81	0.82	-	-	0.27	0.55	0.18	0.83	0.72	0.53	0.86	0.84

Moreover, QBinDiff and QBinDiff$_s$ show different behaviors. Using QBinDiff$_s$ significantly improves performance for most inter and intra-procedural obfuscations compared to QBinDiff, both for general and obfuscated f1-scores. In some cases, it can boost the obfuscated f1-score by as much as 44 points, particularly for the `Virtualization` pass. However, this improvement does not hold for data obfuscations and the `Opaque` pass, where QBinDiff outperforms QBinDiff$_s$. This difference is due to the stable feature set used in QBinDiff$_s$, which was specifically designed for those obfuscations. For data obfuscations, features like assembly mnemonics were removed, reducing the obfuscation noise, but some deleted features may still provide useful information for matching.

In general, Tigress f1-scores are lower than those of OLLVM-14. While OLLVM-14 lacks inter-procedural obfuscations, which would penalize diffing, its intra-procedural and data obfuscations are less severe than Tigress's. Tigress scores tend to be 10 to 20 points lower in terms of f1-score. This is because Tigress includes more advanced obfuscations, such as `Virtualization`, and the obfuscations it shares with OLLVM-14, like `Opaque`, are more aggressive.

Note that the results above include both -O0 and -O2 binaries.[5] In general, -O0 results tend to be higher than -O2, with the f1-score degradation in -O2 being due to inlining caused by the optimization, which causes some functions, both obfuscated and unobfuscated, to disappear. As a result, diffing tools can no longer match those functions, leading to a drop of f1-score.

Overall, these results show that most diffing tools perform well on intra-procedural and data-obfuscated binaries, even when large portions of the binary are obfuscated. This is primarily because these tools rely on the CG structure, which remains unaffected by these types of obfuscations, to start matching functions. However, inter-procedural obfuscations lead to a significant drop in f1-scores, especially for obfuscated f1-scores. As a result, reverse engineers will face more challenges dealing with such obfuscated variants using diffing and should prioritize adaptive diffing tools like QBinDiff, which maintain the highest performances. From a software protection standpoint, these obfuscations should be prioritized if the goal is to counter reverse engineering attacks through diffing.

6   Obfuscated Against Obfuscated

In this Section, we compare two obfuscated versions of a binary to assess the effectiveness of binary diffing and similarity tools in this complex scenario. This use case is valuable for tracking the evolution of software obfuscation techniques over time or for identifying variants that are more vulnerable than others.

For brevity, we only present Tigress results in Table 4, as the OLLVM results follow the same pattern but show higher f1-scores, similar to what we observed in the previous plain-obfuscated experiment. We have chosen to focus on a subset of possible obfuscated-obfuscated pairs, prioritizing intra-intra, inter-inter, and inter-intra pairs when available. This decision is based on the results of the first plain-obfuscated experiment, where data obfuscations yielded less insightful results compared to other obfuscation types. The Mix-Mix pair refers to two binaries obfuscated with the same Mix schema but using different seeds.

Results for QBinDiff$_s$ are not available because when comparing two obfuscated variants, unless both variants are obfuscated using the same type of obfuscation, it is difficult to identify a set of features that are robust across different obfuscation types. Neither the CG nor the CFG are reliable when comparing a variant obfuscated with Split to one obfuscated with Virtualization. Overall, the results reflect a similar trend to previous findings, with QBinDiff, JTrans, and BinDiff generally being the most effective differ tools. In this scenario, JTrans performs worse because the variants are both obfuscated, whereas JTrans was trained on unobfuscated code, making it more reliant on plain binary code.

Although this framework may appear more challenging than the plain-obfuscated one, f1-scores, both obfuscated and unobfuscated, remain acceptable for obfuscated variants without inter-procedural obfuscation, with

---

[5] Due to space limitations, we combined the -O0 and -O2 experiments, even though they exhibit slightly different behavior in terms of diffing results. Specific -O0 and -O2 results will be released as artifacts along with the dataset.

Table 4. Averaged f1-score comparison on Tigress obfuscations, `obfuscated-obfuscated` setting. **First**, <u>second</u> and <u>third</u> best differs are displayed for each obfuscation and obfuscation level.

Tigress obfuscated-obfuscated		General f1-score										Obfuscated f1-score									
		Mix-Mix	Split-Merge	Merge-Copy	Split-Copy	CFF-Split	CFF-Merge	CFF-Copy	CFF-Opaque	Opaque-Enc.A	CFF-Enc.A	Mix-Mix	Split-Merge	Merge-Copy	Split-Copy	CFF-Split	CFF-Merge	CFF-Copy	CFF-Opaque	Opaque-Enc.A	CFF-Enc.A
10%	BinDiff	0.81	0.68	0.79	0.72	0.77	0.80	0.83	0.82	0.83	0.87	0.70	0.08	0.24	0.16	0.35	0.68	0.50	0.75	0.83	0.77
	Diaphora3	0.74	0.65	0.74	0.71	0.73	0.77	0.76	0.77	0.79	0.40	0.09	0.42	0.17	0.18	0.51	0.53	0.71	0.79	0.64	
	GMN	0.62	0.50	0.67	0.56	0.59	0.62	0.68	0.64	0.67	0.70	0.09	0.03	0.30	0.06	0.05	0.16	0.10	0.22	0.33	0.11
	Asm2vec	0.55	0.47	0.55	0.46	0.50	0.37	0.56	0.56	0.58	0.61	0.19	0.05	0.31	0.08	0.11	0.27	0.21	0.40	0.57	0.25
	JTrans	0.79	0.70	0.79	0.71	0.77	0.78	0.82	0.82	0.83	0.86	0.43	0.05	0.37	0.15	0.26	0.49	0.50	0.75	0.83	0.73
	QBinDiff	0.83	0.73	0.82	0.75	0.81	0.83	0.85	0.85	0.86	0.88	0.53	0.15	0.25	0.20	0.33	0.66	0.48	0.69	0.80	0.69
50%	BinDiff	0.55	0.40	0.57	0.38	0.46	0.59	0.60	0.63	0.71	0.80	0.40	0.06	0.18	0.11	0.24	0.43	0.38	0.51	0.66	0.69
	Diaphora3	0.53	0.40	0.60	0.37	0.39	0.52	0.58	0.65	0.69	0.67	0.41	0.07	0.35	0.15	0.13	0.36	0.43	0.64	0.72	0.55
	GMN	0.32	0.25	0.43	0.24	0.27	0.32	0.36	0.34	0.40	0.42	0.09	0.02	0.18	0.04	0.03	0.07	0.06	0.11	0.18	0.05
	Asm2vec	0.32	0.21	0.29	0.19	0.20	0.27	0.29	0.32	0.50	0.39	0.20	0.04	0.16	0.06	0.07	0.20	0.13	0.28	0.53	0.16
	JTrans	*(illegible)*																			
	QBinDiff	0.61	0.48	0.63	0.43	0.52	0.67	0.66	0.72	0.78	0.81	0.43	0.11	0.19	0.17	0.28	0.55	0.43	0.61	0.70	0.64
100%	BinDiff	0.49	0.25	0.36	0.14	0.13	0.38	0.24	0.40	0.39	0.53	0.32	0.04	0.08	0.05	0.06	0.15	0.13	0.21	0.40	0.48
	Diaphora3	0.67	0.23	0.32	0.22	0.14	0.32	0.37	0.55	0.39	0.33	0.61	0.03	0.15	0.14	0.09	0.24	0.36	0.53	0.54	0.42
	GMN	0.30	0.18	0.25	0.10	0.08	0.23	0.13	0.16	0.16	0.13	0.18	0.01	0.04	0.03	0.02	0.04	0.04	0.02	0.06	0.04
	Asm2vec	0.41	0.06	0.08	0.08	0.05	0.16	0.08	0.18	0.33	0.14	0.35	0.03	0.06	0.05	0.05	0.17	0.09	0.18	0.44	0.13
	JTrans	0.71	0.27	0.35	0.19	0.21	0.34	0.43	0.64	0.36	0.41	0.52	0.01	0.06	0.09	0.15	0.12	0.38	0.51	0.29	0.34
	QBinDiff	0.70	0.32	0.41	0.25	0.24	0.61	0.46	0.66	0.48	0.56	0.51	0.07	0.12	0.13	0.18	0.49	0.38	0.49	0.39	0.60

scores reaching at least 0.70 for many pairs at lower obfuscation levels. While this trend diminishes as the number of obfuscated functions reaches 100%, it still demonstrates that diffing obfuscated variants is feasible, yielding satisfactory results, particularly with intra-procedural or data obfuscations that preserve the CG structure for binary diffing. These findings show that data and intra-procedural obfuscations, when used alone, do not fully prevent binary similarity or diffing tools from working, even though they are effective at confusing reverse engineers conducting manual analysis. However, these obfuscations are the most commonly implemented in open-source or free obfuscators [15,18,37]. In contrast, diffing two samples obfuscated with inter-procedural passes is much more challenging, with a significant gap between the f1-scores for unobfuscated and obfuscated pairs, which remain low, typically below 0.50 of f1-score when obfuscation exceeds 50%. Therefore, from a reverse engineering perspective, more focus should be placed on developing new matching algorithms that can better handle transformations that disrupt the relationships between functions. From a software protection standpoint, this suggests a stronger emphasis on inter-procedural obfuscation to limit knowledge transfer between binaries.

7   BinKit Results

As previously mentioned, we expand on our earlier experiment using the BinKit dataset, which includes plain binaries compiled with various compilers, versions, optimizations, and architectures. The obfuscated binaries are exclusively compiled with Clang using OLLVM, with different optimizations and architectures.

We focus only on x64 binaries and limit our analysis to five projects. The obfuscations applied are solely intra-procedural and data obfuscations at the binary level. The results, presented in Table 5, follow the same pattern as our previous dataset experiments, but the effects are even more pronounced. Specifically, binary similarity tools perform notably worse than binary diffing tools, with QBinDiff outperforming both BinDiff and Diaphora.

Table 5. Averaged f1-score comparison for the BinKit obfuscated dataset. **First**, second and third best differs are displayed for each obfuscation and obfuscation level.

Binkit	Plain-obfuscated					Obfuscated-obfuscated				
	bool	cpio	cflow	ccd2cue	a2ps	bool	cpio	cflow	ccd2cue	a2ps
BinDiff	0.9	0.63	0.78	0.94	0.7	0.8	0.42	0.61	0.84	0.44
Diaphora3	0.66	0.6	0.71	0.71	0.63	0.57	0.45	0.4	0.57	0.39
GMN	0.41	0.39	0.30	0.53	0.22	0.40	0.39	0.31	0.53	0.23
Asm2vec	0.37	0.29	0.22	0.55	0.15	0.34	0.25	0.19	0.38	0.13
JTrans	0.86	0.80	0.84	0.90	0.69	0.70	0.55	0.55	0.66	0.42
QBinDiff	0.96	0.92	0.91	0.98	0.82	0.9	0.82	0.82	0.91	0.7
QBinDiff$_s$	**0.97**	**0.94**	**0.93**	**0.99**	**0.87**	**0.92**	**0.86**	**0.86**	**0.91**	**0.80**

These results, tested on a different dataset, further confirm that intra-procedural and data obfuscations do not pose a significant barrier to binary diffing.

8   Real-World Example: XTunnel

This Section aims to extend the previous experiments, conducted on a realistic but yet simulated dataset, to real-world malware samples. Finding any two malware samples is simple, whereas identifying two obfuscated versions of the same malware is more challenging, though still achievable. However, locating two obfuscated malware samples for which establishing a reliable ground truth is practically feasible proves to be significantly more difficult, due to a limited number of functions and limited obfuscation. XTunnel is among the few malware that satisfies these requirements [1]. Establishing a match between different versions eases the reverse engineering of new variant, especially if new functionalities have been added to the obfuscated version. We replicate only the previous `plain-obfuscated` experiment, for brevity, by diffing two XTunnel samples: the plain *42DEE* and the obfuscated *99B45*. The two samples contain 3,196 and 3,693 functions, respectively, with about 400 of them heavily obfuscated using `Opaque Predicates` [1]. The remaining functions appear to be third-party libraries statically linked within the executables.

To create the ground truth for this example, we manually match functions between the plain binary *42DEE* and the obfuscated variant *99B45*, starting with functions that have the same hash and then using manual reverse engineering to identify the rest. Creating such ground truth requires significant effort and may introduce bias, especially due to discrepancies in the number of primary and secondary functions caused by inlining, which can affect its accuracy. As a result, 417 primary functions and 913 secondary functions remain unmatched due to uncertainty in their assignment.

Only QBinDiff$_s$ and BinDiff are used in this experiment. An effective differ should produce high f1-scores for both binaries. The results, shown in Table 6, reveal that while QBinDiff$_s$ and BinDiff are almost identical on the full set of functions, BinDiff performs poorly on the obfuscated functions. Despite potential bias in the ground truth, this example demonstrates that our earlier findings remain true in real-world use cases.

Table 6 f1-scores for the **plain-obfuscated** experiment variant between samples *42DEE* and *99B45*.

	General f1-score	Obfuscated f1-score
BinDiff	0.966	0.303
QBinDiff$_s$	0.97	0.915

Table 7. Obfuscation description

	Passes	Pass type	Description	Unify name
	Copy	Inter	Clone a function	Copy
	Split	Inter	Split a function into chunks	Split
	Merge	Inter	Merge multiple function into one	Merge
	CFF	Intra	The function basic blocks are put at the same level	CFF
Tigress	Virtualize	Intra	Transforms a function into an interpreter	Virtualize
	Opaque	Intra	Insert OpaquePredicates	Opaque
	EncodeArithmetic	Data	Replace arithmetic with complex expressions	Enc.A
	EncodeLiterals	Data	Hide literals with less obvious expressions	Enc.L
	Mix	Intra & Data	Combination of CFF, EncodeArithmetic and Opaque	Mix
	Mix + Split	All	Mix + Split	Mix + Split
	CFF	Intra	The function basic blocks are put at the same level	CFF
OLLVM-14	BogusCF	Intra	Insert basic blocks and OpaquePredicates	Opaque
	InsSub	Data	Substitute operators by more complicated instructions	Enc.A
	Mix	Intra & Data	Combination of CFF, InsSub and BogusCF	Mix

9   Discussions

Limitations and Threats to Validity. First, this research relies on disassembly and functions recovered by IDA-Pro, which may lead to incomplete results due to

Table 8. QBinDiff stable (✔) and unstable (✗) features list depending on the applied obfuscation.

	BBlockNb	SCComponents	BytesHash	Cyclomatic Complexity	MDIndex	JumpNb	SmallPrimeNumbers	MaxParentNb	MaxChildNb	MaxInsNb	MeanInsNb	InsNb	GraphMeanDegree	GraphDensity	GraphNbComponents	Graph Diameter	GraphTransitivity	GraphCommunities	Address	DatName	FuncName	ChildNb	ParentNb	RelativeNb	LibName	ImpName	Constant	StrRefs	MnemonicSimple	MnemonicTyped	GroupsCategory	ReadWriteAccess
Merge Split	✗	✗	✗	✗	✗	✗	✗	✗	✗	✗	✗	✗	✗	✗	✗	✗	✗	✗	✗	✔	✔	✗	✗	✗	✗	✗	✗	✔	✔	✔	✔	✔
Copy	✔	✔	✔	✔	✔	✔	✔	✔	✔	✔	✔	✔	✔	✔	✔	✔	✔	✔	✔	✔	✔	✗	✗	✗	✗	✗	✗	✔	✔	✔	✔	✔
Intra	✗	✗	✗	✗	✗	✗	✗	✗	✗	✗	✗	✗	✗	✗	✗	✗	✗	✗	✗	✔	✔	✔	✔	✔	✔	✔	✔	✔	✔	✔	✔	✔
Data	✔	✔	✗	✔	✔	✔	✔	✔	✔	✗	✗	✗	✔	✔	✔	✔	✔	✔	✔	✔	✔	✔	✔	✔	✔	✔	✗	✗	✗	✗	✗	✗

the complexity of these problems [28]. Additionally, some obfuscation techniques are specifically designed to hinder disassemblers from working correctly.

Second, most differ tools are limited to one-to-one matching, which may not be enough for obfuscated or optimized functions where a one-to-many approach would be more appropriate. This is a complex issue, and only a few differ tools offer solutions, with no clear way to assess their effectiveness [20].

Third, obfuscation may not achieve the desired effect in the final binary, as the compiler can interfere with or even reverse obfuscation, given their conflicting purposes. The higher the optimization level, the greater the risk that the obfuscation will be altered. This is especially true for optimization level -O2, which is more realistic but can remove obfuscations or apply inlining. This issue is noticeable with newer compiler versions, such as clang-14, when using older obfuscation techniques like OLLVM. Even at optimization level -O0, this behavior can still occur, though to a lesser extent (e.g., OLLVM EncodeArith may be simplified through constant propagation). Preventing these optimizations is compiler-dependent and often requires tweaking internals, if even possible. For example, constant propagation occurs both in the clang front-end and optimization phase, making it difficult to disable. Source-to-source obfuscators, like Tigress, cannot directly interact with the compiler, and most OLLVM passes are applied before any optimization, meaning both are susceptible to simplification. On the other hand, other compilation-pass-based obfuscators apply their transformations within an optimization chain, not at the beginning, to avoid slowing down the final binary, and not at the end, to slightly optimize the obfuscated code for performance [2]. Determining the best way to combine optimization and obfuscation is beyond the scope of this paper.

Fourth, this study primarily focuses on single obfuscation types, except the Mix and Mix + Split combinations. In the Mix configuration, CFF is followed by another intra-procedural pass and a data obfuscation. The pass order here was fixed, and it may not be the most efficient for resisting an attacker. Similarly,

`Mix` combines `CFF`, shared by both Tigress and OLLVM-14, with slightly different passes that modify control flow (`BogusCF` and `Opaque`) and data (`InsSub` and `Enc.A`). This comparison evaluates an obfuscator's effectiveness based on its different obfuscation types, rather than the overall robustness of each pass.

Fifth, to level the playing field for all the differs, Diaphora3's decompiler-based features were disabled. Including them will be considered in future work.

Future Work. Studying obfuscation in all its facets remains an ongoing research challenge, as functions, optimizations, obfuscation passes, and obfuscators are all intertwined. There is still much work to be done. Specifically, the dataset that we created is an initial effort to address a gap in obfuscation research. Expanding it by incorporating additional obfuscators or more projects would be a valuable step toward improving generalization. We plan to gradually enhance it over time.

Additionally, the relationship between an obfuscator and a differ is similar to the dynamic between a defender and an attacker. This opens the door for adversarial refinement on both sides. A defender could use diffing tools to identify and analyze weaknesses in their obfuscations, improving their resilience against attackers. This line of research could ultimately lead to more robust and resilient obfuscation schemes.

10   Conclusion

This paper presents a comprehensive study of binary diffing in both `plain-obfuscated` and `obfuscated-obfuscated` settings. We evaluate various binary differ and similarity tools, revealing that, surprisingly, standard binary diffing methods perform well against intra-procedural and data obfuscations. On the other hand, the most effective differ tools, which rely heavily on the CG, are susceptible to disruption from inter-procedural obfuscations. The modularity offered by QBinDiff proves useful in obtaining resilient diffing results based on the type of obfuscation applied, showing promising outcomes. These findings demonstrate that diffing obfuscated binaries is feasible and can yield satisfactory results in certain cases. From a software protection perspective, they also highlight the advantages of using more extensive inter-procedural obfuscation.

Acknowledgments. We would like to thank Bruno Mateu for porting OLLVM-4 to OLLVM-14 and the Agence Innovation Defense (AID) that supports this research.

Data Availability Statement. The dataset of obfuscated binaries and the artifacts are publicly released. (https://github.com/quarkslab/diffing_obfuscation_dataset)

References

1. Bardin, S., David, R., Marion, J.Y.: Backward-bounded DSE: targeting infeasibility questions on obfuscated codes. In: 2017 IEEE Symposium on Security and Privacy (SP), pp. 633–651 (2017). https://doi.org/10.1109/SP.2017.36

2. Brunet, P., Creusillet, B., Guinet, A., Martinez, J.M.: Epona and the obfuscation paradox: transparent for users and developers, a pain for reversers. In: Proceedings of the 3rd ACM Workshop on Software Protection. SPRO'19, New York, NY, USA, pp. 41–52. Association for Computing Machinery (2019)
3. Capozzi, G., D'Elia, D.C., Di Luna, G.A., Querzoni, L.: Adversarial attacks against binary similarity systems. arXiv preprint arXiv:2303.11143 (2023)
4. Cohen, R., David, R., Mori, R., Yger, F., Rossi, F.: Improving binary diffing through similarity and matching intricacies. In: Conference on Artificial Intelligence for Defense (CAID) (2024)
5. Cohen, R., David, R., Yger, F., Rossi, F.: Identifying obfuscated code through graph-based semantic analysis of binary code. In: The 13th International Conference on Complex Networks and their Applications (2024)
6. Collberg, C.: The tigress C obfuscator (2016). https://tigress.wtf/index.html
7. David, R., Coniglio, L., Ceccato, M.: Qsynth-a program synthesis based approach for binary code deobfuscation. In: BAR 2020 Workshop (2020)
8. De Ghein, R., Abrath, B., De Sutter, B., Coppens, B.: Apkdiff: matching android app versions based on class structure. In: Proceedings of the 2022 ACM Workshop on Research on Offensive and Defensive Techniques in the Context of Man At The End (MATE) Attacks. Checkmate '22, New York, NY, USA, pp. 1–12. Association for Computing Machinery (2022)
9. Ding, S.H., Fung, B.C., Charland, P.: Asm2vec: boosting static representation robustness for binary clone search against code obfuscation and compiler optimization. In: 2019 IEEE Symposium on Security and Privacy (SP). IEEE (2019)
10. Duan, Y., Li, X., Wang, J., Yin, H.: Deepbindiff: learning program-wide code representations for binary diffing. In: Network and Distributed System Security Symposium (2020)
11. Dullien, T., Rolles, R.: Graph-based comparison of executable objects (English version). SSTIC 5(1), 3 (2005)
12. Flake, H.: Structural comparison of executable objects. DIMVA 2004, July 6-7, Dortmund, Germany (2004)
13. Gao, H., Zhang, T., Chen, S., Wang, L., Yu, F.: Fusion: measuring binary function similarity with code-specific embedding and order-sensitive GNN. Symmetry (2022)
14. Greco, C., Ianni, M., Guzzo, A., Fortino, G.: Explaining binary obfuscation, pp. 22–27 (2023). https://doi.org/10.1109/CSR57506.2023.10224825
15. Hikari: Hikari-llvm15 (2019). https://github.com/61bcdefg/Hikari-LLVM15
16. Hosseinzadeh, S., et al.: Diversification and obfuscation techniques for software security: a systematic literature review. Inf. Software Technol. (2018)
17. Hu, Y., Zhang, Y., Li, J., Wang, H., Li, B., Gu, D.: Binmatch: a semantics-based hybrid approach on binary code clone analysis. In: 2018 IEEE International Conference on Software Maintenance and Evolution (ICSME). IEEE (2018)
18. Junod, P., Rinaldini, J., Wehrli, J., Michielin, J.: Obfuscator-LLVM–software protection for the masses. In: Wyseur, B. (ed.) Proceedings of the IEEE/ACM 1st International Workshop on Software Protection, SPRO'15, Firenze, Italy, May 19th, 2015 (2015)
19. Kim, D., Kim, E., Cha, S.K., Son, S., Kim, Y.: Revisiting binary code similarity analysis using interpretable feature engineering and lessons learned. IEEE Trans. Software Eng. 1–23 (2022)
20. Koret, J.: Diaphora (2015). https://github.com/joxeankoret/diaphora

21. Kostakis, O., Kinable, J., Mahmoudi, H., Mustonen, K.: Improved call graph comparison using simulated annealing. In: Proceedings of the 2011 ACM Symposium on Applied Computing, pp. 1516–1523 (2011)
22. Kuhn, H.W.: The hungarian method for the assignment problem. Naval Res. Logist. Quart. **2**(1–2), 83–97 (1955)
23. Li, Y., Gu, C., Dullien, T., Vinyals, O., Kohli, P.: Graph matching networks for learning the similarity of graph structured objects. In: International Conference on Machine Learning, pp. 3835–3845. PMLR (2019)
24. Luo, L., Ming, J., Wu, D., Liu, P., Zhu, S.: Semantics-based obfuscation-resilient binary code similarity comparison with applications to software plagiarism detection. In: Proceedings of the 22nd ACM SIGSOFT International Symposium on Foundations of Software Engineering, pp. 389–400 (2014)
25. Marcelli, A., Graziano, M., Ugarte-Pedrero, X., Fratantonio, Y., Mansouri, M., Balzarotti, D.: How machine learning is solving the binary function similarity problem. In: 31st USENIX Security Symposium (USENIX Security 22) (2022)
26. Massarelli, L., Di Luna, G.A., Petroni, F., Querzoni, L., Baldoni, R., et al.: Investigating graph embedding neural networks with unsupervised features extraction for binary analysis. In: 2nd Workshop on Binary Analysis Research (BAR) (2019)
27. Massarelli, L., Di Luna, G.A., Petroni, F., Querzoni, L., Baldoni, R.: Safe: self-attentive function embeddings for binary similarity. In: Proceedings of 16th Conference on Detection of Intrusions and Malware and Vulnerability Assessment (DIMVA) (2019)
28. Meng, X., Miller, B.P.: Binary code is not easy. In: Proceedings of the 25th International Symposium on Software Testing and Analysis, pp. 24–35 (2016)
29. Mengin, E., Rossi, F.: Binary diffing as a network alignment problem via belief propagation. In: 2021 36th IEEE/ACM International Conference on Automated Software Engineering (ASE), pp. 967–978. IEEE (2021)
30. Mengin, E., Rossi, F.: Improved algorithm for the network alignment problem with application to binary diffing. Procedia Comput. Sci. **192**, 961–970 (2021)
31. Mikolov, T., Chen, K., Corrado, G., Dean, J.: Efficient estimation of word representations in vector space. arXiv preprint arXiv:1301.3781 (2013)
32. Ming, J., Xu, D., Jiang, Y., Wu, D.: BinSim: Trace-based semantic binary diffing via system call sliced segment equivalence checking. In: 26th USENIX Security Symposium (USENIX Security 17), pp. 253–270 (2017)
33. Ming, J., Xu, D., Wu, D.: Memoized semantics-based binary diffing with application to malware lineage inference. In: Federrath, H., Gollmann, D. (eds.) ICT Systems Security and Privacy Protection. Springer, Cham (2015)
34. Nagra, J., Collberg, C.: Surreptitious Software: Obfuscation, Watermarking, and Tamperproofing for Software Protection: Obfuscation, Watermarking, and Tamperproofing for Software Protection. Pearson Education (2009)
35. Pei, K., Xuan, Z., Yang, J., Jana, S., Ray, B.: Trex: learning execution semantics from micro-traces for binary similarity. arXiv preprint arXiv:2012.08680 (2020)
36. Shirani, P., Wang, L., Debbabi, M.: Binshape: scalable and robust binary library function identification using function shape. In: International Conference on Detection of Intrusions and Malware, and Vulnerability Assessment, pp. 301–324. Springer (2017)
37. Thomas, R.: O-mvll (2022). https://github.com/open-obfuscator/o-mvll
38. Tofighi-Shirazi, R., Asavoae, I.M., Elbaz-Vincent, P., Le, T.H.: Defeating opaque predicates statically through machine learning and binary analysis. In: Proceedings of the 3rd ACM Workshop on Software Protection, pp. 3–14 (2019)

39. Ullah, S., Oh, H.: Bindiff NN: learning distributed representation of assembly for robust binary diffing against semantic differences. IEEE Trans. Software Eng. **48**(9), 3442–3466 (2021)
40. Wang, C.: A Security Architecture for Survivability Mechanisms. University of Virginia (2001)
41. Wang, H., et al.: Jtrans: jump-aware transformer for binary code similarity detection. In: Proceedings of the 31st ACM SIGSOFT International Symposium on Software Testing and Analysis, pp. 1–13. ISSTA 2022, New York, NY, USA (2022)
42. Zhang, P., et al.: Khaos: the impact of inter-procedural code obfuscation on binary diffing techniques. In: Proceedings of the 21st ACM/IEEE International Symposium on Code Generation and Optimization. CGO '23, New York, NY, USA, pp. 55–67. Association for Computing Machinery (2023)
43. Zhao, L., Zhu, Y., Ming, J., Zhang, Y., Zhang, H., Yin, H.: Patchscope: memory object centric patch diffing. In: Proceedings of the 2020 ACM SIGSAC Conference on Computer and Communications Security, pp. 149–165 (2020)
44. Zhou, Y., Main, A., Gu, Y.X., Johnson, H.: Information hiding in software with mixed boolean-arithmetic transforms. In: International Workshop on Information Security Applications, pp. 61–75. Springer (2007)

Quantifying and Mitigating the Impact of Obfuscations on Machine-Learning-Based Decompilation Improvement

Luke Dramko[1]([✉]), Deniz Bölöni-Turgut[2], Claire Le Goues[1], and Edward Schwartz[3]

[1] Carnegie Mellon University, Pittsburgh, PA 15213, USA
{lukedram,clegoues}@cs.cmu.edu
[2] Cornell University, Ithaca, NY 14850, USA
db823@cornell.edu
[3] Carnegie Mellon University Software Engineering Institute,
Pittsburgh, PA 15213, USA
eschwartz@cert.org

Abstract. Decompilers are tools that reverse the process of compilation, converting executable binaries into a high-level language like C. They are useful in situations where the original source code is unavailable, such as when analyzing malware, doing vulnerability research, and patching legacy software. Unfortunately, decompilation is necessarily incomplete, because the compiler discards many of the abstractions that make source code readable, like identifier names and types. A large body of existing work uses machine learning to predict missing names, types, and other abstractions in decompiled code. However, little of this work considers *obfuscations*: semantics-preserving transformations that obscure the functionality and design of a program. At the same time, obfuscations are common in practice, especially in malware. In this work, we perform a quantitative analysis of the impact that obfuscations have on decompiled code. Further, we investigate the degree to which training on obfuscated code mitigates the impact of obfuscations. We perform our experiments on three different models from the literature: DIRTY, HexT5, and VarBERT. We find that obfuscations do negatively impact machine learning models, but training on obfuscations can partially help recover lost accuracy.

Keywords: Decompilation · Reverse Engineering · Machine Learning

1 Introduction

A *decompiler* is a tool that reverses the process of compilation, converting executable binary programs into a high level language such as C. Decompilers are useful for a variety of security related tasks, including malware analysis, vulnerability research, and patching legacy software [37,38]. Source code is a dual-channel medium, containing a formal channel that specifies execution semantics,

M. Egele et al. (Eds.): DIMVA 2025, LNCS 15747, pp. 244–266, 2025.
https://doi.org/10.1007/978-3-031-97620-9_14

and a natural language channel that communicates information to developers who read and write the code [5].

```
1  int uv_exepath(char* buffer, size_t* size){        1  __int64 __fastcall <func>(char *buf,
2    if (!buffer || !size) {                                  ssize_t *a2){
3      return -1;                                       2    ssize_t v2;
4    }                                                  3    if (!buf) return 0xFFFFFFFFLL;
5                                                       4    if (!a2) return 0xFFFFFFFFLL;
6    *size = readlink("/proc/self/exe", buffer         5    v2 = readlink("/proc/self/exe", buf, *
        , *size - 1);                                          a2 - 1);
7    if (*size <= 0)                                    6    *a2 = v2;
8      return -1;                                       7    if (!v2) return 0xFFFFFFFFLL;
9    buffer[*size] = '\0';                              8    buf[v2] = 0;
10   return 0;                                          9    return 0LL;
11 }                                                   10  }
```

(a) A function that returns the filename of the currently running process.

(b) The same function as 1a, but after being compiled and then decompiled by IDA Pro.

Fig. 1. Decompiled code is harder to read than original source code.

Computers, however, only require the formal channel, and so compilers discard the abstractions in the natural channel during compilation. As a result, traditional decompilers struggle to recover many of the natural abstractions that make source code readable, such as variable names, types, comments, and some aspects of code structure [12]. This makes reverse engineering slow and painstaking [37,38].

To make matters worse, some targets of decompilation—especially malware—are intentionally *obfuscated*: they are transformed to obscure the functionality and design of a program (in other words, making it more difficult to comprehend) without changing the program's behavior. Figure 1a shows a simple function, and Fig. 1b shows the function after being compiled and then decompiled. Figure 2a –2d show the same function after applying an obfuscation, compilation, and decompilation. For instance, in Fig. 2a, the string literal "/proc/self/exe", which provides a key clue to what the function does, has been replaced with a sequence of operations on a collection of variables that obscure the content of the string. In Fig. 2c, control flow has been completely restructured so that it is difficult to tell what statements are executed in what order. As these examples demonstrate, decompilers typically do not undo obfuscations; they simply propagate the obfuscation from the binary level to the source level.

Recently, researchers have turned to machine learning models to *probabilistically predict* missing abstractions in decompiled code [2,6,21,26,32,41,43], such as variable names and types. These techniques are based on the principle that software is *natural*, or predictable given context [17]. For example, it is possible to predict a variable's name based on how it is used. These models take as input an executable or a representation of the executable that can be deterministically derived from it—such as disassembly or the output of a deterministic decompiler—and output one or more natural-channel abstractions. We collectively refer to these as *decompilation improvement* tasks.

```
1  __int64 __fastcall <func>(char *buf,
       ssize_t *a2) {
2    __int64 i, len, v9;
3    ssize_t v3;
4    char path[8], v8[8], v11[8];
5    int v7, v10;
6    if (!buf) return 0xFFFFFFFFLL;
7    if (!a2) return 0xFFFFFFFFLL;
8    len = *a2 - 1;
9    v9 = 0x4958054A4755560ALL;
10   v10 = 1426081857;
11   strcpy(v11, "IW");
12   *(_QWORD *)path = 0x4958054A4755560ALL;
13   v7 = 1426081857;
14   strcpy(v8, "IW");
15   for (i = 0LL; i != 14; ++i)
16       path[i] ^= (_BYTE)i + 37;
17   v8[2] = 0;
18   v3 = readlink(path, buf, len);
19   *a2 = v3;
20   if (!v3) return 0xFFFFFFFFLL;
21   buf[v3] = 0;
22   return 0LL;
23 }
```

(a) Figure 1a, obfuscated with ADVobfuscator [1] string obfuscation (str). The string literal is represented as integers that are transformed by a convoluted series of operations.

```
1  __int64 __fastcall <func>(char *buf,
       ssize_t *a2) {
2    unsigned int v2;
3    ssize_t v3;
4    v2 = -1;
5    if (buf != 0LL && a2 != 0LL) {
6        v3 = readlink("/proc/self/exe", buf
           ,
7            *a2 - 0x3EBD892878945E8LL +
                 0x3EBD892878945E7LL);
8        *a2 = v3;
9        if (v3) {
10           buf[v3] = 0;
11           return 0;
12       }
13   }
14   return v2;
15 }
```

(b) Figure 1a, obfuscated with instruction substitution (sub) [22]. The subtraction instruction on line 5 of Figure 1b is replaced with an
equivalent but more convoluted sequence of
operations above.

```
1  __int64 __fastcall <func>(char *buf,
       ssize_t *a2) {
2    unsigned int v2;
3    int i, v4;
4    ssize_t v6;
5    for (i = -379896799;; i = 1196796914) {
6        while (i <= -157289568) {
7            v4 = -2108226211;
8            if (a2) v4 = -1403965279;
9            if (!buf) v4 = -2108226211;
10           if (v4 == -1403965279) {
11               v6 = readlink("/proc/self/
                   exe", buf, *a2 - 1);
12               *a2 = v6;
13               i = -157289567;
14               if (!v6) i = 1558099169;
15           } else {
16           LABEL_5:
17               i = 1196796914;
18               v2 = -1;
19           }
20       }
21       if (i != -157289567) break;
22       buf[v6] = 0;
23       v2 = 0;
24   }
25   if (i != 1196796914) goto LABEL_5;
26   return v2;
27 }
```

(c) Figure 1a, obfuscated with control flow flattening (fla) [22]. The functions' control flow
is rearranged so that basic blocks are arranged
in a loop and a control variable (i, above) determines which blocks are executed on each iteration.

```
1  __int64 __fastcall <func>(char *buf,
       ssize_t *a2) {
2    ssize_t v2;
3    __int64 result;
4    if (buf && a2) {
5        v2 = readlink("/proc/self/exe", buf
           , *a2 - 1);
6        *a2 = v2;
7        if (v2) {
8            buf[v2] = 0;
9            result = 0LL;
10           if (y_26 < 10) return result;
11           goto LABEL_11;
12       }
13       result = 0xFFFFFFFFLL;
14   } else {
15       result = 0xFFFFFFFFLL;
16       if (y_26 >= 10 && (((_BYTE)x_25 *
           ((_BYTE)x_25 - 1)) & 1) != 0)
           {
17           while (1);
18       }
19   }
20   if (y_26 < 10) return result;
21 LABEL_11:
22   if ((((_BYTE)x_25 * ((_BYTE)x_25 - 1))
       & 1) != 0) {
23       while (1);
24   }
25   return result;
26 }
```

(d) Figure 1a, obfuscated with bogus control
flow (bcf) [22]. Extra control-flow constructs
like if-statements, while loops, and gotos are
added.

Fig. 2. Figure 1a, obfuscated in different ways, then decompiled. The obfuscated versions are more difficult to read than without obfuscations Fig. 1b. Function names are
normalized to <func>, as is the convention in HexT5 [41].

In this paper, we analyze the impact that obfuscations have on the accuracy of decompilation-improvement machine-learning models. Further, we quantify the impact that training on obfuscated code has on a model's ability to handle obfuscations. While obfuscation is commonly employed by malware in practice, virtually no existing work in ML-based decompilation improvement considers obfuscations in training or evaluation. This is concerning because obfuscation undermines the *naturalness* assumption on which these techniques are based. By its nature, obfuscation changes the context under which variable names, types, and other abstractions occur. For instance, on Fig. 1a, line 6, the size of the buffer is decremented by 1 to accommodate the null terminator that must be present at the end of all C strings. Decrementing a string's length for the null terminator is common in C. In contrast, subtracting `0x3EBD892878945E8` and adding `0x3EBD892878945E7`, as occurs under the instruction substitution obfuscation in Fig. 2b, while semantically equivalent, is syntactically unusual; it is not what would normally be predicted given the surrounding context, and thus can be considered *unnatural*. We expect that this should undermine machine-learning-based tools whose models are only "familiar with" unobfuscated code. However, it may also be possible to learn a model that is robust to the presence of obfuscations by training on those obfuscations, making the obfuscations an expected, or at least not unexpected, part of the context.

In this paper, we answer four research questions:

1. How much does the presence of obfuscations impact the accuracy of ML-based decompilation improvement models?
2. How difficult is decompilation improvement under each type of obfuscation?
3. How well does learning transfer from one type of obfuscation to another?
4. How does varying the amount of obfuscata in training affect model performance?

In particular, we perform our experiments on three models covering three different decompilation improvement tasks: DIRTY [6], a model that predicts variable names and types, VarBERT [32], a model which only predicts variable names, and HexT5 [41], a language model which can solve a variety of tasks, but which we use to predict variable and function names. We answer our research questions by performing a series of experiments in which we train and evaluate machine learning models on unobfuscated and obfuscated code from a novel, large-scale dataset of obfuscated and unobfuscated executable binaries.

In short, we contribute:

- Four experiments answering our research questions involving 30 trained machine learning models which quantify the impact that obfuscations have on machine-learning-based decompilation improvement.
- A tool for building datasets including obfuscations at scale.
- A novel dataset consisting of unobfuscated and obfuscated binaries, with up to four obfuscations per binary.

We make available the code and data used in our experiments.

2 Research Questions and Findings

Our high level research goal is to understand and mitigate the effect of intentional software obfuscation on decompilation improvement models. In machine learning parlance, unobfuscated code and obfuscated code represent different distributions. Using a neural model trained on unobfuscated code to decompile obfuscated code represents a *covariate shift* [33]. Note that naturalness [17] is defined with respect to a distribution, because what is "natural" is context dependent: a natural-sounding sentence in American English may sound unnatural in British English. Meanwhile, some distributions may be closer to each other than others, i.e., two English dialects may be closer to one another than either is to Spanish. Here, different obfuscations may transform code in similar ways, and distributions for code obfuscated by these obfuscations might be closer to one another than they are to that of a dissimilar obfuscation. While it is difficult to measure the distances between distributions directly, we can instead indirectly measure it by quantifying the performance of a model trained on one distribution when evaluated on another. This idea underlies our high-level approach: we train and evaluate models on different code distributions, represented by different obfuscations, to understand their impact on decompilation improvement models.

In RQ1, (Sect. 5.1), we ask *"How much does the presence of obfuscations impact the accuracy of decompilation improvement models?"* To answer this question, we measure the performance of models of unobfuscated and obfuscated code both with and without covariate shift. We do find empirical evidence of a covariate shift that harms the performance of the three models. In particular, training solely on unobfuscated code, as virtually all decompilation improvement models do today, leads to poorer performance on obfuscated code, which is found in many real-world reverse-engineering scenarios. Fortunately, training on obfuscated code alleviates most or all of the impact by eliminating the covariate shift.

Code obfuscated in different ways may represent different distributions. In RQ1, we trained the obfuscated-code model to learn a combined distribution of all obfuscations. But some obfuscations' individual distributions may be farther away from the distribution of unobfuscated code than others, leading to poorer performance. Driven by this intuition, in RQ2 (Sect. 5.2), we ask *"How difficult is decompilation improvement under each type of obfuscation?"* We find that control flow flattening is the most difficult obfuscation for a model trained on unobfuscated code.

In RQ3 (Sect. 5.3), we further quantify the differences between individual obfuscation's distributions. We ask *"How well does learning transfer from one type of obfuscation to another?"* It is possible that the distributions for obfuscations that are similar are close to one another, and that training on one obfuscation means the performance on a related one may be relatively good. In other words, the learning transfers well between the two obfuscations. Unfortunately, we find that learning usually transfers poorly amongst the different obfuscations in our dataset. These results imply that models may perform poorly in practice when they encounter a new obfuscation.

If in-distribution data is required to learn obfuscations, then a natural question is *how much data is required?* In RQ4 (Sect. 5.4), we ask *"How does varying the amount of obfuscated data in training affect model performance?"* Neural networks require a large amount of data to be fit well; for novel obfuscations, there will likely be little data available. However, all decompiled C code—obfuscated or not—is similar in many ways: it obeys the same syntactic rules, and the semantics assigned to that syntax is the same. This suggests that there are at least some common parts of code, regardless of obfuscation, that can be used to at least partially inform predictions on unseen obfuscations, though our results from RQ3 suggest that at least some data of a particular obfuscation is required for good performance on that obfuscation. This is reminiscent of the popular pretrain/finetune paradigm in machine learning, where a model is (pre)trained on one task for which there is much data, and then trained a little more (finetuned) on data for a related task. To answer RQ4, then, we start by building a dataset of unobfuscated code, the purpose of which is to provide the model with information about the syntax and semantics of C code. Then we vary the amount of obfuscated code, measuring model performance for several different base-2 orders of magnitude sizes. Model performance gains increase rapidly after the first obfuscated data are added, but drop off rapidly as more are added. This suggests that large performance gains on obfuscated code requires a substantial amount of obfuscated data.

3   Datasets

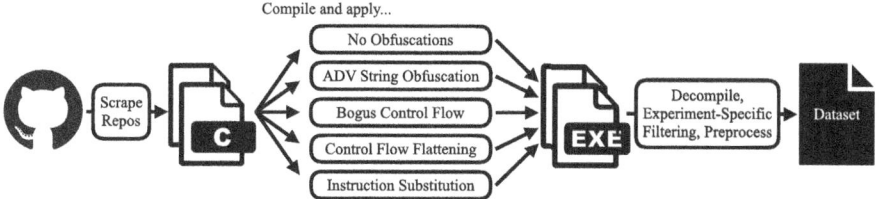

Fig. 3. An overview of our approach. We download open-source software from GitHub, then compile each project five times: once without obfuscations, once with each of our four obfuscations. In doing so, we ensure that each executable is compiled using debug information. Next, we use IDA Pro's decompiler through DIRTY's [6] dataset generation scripts to produce labeled training data. Finally, we select and preprocess data, building datasets to answer each research question.

To answer our research questions, we needed a large collection of obfuscated binaries compiled with debug information. An overview of our approach for constructing datasets with respect to the desired experiments is shown in Fig. 3. We generated training data by cloning and compiling a large collection of open-source software, in line with prior work [2,6,9,18,24,26,31,32,41]. In particular, we targeted majority C-language repositories, though these may occasionally contain a minority of C++. We collected data from 19,552 such repositories. Unlike in prior work, however, we compiled each project five times: once with no obfuscations applied, then once with each of four obfuscations. To do this, we adapted the GitHub Cloner and Compiler (GHCC) tool [20]. GHCC automatically clones and compiles a given list of GitHub repositories, first by executing configuration scripts if they exist, then executing each Makefile found in the project. Our adapted tool, GHCC-Obfuscator, first clones, then compiles the repository without any obfuscations using the repository's original Makefile(s). It then repeats the process, applying one obfuscation at a time, resetting the repository to a freshly-cloned state in between. In performing each compilation, GHCC-Obfuscator intercepted the calls to the compiler and added the -g flag for debug information to each compilation command. This feature was also used to add obfuscation-specific flags as necessary. This process produces 8,081,059 unique binaries, from which we sampled to build the datasets in the experiments.

With the binaries compiled, we used DIRTY [6]'s dataset generation scripts to extract labeled data from the binaries compiled with debug information. Each binary is decompiled using IDA Pro in batch mode. Because the binaries contains debug information, its developer-provided identifier names and types are present in the decompiled code. Next, the binaries are stripped of debug symbols. The binaries are decompiled again, and this time they are missing developer-provided names and types. The two decompilations form input-output pairs for supervised training: the second decompilation is the input, and the first, the output. Offsets within the binary are used to map the variables in the decompiled code and original code together. The obfuscations we use preserve debug information, making it possible to establish ground truth in this way for them as well. Obfuscations make the functions larger on average, 297 as opposed for 249 tokens for unobfuscated code.

3.1 Obfuscations

We chose a diverse collection of obfuscations that modify the code in a variety of ways. We describe each below.

String Obfuscation with ADVobfuscator. ADVobfuscator [1] is a library of C++ header files with macros that are applied by modifying the source code. In particular, used it to obfuscate string literals found in the source code. Obfuscated strings are either encoded as an integer or as another string which is transformed by a series of operations into the original string. See Fig. 2a for an example of ADVobfuscator applied to a function. Since ADVobfuscator depends on C++ macros, we compiled any programs obfuscated with ADVobfuscator using the C++ compiler g++ (as opposed to gcc). ADVobfuscator only produced obfuscated binaries when the C files were compiled with at least an O1 optimization level. (For consistency, we also compiled all other binaries in the experiments at O1 as well). Prior work has shown that optimization levels have a small-to-negligible impact on model performance of at most a few percentage points [6,32], likely because the deterministic decompiler undoes the optimizations when generating decompiled code. We abbreviate string obfuscation as `str` when presenting results elsewhere in the remainder of this paper.

Prior work has shown that string literals are a helpful feature of code for DIRE [13], the predecessor of DIRTY, which predicts variable names in decompiled code (but not types). It seems likely that this is a consequence of how language models (such as transformers) represent text. To input code to a language model, the input is split into a sequence of discrete tokens, each of which maps to a learned vector that encodes the semantics of that token. In general, there are more tokens allocated to natural language words and subwords then there are to C-language syntatic symbols like *, (, and { . As a result, natural language words carry an inflated importance in helping models reason about code; misleading natural language can substantially confuse even powerful models, even when the syntax is otherwise identical, as Miceli-Barone et al. show with identifier names [30]. Because identifier names are discarded during compilation, string literals are one of the sole sources of natural language in non-obfuscated decompiled code.

Obfuscator-LLVM Compile-Time Obfuscations. Obfuscator-LLVM [22] is a compile-time obfuscation tool based on LLVM that includes three different forms of obfuscation: instruction substitution (abbreviated `sub`), control flow flattening (`fla`), and bogus control flow (`bcf`).

Instruction substitution (`sub`) replaces arithmetic and binary operations on integers with a more complicated—but equivalent—series of operations. For example, this process may introduce random numbers into computations which cancel out because of mathematical identities. Figure 2b shows an example of a function obfuscated with instruction substitution.

Control flow flattening (`fla`) [27,39] is a form of obfuscation that implements control flow without using the traditional control structures of a programming language. Specifically, it creates a new control variable that represents which block should be executed next, and transforms each function's control flow into a loop that uses conditional statements to dispatch to the code that corresponds to the current value of the control variable. Each conditional statement body represents a basic block from the original, unobfuscated version of the code. The value of the variable determines which conditional statement body is executed; at the end of each statement body, the control variable is reset such that control flow mimics that of the original function. Figure 2c shows an example a function obfuscated with control flow flattening.

The bogus control flow (`bcf`) obfuscation inserts additional conditional statements with complex, opaque predicates that ultimately end up having no impact on the dynamic control flow of a program. In doing so, `bcf` introduces many irrelevant lines of code into the function. Figure 2d shows an example of a function obfuscated with bogus control flow.

To apply Obfuscator-LLVM to our dataset, we built the code using its original Makefiles, but forced the Obfuscator-LLVM version of the `clang` compiler to be used and specified the obfuscation's compiler flag.

3.2   Dataset Preprocessing

A particularly difficult problem in dataset preparation for decompilation-based models is data leakage. Machine learning models have the tendency to "memorize" their training data; they perform unrealistically well when evaluated on their training data. Data leakage occurs when a model is evaluated on data on which it was trained [23]. Duplicate copies of software projects can often result in data leakage when one is added to the training set and others are added to the evaluation (test) set. Because duplicate copies (e.g. forks) of software projects are extremely common on open source hosting services like GitHub [34], careful attention must be paid to data leakage. Identifying duplicate repositories is a difficult problem, which in turn makes data leakage hard to prevent. We use the following measures to prevent data leakage:

- *By-project splitting*: All three models operate at the function level. Putting some functions from a given project in the training set and others in the test set allows for the leakage of project-specific details. As a result, data from each project are placed exclusively in either the training set or test set. Xiong et al. [41] evaluate both with and without by-project splitting; by-project splitting causes the accuracy to decrease by more than two thirds, highlighting the importance of this data leakage prevention strategy.
- *MinHashing* [4]: This is a technique for efficiently approximating the Jaccard similarity between words or sequences of words in documents. Here, we treat all of the C code in a software project as a "document." MinHashing is often used with locality-sensitive hashing (LSH) to group similar documents into "buckets"; we consider software projects that end up in the same "buckets" to be duplicates. We ensure that each project on which we evaluate is not a duplicate of any project in the training set. The Stack [25], a popular dataset of source code for training machine learning models, also uses minhashing with LSH for deduplication.
- *Binary Hashing*: Following Chen et al. [6], we hash each executable file produced by the model and ensure that models with the same binary hash do not end up in both the train and the test sets.

We control for dataset composition for each model across each experiment; that is, for each trial in each experiment, all three models are trained on the same datasets. To ensure that this is the case, we perform dataset preprocessing and splitting once (using a modified version of the DIRTY dataset preprocessor) then convert the prepared train and test sets into formats suitable for VarBERT and HexT5.

4   Models

Here, we describe the three different model types on which we perform experiments and reproduce the relevant experiments from the original papers, though on our dataset, to establish a baseline.

4.1   Architecture and Training Practices

We perform all of our experiments on three different models: DIRTY [6], VarBERT [32], and HexT5 [41].

DIRTY [6] is a decompilation improvement model with a transformer-based encoder-decoder architecture [36] that predicts variable names and types in decompiled C code. DIRTY models are trained from scratch (that is, from randomly initialized parameters).

VarBERT [32] is a decompilation improvement model based on a transformer encoder that predicts variable names in decompiled C code. VarBERT is pre-trained on both a masked language modeling objective [10] and a constrained masked-language-modeling objective, before being trained on variable-name-prediction data, in line with Gu et al. [15]. We make use of the pretrained checkpoints provided by the authors, but fine-tune on our own variable-prediction data.

HexT5 [41] is emblematic of the modern trend of representation learning in natural language processing, whereby large neural networks, often transformer-architecture sequence-to-sequence models, are trained to predict parts of a sequence that are artificially hidden from the model. HexT5 is pretrained to learn representations of source code, decompiled code, assembly code, and intermediate representations. The authors evaluate it on four different tasks: summarization, function name prediction, variable name prediction, and code similarity. We evaluate only on the function name and variable name prediction tasks because our dataset does not have an oracle for textual summaries or code similarity.

Note that the choice of model is orthogonal to our research questions. While it is possible that obfuscations may affect different models in different ways, in general, we have reason to believe our findings are likely to generalize beyond the three models we evaluate here. First, like DIRTY, VarBERT, and HexT5, virtually all modern decompilation improvement models are transformer-architecture models [2,6,18,19,24,31,32,41]. Further, they are almost invariably trained on large corpora of open source code downloaded from internet repositories [2,6,9,18,24,26,31,32,41]. Finally, theory predicts that a covariate shift is expected to decrease the performance of any machine learning model, independent of architecture, so the trends in performance degredation (if not their magnitude) should generalize widely.

4.2   Baseline

(a) DIRTY			(b) VarBERT		(c) HexT5		
Retyping		Renaming		Renaming		Renaming	
overall	structs	variables		variables		variable	function
DIRT [6]	56.4	54.6	36.9	VarCorpus [32]	42.6	NSP [41][1]	25.7	–
Ours	54.5	48.8	27.7	Ours	28.7	Ours	25.2	30.6

Fig. 4. Baseline performance numbers taken from the original papers for the three models compared with performance when trained and evaluated on our dataset. DIRT was the dataset originally used to train DIRTY; VarCorpus was originally used to train VarBERT, and NSP was the dataset used to train HexT5. All values are in percent accuracy.

To set a baseline, we reproduce results from the original DIRTY [6], Var-BERT [32] and HexT5 [41] papers, but using our dataset. That is, we train and evaluate them on a subset of the full dataset consisting only of unobfuscated code.

In all cases, the scores we obtain in our reproduction are lower than the original works, sometimes significantly. Because machine learning is a "black box" method, it is difficult to conclusively determine the cause for the difference. We suspect it is due to our data-leakage-prevention measures outlined in Sect. 3.2; with less data leakage; there are fewer examples in the test set which the trained models have memorized.

DIRTY is a variable name and type prediction model. We report three metrics, as shown in Fig. 4a: the percentage of correctly predicted variable names, the percentage of correctly predicted variable types, and the percentage of correctly predicted types that are `structs` in the original code. Structures make up a minority of variables' types in source code yet are often more important for understanding the functionality of code than primitive types.

VarBERT is a variable name prediction model. We use their IDA-O1 accuracy number because we use the IDA Pro decompiler at optimization level O1, as discussed in Sect. 3. The results are shown in Fig. 4b.

HexT5 is a language model fine-tuned on several tasks; we use it for variable name and function name prediction here. The results are shown in Fig. 4c[1].

---

[1] The HexT5 results reported here are imprecise estimates based on information in a bar graph in the paper; exact numbers are not provided. The original HexT5 paper reports function name prediction efficacy in terms of precision and recall. (We reached out to the authors for exact numbers but did not hear back).

5 Experiments

We evaluated the effect of introducing obfuscations to the decompilation problem over four separate experiments. For each of these experiments, we constructed several different datasets which we used to train DIRTY [6], VarBERT [32], and HexT5 [41] models. We performed model training and evaluation on NVIDIA A100 and Titan X (Pascal) GPUs and other tasks on 64-bit Linux with Intel Xenon CPUs. All three models are implemented as python programs using the `pytorch` library. For DIRTY and VarBERT, we used the dataset preprocessing, training, and evaluation scripts and environment files provided by the authors. For HexT5, these were not released, so we wrote our own, available in the replication package, using the `transformers` API [40]. For each model in each experiment, we selected training data (in compiled binary form) according to the experiment's aims, then ran DIRTY's decompilation and preprocessing scripts. We converted this data into the formats required by HexT5 and VarBERT, then ran the corresponding model-specific preprocessing scripts. With the datasets prepared, we train each model. We dedicated the bulk of the data to training but ensured that each test set, derived using the same process, contained at least 5000 examples that are drawn from repositories that do not overlap the training set. We evaluate each model with each test set as dictated by the experiments' goals.

5.1 RQ1: How much does the Presence of Obfuscations Impact the Accuracy of Decompilation Improvement Models?

Table 1. RQ1: Impact of Obfuscations on Accuracy. Results displayed in terms of percent accuracy, along with a relative percent change compared with the baseline (first row). Higher is better.

| Train | Test | DIRTY | | | VarBERT | HexT5 | |
| | | Retyping | | | Renaming | | |
		overall	structs	variables	variables	variables	functions
Unobf	Unobf	54.5 –	48.8 –	27.7 –	28.7 –	25.2 –	30.6 –
Unobf	Obf	51.7 (-5.1%)	44.1 (-9.6%)	21.0 (-24.0%)	20.6 (-28.4%)	23.2 (-8.0%)	26.8 (-12.3%)
Obf	Obf	56.2 (+3.3%)	46.3 (-5.2%)	27.8 (+0.3%)	25.8 (-10.1%)	27.0 (+6.9%)	29.0 (-5.3%)
Obf	Unobf	50.1 (-8.1%)	44.6 (-8.5%)	24.9 (-10.1%)	23.8 (-17.1%)	22.3 (-11.8%)	28.5 (-6.9%)

In our first experiment, we investigate the impact that obfuscations have on the performance of the three model types, and to what degree training on obfuscated data can mitigate the impact of obfuscations.

Methodology. We use our reproductions of the three models from Sect. 4.2 as a baseline against which we evaluate the impact of obfuscations. We reuse the dataset from the reproduction, but also build a dataset of obfuscated code of the same size. To provide an additional control, we attempt to use data from the same projects as in the reproduction where possible. (This may not be possible if the project failed to build with obfuscations applied.) Because we compile each repository with four different obfuscations, including all obfuscated data would result in a training set that is several times larger than the unobfuscated training set. To control for training set size, instead, we select an obfuscation uniformly at random and use data from that obfuscation, up to the amount used in the unobfuscated training set. If there is insufficient data (perhaps due to an early compilation failure on that obfuscation), we continue sampling from other obfuscations. The final sizes of the unobfuscated and obfuscated training sets are 1,627,991 and 1,578,083 functions, respectively. We train each type of model using the obfuscated dataset. Finally, we evaluate each model trained on the unobfuscated and obfuscated training sets against both of the unobfuscated and obfuscated test sets, leading to a total of four different combinations.

Results. The results are shown in Table 1. The first row contains our baseline reproductions. The second row represents what happens when these same models—trained only on unobfuscated code—are exposed to obfuscations in evaluation. This simulates what would happen if these models, as released, were applied to obfuscated code, as is commonly found in practice. In all cases, the accuracy drops, but the magnitude of the drop varies considerably. The relative drop in accuracy varies between 5.1% for overall retyping with DIRTY and 28.4% for variable renaming with VarBERT. However, training on obfuscations can help mitigate the loss in accuracy. The third row of Table 1 illustrates this. Training on obfuscated code substantially improves prediction accuracy on obfuscated code. However, there is no free lunch: a model trained solely on obfuscated code performs worse evaluated on *un*obfuscated code (row 4) than a model trained on unobfuscated code. This is perhaps because unobfuscated code is not "natural" from the perspective of an obfuscated-code-only model; unobfuscated code represents a covariate shift with respect to a model trained on obfuscated code.

> **Answer to RQ1**: Training a model on unobfuscated code leads to poor performance on obfuscated code, but training on obfuscated code can mitigate the performance loss, at the cost of poorer performance on unobfscated code.

5.2 RQ2: How Difficult is Decompilation Improvement Under Each Type of Obfuscation?

Some obfuscations may create a more challenging context than others for name and type prediction. In this experiment, we benchmark the difficulty of each obfuscation.

Methodology. For this experiment, we used the baseline model trained on unobfuscated code from Sect. 4.2. The model was then tested on 4 disjoint subsets of the obfuscated test set from RQ1 (Sect. 5.1), where each subset contained binaries with a particular obfuscation applied. In partitioning the test set, we excluded examples that appeared under multiple obfuscations. This happens for simple functions where the obfuscations do not apply or where obfuscations fail (the latter primarily in the dataset's C++ minority). Excluding simple functions has the effect of making the task slightly more challenging; filtering C++ has the effect of increasing the difficulty of the struct-prediction task in particular because structs are more common in C++ code, including easier-to-predict standard-library features like iterators and strings.

We use the same model's performance on the unobfuscated test set from Sect. 4.2 as a baseline.

Table 2. RQ2: Accuracy on Individual Obfuscations. Results displayed in terms of percent accuracy, along with a relative percent change compared with the baseline (first row). Higher is better.

Obfuscation	DIRTY			VarBERT	HexT5	
	Retyping		Renaming			
	overall	structs	variables	variables	variables	functions
none	54.5 –	48.8 –	27.7 –	28.7 –	25.2 –	30.6 –
fla	51.6 (-5.2%)	15.5 (-68.2%)	8.9 (-67.9%)	9.1 (-68.5%)	12.5 (-50.5%)	13.2 (-56.7%)
sub	50.9 (-6.5%)	37.0 (-24.2%)	17.1 (-38.3%)	24.8 (-13.5%)	19.0 (-24.9%)	19.2 (-37.3%)
bcf	45.6 (-16.2%)	21.7 (-55.5%)	10.6 (-61.8%)	18.6 (-35.4%)	13.5 (-46.7%)	14.8 (-51.7%)
str	52.8 (-3.1%)	25.6 (-47.4%)	28.3 (+2.4%)	28.4 (-0.1%)	24.2 (-4.2%)	21.5 (-29.6%)

Results. The results are summarized in Table 2. Control flow flattening generally provides the model with the most difficulty across all three tasks and models. Instruction substitution and ADVobfuscator's string obfuscation provide the least difficulty.

These results are not surprising. Control flow flattening provides the largest textual and syntactic differences from unobfuscated code, and is thus most likely to be the least natural. Conversely, instruction substitution only affects operations on integers, leaving parts of the code that don't deal with integer arithmetic unaffected. ADVobfuscator only affects string literals. While string literals are important, they exist in a minority of functions; functions with no string literals are unaffected.

> **Answer to RQ2**: Different obfuscations may vary widely in difficulty. Difficulty is correlated with the amount of textual changes made to the code.

5.3   RQ3: How well does Learning Transfer from One Type of Obfuscation to Another?

In this experiment, we measure how well learning models trained on one obfuscation perform on binaries compiled with a different obfuscation. Since there are an infinite number of possible obfuscations, with malware authors often inventing new obfuscations as well, it is impossible for a model to be trained on every possible obfuscation. Therefore, it is important to measure a model's ability to generalize to new obfuscations unseen during training.

Methodology. We trained models on two obfuscations—control flow flattening (fla) and instruction substitution (sub)—and evaluated their performance on all other obfuscations in our dataset individually. These two obfuscations complement each other: one affects the control flow while largely leaving basic blocks intact, while instruction substitution involves modifying computations within basic blocks while leaving control flow unchanged. We choose these obfuscations because they are substantially different so we can better make general claims about the transferability of learning on obfuscations. Training a model and evaluating a model on the same obfuscation serves as a baseline (fla on fla and sub on sub). We trained the fla models on 4,394,527 functions and the sub models on 5,171,901 functions.

Results. The results are shown in Table 3. We note that in both cases, the baseline trial performance, where the train and test sets are drawn from the same obfuscations, generally perform better than other trials. Both bogus control flow and control flow flattening are control flow related obfuscations, but training on one and evaluating on the other (Table 3 line 2) does not produce results that are consistently or meaningfully better than other obfuscations. Similarly, the model trained on instruction-substitution for the most part does not generalize well to other obfuscations, though it does for retyping with DIRTY on string obfuscation, the other non-control flow obfuscation. These results illustrate the magnitude of the shift between even superficially similar obfuscations. Further, they suggest that training on one obfuscation will not necessarily transfer to other obfuscations.

Table 3. RQ3: Transferability of Learning. Results displayed in terms of percent accuracy, along with a relative percent change compared with the baselines (first row of each table section). Higher is better.

		DIRTY			VarBERT	HexT5	
Train	Test	Retyping			Renaming		
		overall	structs	variables	variables	variables	functions
fla	fla	59.3	– 30.5	– 19.8	– 22.5	– 20.1	– 18.6 –
fla	bcf	41.0 (-30.9%)	15.7 (-48.3%)	10.6 (-46.1%)	20.1 (-10.6%)	14.1 (-29.8%)	13.5 (-27.3%)
fla	sub	48.6 (-18.1%)	24.4 (-19.9%)	12.3 (-38.0%)	21.1 (-6.2%)	14.9 (-26.2%)	15.3 (-17.7%)
fla	str	56.5 (-4.8%)	28.0 (-8.0%)	10.1 (-48.9%)	10.1 (-55.0%)	12.6 (-37.4%)	11.9 (-36.2%)
sub	sub	50.5	– 25.1	– 14.6	– 30.1	– 20.3	– 19.0 –
sub	str	56.0 (+11.0%)	28.1 (+11.8%)	7.6 (-47.8%)	12.1 (-59.6%)	14.0 (-31.1%)	11.9 (-37.6%)
sub	bcf	41.1 (-18.6%)	11.4 (-54.6%)	13.4 (-8.4%)	22.3 (-25.9%)	17.4 (-14.4%)	14.3 (-24.8%)
sub	fla	51.7 (+2.4%)	22.1 (-12.0%)	8.0 (-45.0%)	10.3 (-65.6%)	12.6 (-38.0%)	16.1 (-15.4%)

> **Answer to RQ3**: Learning is not easily transferred between obfuscations, even those which are similar.

5.4 RQ4: How Does Varying the Amount of Obfuscated Data in Training Affect Model Performance?

Table 4. RQ4: Effect of the Quantity of Obfuscated Training Data on Accuracy

	DIRTY			VarBERT	HexT5	
Train Set	Retyping			Renaming		
	overall	structs	variables	variables	variables	functions
128-none	47.1	– 18.9	– 12.9	– 14.9	– 14.4	– 12.1 –
1024-none	44.5 (-5.5%)	19.6 (+4.0%)	13.4 (+3.7%)	15.1 (+2.0%)	14.8 (+3.0%)	12.2 (+1.1%)
8192-none	45.2 (-3.9%)	18.3 (-3.2%)	13.0 (+0.8%)	15.1 (+1.7%)	15.4 (+7.0%)	12.7 (+5.1%)
128-obf	49.8 (+5.7%)	18.2 (-3.6%)	16.2 (+25.4%)	15.7 (+5.4%)	14.9 (+3.1%)	14.4 (+19.4%)
1024-obf	50.6 (+7.6%)	17.4 (-7.9%)	14.3 (+10.7%)	15.7 (+6.0%)	15.1 (+4.4%)	14.8 (+22.4%)
8192-obf	50.5 (+7.2%)	17.5 (-7.2%)	16.8 (+30.1%)	17.4 (+16.9%)	16.6 (+14.9%)	14.9 (+23.1%)

Adding small amounts of obfuscated data improves performance on obfuscated data, while adding small amounts of unobfuscated data does not. VarBERT's 1024-none accuracy is slightly greater than its 8192-none accuracy, though they round to the same value.

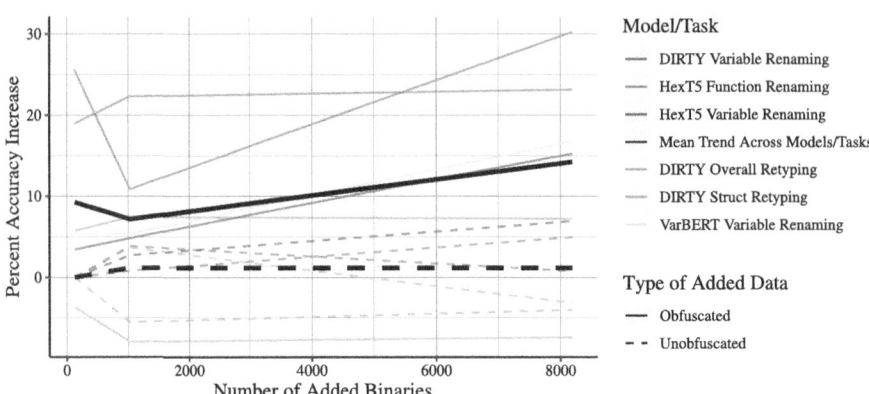

Fig. 5. Accuracy gains relative to the 128-none trials for each model/task combination (background colored lines) and the mean across all models and tasks (the black lines). Adding even a small amount of obfuscated data leads to accuracy gains, though the gains are very noisy. The results suggest steeply diminishing returns for additional data, possibly following a power law error curve, as predicted by theory [16].

Generating obfuscated training data at scale is a nontrivial task. Further, one may encounter new obfuscations for which no commercial generation tools are available. In these cases, there may be only a limited amount of labeled training data available (perhaps produced by malware analysts as they encounter new examples). This is especially important, because, as discussed in Sect. 5.3, learning does not transfer well between obfuscations.

Methodology. We designed an experiment to quantify the relationship between model performance and the number of obfuscated examples in its training set. Neural networks, including almost all state-of-the-art techniques, require large datasets to be fit well. If the amount of obfuscated training data is limited, as is likely the case for newly discovered obfuscations in practice, then training only on obfuscated training data would be insufficient. On the other hand, generating unobfuscated training samples is straightforward using the process described by Chen et al. [6] and Pal et al. [32]. Therefore, we build datasets consisting primarily of unobfuscated data with a limited amount of obfuscated data. In principle, this is very similar to the pretrain/finetune paradigm. The two steps are typically separated so that a pretrained model can be copied and used for multiple different finetuning tasks, but for our experiments, that is not relevant, so we don't separate the steps to simplify the training process.

Our training sets were created by combining a fixed set of unobfuscated data with varying numbers of obfuscated binaries. We created 3 training sets each with 128, 1024, and 8192 obfuscated binaries respectively. For each obfuscated training set, there are an equal number of binaries of each of the four obfuscation types. These had 868,968, 883,793, and 1,013,980 functions, respectively.

Our goal is to measure the relationship between amount of obfuscated data used in training and model accuracy on obfuscated code. However, training set size is a factor in model accuracy; typically, the greater the size of the training set, the higher the accuracy. To account for this, however, we also control for training set size by running separate trials with the same amount of purely unobfuscated training data. We created training sets consisting purely of unobfuscated binaries which had the same total number of decompiled executable files as the 128, 1024, and 8192 obfuscated training sets. That is, the training set of equal size to the 8192 obfuscated executable training set, contained the same unobfuscated executables as the 1024 unobfuscated train set along with 7168 more unobfuscated binaries. The sizes of these control training sets are 870,089, 885,795, and 1,030,167 functions, respectively.

We also created shared validation and test sets, with contents evenly split between no obfuscations, `fla`, `bcf`, `sub`, and `str`. We then evaluated all six models on our single test set.

Results. The results are shown in Table 4. As expected, adding a small amount of unobfuscated data (Table 4, rows 1–3) has very little impact on performance. On the other hand, adding obfuscated training data does increase performance. Most of the performance gain happens with the first 128 binaries worth of functions. Accuracy gains after this point are more muted and are very noisy. The results suggest sharply diminishing returns after adding a relatively small amount of data, as illustrated in Fig. 5. It is possible that the results follow a power law curve as well; there are diminishing returns as the amount of obfuscated data added increases. Unfortunately, the scale of the curve is such that only a few examples is insufficient to achieve a meaningful improvement in performance.

> **Answer to RQ4**: Model performance gains increase rapidly after the first obfuscated data are added, but drop off rapidly as more obfuscated data are added.

6   Threats to Validity

Internal validity is the degree to which an experiment establishes an causal relationship. A source of internal validity was that during data generation, different obfuscators can cause build failures at different stages. The automated compilation tool we used, GHCC, collects all products of compilation, even if there are failures on subsequent steps. Thus, in some cases, compiling the same project with different obfuscations may yield different collections of binaries if an obfuscation causes a compilation error. In Sect. 5.1, we account for this by sampling from other obfuscations from the same project to ensure that the training set size is consistent among the trials. Another option would have been to include only repositories which built completely, but this would have substantially decreased the available pool of training data and biased the dataset against complex software projects that are difficult to build.

External Validity is the degree to which an experiment's results generalize beyond the population in the experiment. We expect our results to generalize well across other real-world software projects because we train on a variety of different real world software and evaluate on a randomly selected subset of them. However, our dataset is necessarily biased towards some projects which we could build automatically; it is possible that binaries from unbuilt projects might systematically present some unique challenges that may impact the results. Similarly, our dataset is biased towards open-source software, which may not be representative of all software. Malware may include other techniques to defend against reverse engineering, such as packing and encryption; however, there exist a wide variety of unpacking and decryption techniques and commercial services. From the perspective of neural variable name and type prediction, these are pre-processing steps that are orthogonal to our problem. Finally, it is possible that there are other obfuscations to which our findings do not generalize.

Construct Validity is the degree to which the experiment design supports the claims. We only claim to measure the accuracy of the models with respect to the original names and types in the source code, a common strategy in virtually all existing work in ML-based decompilation improvement. Notably, however, we cannot directly make claims that the names and types predicted are *helpful* to reverse engineers. However, we expect that the names and types in the original source code are typically more helpful than those in the decompiled code, and thus predicting more variable names and type names correctly is, in general, a positive sign.

Following Chen et al. [6], we filter away variables for which the decompiled variable name is the same as what the dataset-generation technique identifies as an original variable name. This usually happens when a variable is decompiler-generated, but also for loop indices and a few other types of variables that have predictable names. There are also other variables, about 10% of the total, that look decompiler-generated but have original variable names different from the original (e.g. v2 vs v5). For consistency with prior work [6], our results include predictions on these variables. Excluding these causes numbers to shift (in either direction) a small amount, though the trends reported in each experiment remain the same.

7   Related Work

ML-Based Decompilation Improvement There is a significant body of existing work in leveraging machine learning to improve and complement decompilation. Much of the existing work focuses on predicting one particular type of information lost during compilation: variable names [26, 31, 32], variable types [28, 43, 44], function names [9, 21, 24], or the original syntactic structure of the original source code [2, 14, 18, 19].

Handling Obfuscations Despite significant work in learning for decompilation, very little existing work considers the impact that obfuscations have on these techniques. The authors of SymLM [21], which predicts function names, do evaluate their work on binaries with four obfuscations (bogus control flow, control-flow-flattening, instruction substitution, and basic block splitting), similar to RQ2 in our work (Sect. 5.2). We use three of the same obfuscations, but instead of basic block splitting we use string literal obfuscation. They find that obfuscations decrease the accuracy of their tool by 2–10%, depending on the obfuscation.

Deobfuscation There is also work on undoing obfuscations. Instruction substitution can be undone using common compiler optimization passes. Other work focuses on eliminating bogus control flow constraints[42] and eliminating dead or bogus code [7]. There is also work on undoing control-flow flattening obfuscations [11]. Symbolic execution can be used in conjunction with techniques similar to those in compiler optimization to simplify functions [29,35], sometimes in conjunction with program synthesis [8]. Deobfuscation techniques generally produce code that is typically equivalent to the original but may be syntactically very different. We leave a study on how deterministic deobfuscation techniques affect neural decompilers to future work.

Neural decompilers may be used on obfuscated code in scenarios where deterministic deobfuscation is not integrated with the security researchers' toolchains or when deterministic deobfuscation fails. In addition, new obfuscations may be created at any time; while neural models only need examples, deterministic deobfuscation may require the design and implementation of new algorithms to handle them if they exploit the limitations in existing techniques.

8 Conclusion

Neural decompilation improvement models predict missing abstractions, like variable names and types, in decompiled code. Little existing work in neural decompilation improvement considers obfuscated code, despite obfuscations being widespread in practice. In this work, we quantified the impacts that four obfuscations have on three prominent decompilation improvement models.

We find that obfuscations do negatively impact the performance these models, though training on obfuscated code largely mitigates the impact of obfuscations. Unfortunately, as we show in Sects. 5.1, 5.2 and 5.3, each obfuscation we tested produced its own substantially different distribution of decompiled code. Practically, this means that if an attacker creates malware using their own secret obfuscation, a decompilation improvement model will likely perform poorly. However, there is a silver lining: as we show in Sect. 5.4, a model can see gains in performance on obfuscations when trained on only a few hundred examples, which means that models can be adapted for *known* obfuscations.

In this work, we focus on predictions at the function level, following DIRTY [6], VarBERT [32], and HexT5 [41]; it may also be interesting to examine prediction at different levels of granularity (e.g. partial function or full-program level), though function-level remains the most common approach.

Acknowledgments. This material is based upon work funded and supported by the Department of Defense under Contract No. FA8702-15-D-0002 with Carnegie Mellon University for the operation of the Software Engineering Institute, a federally funded research and development center. Additionally, this material is based upon work supported by the National Science Foundation Graduate Research Fellowship Program under Grant No DGE2140739. This work used GPU nodes at Purdue Anvil through allocation CIS240492 from the Advanced Cyberinfrastructure Coordination Ecosystem: Services & Support (ACCESS) program [3], which is supported by U.S. National Science Foundation grants #2138259, #2138286, #2138307, #2137603, and #2138296. Any opinions, findings, and conclusions or recommendations expressed in this material are those of the authors and do not necessarily reflect the views of the National Science Foundation.

References

1. Andrivet, S.: ADVobfuscator (2020). https://github.com/andrivet/ADVobfuscator
2. Armengol-Estapé, J., Woodruff, J., Cummins, C., O'Boyle, M.F.: Slade: a portable small language model decompiler for optimized assembly. In: CGO, pp. 67–80. IEEE (2024)
3. Boerner, T.J., Deems, S., Furlani, T.R., Knuth, S.L., Towns, J.: Access: advancing innovation: NSF's advanced cyberinfrastructure coordination ecosystem: services & support. In: Practice and Experience in Advanced Research Computing 2023: Computing for the Common Good, pp. 173–176. PEARC '23, Association for Computing Machinery, New York, NY, USA (2023). https://doi.org/10.1145/3569951.3597559
4. Broder, A.Z.: On the resemblance and containment of documents. In: Proceedings. Compression and Complexity of SEQUENCES 1997 (Cat. No. 97TB100171), pp. 21–29. IEEE (1997)
5. Casalnuovo, C., Barr, E.T., Dash, S.K., Devanbu, P., Morgan, E.: A theory of dual channel constraints. In: ICSE-NIER, pp. 25–28 (2020)
6. Chen, Q., Lacomis, J., Schwartz, E.J., Le Goues, C., Neubig, G., Vasilescu, B.: Augmenting decompiler output with learned variable names and types. In: 31st USENIX Security Symposium, pp. 4327–4343 (2022)
7. Coogan, K., Lu, G., Debray, S.: Deobfuscation of virtualization-obfuscated software: a semantics-based approach. In: CCS, pp. 275–284 (2011)
8. David, R., Coniglio, L., Ceccato, M., et al.: QSynth-a program synthesis based approach for binary code deobfuscation. In: BAR 2020 Workshop (2020)
9. David, Y., Alon, U., Yahav, E.: Neural reverse engineering of stripped binaries using augmented control flow graphs 4(OOPSLA), 1–28 (2020)
10. Devlin, J., Chang, M.W., Lee, K., Toutanova, K.: BERT: Pre-training of deep bidirectional transformers for language understanding. In: Burstein, J., Doran, C., Solorio, T. (eds.) NAACL, pp. 4171–4186. Association for Computational Linguistics, Minneapolis, Minnesota (2019)
11. Dong, W., Lin, J., Chang, R., Wang, R.: CaDeCFF: compiler-agnostic deobfuscator of control flow flattening. In: Proceedings of the 13th Asia-Pacific Symposium on Internetware (2022)
12. Dramko, L., Lacomis, J., Schwartz, E.J., Vasilescu, B., Le Goues, C.: A taxonomy of C decompiler fidelity issues. In: 33rd USENIX Security Symposium (2024)

13. Dramko, L., et al.: Dire and its data: neural decompiled variable renamings with respect to software class. ACM Trans. Softw. Eng. Methodol. **32**(2), 1–34 (2023)
14. Fu, C., et al.: Coda: an end-to-end neural program decompiler. NeurIPS **32** (2019)
15. Gu, Y., Zhang, Z., Wang, X., Liu, Z., Sun, M.: Train no evil: selective masking for task-guided pre-training. In: Webber, B., Cohn, T., He, Y., Liu, Y. (eds.) EMNLP (2020)
16. Hestness, J., et al.: Deep learning scaling is predictable, empirically. arXiv preprint arXiv:1712.00409 (2017)
17. Hindle, A., Barr, E.T., Gabel, M., Su, Z., Devanbu, P.: On the naturalness of software. Commun. ACM **59**(5), 122–131 (2016)
18. Hosseini, I., Dolan-Gavitt, B.: Beyond the c: retargetable decompilation using neural machine translation. In: NDSS (2022)
19. Hu, P., Liang, R., Chen, K.: DeGPT: optimizing decompiler output with LLM. In: NDSS (2024)
20. Hu, Z.: GHCC (2021). https://github.com/huzecong/ghcc
21. Jin, X., Pei, K., Won, J.Y., Lin, Z.: SymLM: predicting function names in stripped binaries via context-sensitive execution-aware code embeddings. In: CCS, pp. 1631–1645 (2022)
22. Junod, P., Rinaldini, J., Wehrli, J., Michielin, J.: Obfuscator-LLVM–software protection for the masses. In: SPRO, pp. 3–9. IEEE (2015)
23. Kapoor, S., Narayanan, A.: Leakage and the reproducibility crisis in machine-learning-based science. Patterns **4**(9), 100804 (2023)
24. Kim, H., Bak, J., Cho, K., Koo, H.: A transformer-based function symbol name inference model from an assembly language for binary reversing. In: Asia-CCS, pp. 951–965 (2023)
25. Kocetkov, D., et al.: The stack: 3 TB of permissively licensed source code. arXiv preprint (2022)
26. Lacomis, J., et al.: DIRE: a neural approach to decompiled identifier naming. In: ASE, pp. 628–639. IEEE (2019)
27. László, T., Kiss, Á.: Obfuscating C++ programs via control flow flattening (2009)
28. Lehmann, D., Pradel, M.: Finding the Dwarf: recovering precise types from WebAssembly binaries. In: PLDI, pp. 410–425 (2022)
29. Liang, M., Li, Z., Zeng, Q., Fang, Z.: Deobfuscation of virtualization-obfuscated code through symbolic execution and compilation optimization. In: ICICS 2017. Springer (2018)
30. Miceli-Barone, A.V., Barez, F., Cohen, S.B., Konstas, I.: The larger they are, the harder they fail: language models do not recognize identifier swaps in Python. In: ACL 2023 (2023)
31. Nitin, V., Saieva, A., Ray, B., Kaiser, G.: DIRECT: a transformer-based model for decompiled variable name recovery. In: NLP4Prog 2021, p. 48 (2021)
32. Pal, K.K., et al.: Len or index or count, anything but v1: predicting variable names in decompilation output with transfer learning. In: 2024 IEEE Symposium on Security and Privacy (SP), p. 152 (2024)
33. Quionero-Candela, J., Sugiyama, M., Schwaighofer, A., Lawrence, N.D.: Dataset Shift in Machine Learning. The MIT Press (2009)
34. Spinellis, D., Kotti, Z., Mockus, A.: A dataset for GitHub repository deduplication. In: International Conference Mining Software Repositories, pp. 523–527 (2020)
35. Tofighi-Shirazi, R., Christofi, M., Elbaz-Vincent, P., Le, T.H.: DoSE: deobfuscation based on semantic equivalence. In: SSPREW, pp. 1–12 (2018)
36. Vaswani, A., et al.: Attention is all you need. NeurIPS **30** (2017)

37. Votipka, D., Rabin, S., Micinski, K., Foster, J.S., Mazurek, M.L.: An observational investigation of reverse engineers' process and mental models. In: Extended Abstracts of the 2019 CHI Conference on Human Factors in Computing Systems, pp. 1–6 (2019)
38. Votipka, D., Rabin, S., Micinski, K., Foster, J.S., Mazurek, M.L.: An observational investigation of reverse engineers' processes. In: 29th USENIX Security Symposium, pp. 1875–1892 (2020)
39. Wang, C., Davidson, J., Hill, J., Knight, J.: Protection of software-based survivability mechanisms. In: DSN, pp. 193–202. IEEE (2001)
40. Wolf, T., et al.: Transformers: state-of-the-art natural language processing (2020).https://doi.org/10.5281/zenodo.7391177
41. Xiong, J., Chen, G., Chen, K., Gao, H., Cheng, S., Zhang, W.: Hext5: unified pre-training for stripped binary code information inference. In: ASE, pp. 774–786. IEEE (2023)
42. You, G., Kim, G., Han, S., Park, M., Cho, S.J.: Deoptfuscator: defeating advanced control-flow obfuscation using android runtime (ART). IEEE Access 10, 61426–61440 (2022)
43. Zhang, Z., et al.: OSPREY: recovery of variable and data structure via probabilistic analysis for stripped binary. In: 2021 IEEE Symposium on Security and Privacy (SP), pp. 813–832. IEEE (2021)
44. Zhu, C., et al.: TYGR: type inference on stripped binaries using graph neural networks. In: 33rd USENIX Security Symposium, pp. 4283–4300 (2024)

Exploring the Potential of LLMs for Code Deobfuscation

David Beste[1]([✉]) [ID], Grégoire Menguy[2] [ID], Hossein Hajipour[1] [ID], Mario Fritz[1] [ID], Antonio Emanuele Cinà[3] [ID], Sébastien Bardin[2] [ID], Thorsten Holz[1] [ID], Thorsten Eisenhofer[4] [ID], and Lea Schönherr[1] [ID]

[1] CISPA Helmholtz Center for Information Security, Saarbrücken, Germany
{david.beste,hossein.hajipour,mario.fritz,thorsten.holz,
lea.schonherr}@cispa.de
[2] Université Paris-Saclay, CEA, List, Gif-sur-Yvette, France
{gregoire.menguy,sebastien.bardin}@cea.fr
[3] University of Genoa, Genoa, Italy
antonio.cina@unige.it
[4] BIFOLD and TU Berlin, Berlin, Germany
thorsten.eisenhofer@tu-berlin.de

Abstract. Code obfuscation alters software code to conceal its logic while retaining functionality, aiding intellectual property protection but hindering security audits and malware analysis. To address this, automated deobfuscation techniques have been developed, though existing approaches remain constrained by limited scope and specificity. Motivated by these challenges, this paper explores a novel approach for code deobfuscation based on Large Language Models (LLMs). First, we investigate the general capabilities of LLMs in reducing code complexity by choosing five different source-to-source obfuscation methods. Despite challenges regarding semantical correctness, our findings indicate that LLMs can be very effective in this task. Building on this, we fine-tune two versatile models capable of simplifying code obfuscated through up to seven different chained obfuscation transformations while consistently outperforming deobfuscation based on compiler optimizations and general-purpose LLMs. Our best model demonstrates an average Halstead metric program length reduction of 89.21% for our most challenging scenario. Finally, we conduct a memorization test to assess if performance stems from memorized code rather than true deobfuscation capabilities, which our models pass.

1 Introduction

Code obfuscation [8,9] refers to various methods to disguise a program's functionality, making its source or machine code harder for a human analyst to comprehend. Malicious actors often use obfuscation techniques to complicate malware analysis, thus impeding the development of effective detection methods and countermeasures [21,35]. For these reasons, deobfuscation methods have been developed to recover the original structure of the code. These methods

© The Author(s), under exclusive license to Springer Nature Switzerland AG 2025
M. Egele et al. (Eds.): DIMVA 2025, LNCS 15747, pp. 267–286, 2025.
https://doi.org/10.1007/978-3-031-97620-9_15

use sophisticated approaches such as advanced static analysis [31,35], dynamic analysis [37,41], and, more recently, program synthesis [4,18,29]. Despite their potential, the effectiveness of these techniques often encounters significant limitations. Many focus on specific obfuscation strategies, such as mixed Boolean arithmetic (MBA) [31,32] or opaque predicates [3,30], which limits their applicability to broader obfuscation scenarios. Furthermore, although program synthesis offers a promising approach, it is so far only suitable for deobfuscating small and simple pieces of code without complex structures [4,29]. This limitation significantly diminishes its practical utility for broad-scale deobfuscation tasks and emphasizes the need for further progress in this area.

In this paper, we explore the capabilities of LLMs in code deobfuscation tasks. LLMs have demonstrated remarkable abilities in generating code, as evidenced by applications such as writing code from natural language descriptions [33], code summarization [39], and automatic code repair [14,22]. Based on these advances, we investigate whether LLMs provide enough inherent code comprehension to integrate specialized knowledge with broad applicability, which is essential for deobfuscating complex code. This focus allows us to address two main challenges:

First, the LLM needs to identify the transformation(s) applied to the code to remove the obfuscation structures entirely. Second, the LLM needs to develop a nuanced understanding of the context in which the obfuscated code operates so that it can reconstruct the code without breaking functionality.

To better understand the potential of LLMs, our first step involves evaluating the foundational capabilities of recent code models, including DeepSeek Coder [16], Code Llama [33], and GPT-4 [1], for this task. To this end, we consider a variety of source-to-source obfuscation techniques such as control-flow flattening [38], opaque predicates [8], and MBA encoding [4] as representative examples of different code obfuscation strategies. In total, we select five different transformation methods to construct a dataset that contains pairs of original and obfuscated codes. The data set contains 30,000 training samples and 2,400 test samples for the single transformation and multi-chain deobfuscation scenarios.

By controlling the construction of the dataset, we can generate code pairs that undergo one or more transformations. Using this constructed dataset, we conduct a series of systematic experiments to evaluate the capabilities of LLMs in deobfuscating codes obfuscated by different sets of transformations. In our study, we examine how base models and instruction-tuned code models perform in this task, both in zero-shot scenarios and through fine-tuning. Our key metrics are the complexity reduction of the code and the preservation of its functionality.

Main Findings. Our experiments result in the following observations:

- While LLMs can effectively reduce the complexity of obfuscated codes, they sometimes break the functionality of the code.
- Fine-tuned models show significant improvements over general-purpose LLMs such as GPT-4 and existing compiler optimizations, which we used as a baseline to compare against.
- The models exhibited very good syntactical correctness even in our most complex scenarios.

- In fact, in the most demanding scenarios, our best model achieved an average program length reduction of 89.21% according to the Halstead metric [20].
- With increasing complexity of the obfuscations, the semantical correctness declines.
- After performing a memorization test, we find that the LLMs do not make use of memorized samples but rather truly deobfuscate the code.

Our research highlights the potential of LLMs to complement existing deobfuscation methods.

Contribution. In summary, we systematically analyze the potential of LLMs for code deobfuscation and compare general-purpose models, specialized code models, and instruction-tuned models. This analysis not only highlights the strengths of LLMs in tackling code obfuscation, but it also shows their current limitations, such as generating semantically incorrect code. We build the first scalable dataset for training and evaluating the performance of LLMs for the deobfuscation task that can be used with arbitrary C programs. Our code is available at https://github.com/DavidBeste/llm-code-deobfuscation.

2   Obfuscation of Code

Obfuscation refers to the process of making software code difficult to understand. To protect a program P from reverse engineering, obfuscation translates it into a program P', which is harder to analyze but semantically equal. Figure 1 shows an example where three transformations (control-flow flattening, arithmetic encoding, and argument randomization) have been applied to obfuscate the code.

Formally, we define this *obfuscation transformation* T as a function

$$T \colon P \mapsto P', \tag{1}$$

which maps a program P into an obfuscated version P' subjected to semantic constraints. The *obfuscator* is assumed to be equipped with a set of such transformations $\mathcal{T}$, and obfuscation is done for a *chain* of transformations $[T_1, \ldots, T_k] \in \mathcal{T}^k$ by iteratively applying the transformations to the program, i.e.,

$$P' = (T_1 \circ \ldots \circ T_k)(P). \tag{2}$$

The result P' then represents the obfuscated program. In this work, we focus on the *Tigress* obfuscation toolkit [7], which represents a state-of-the-art C code obfuscator and includes a wide range of configurable obfuscation schemes [9, 25, 26, 35], making it well suited for scientific investigations.

Code Deobfuscation. Similarly, efforts have been devoted to deobfuscate programs in an automated way. Approaches that rely on static analysis [31, 32, 35], dynamic analysis [11, 37, 41] or program synthesis [4, 29] have been shown to be very efficient. These approaches aim to be *obfuscator-independent* and see each

```
 1   __inline static void strtoupper(char *s) {
 2     char *c;
 3     c = s;
 4     while (*c) {
 5       if ((int )*c >= 97) {
 6         if ((int )*c <= 122) {
 7           *c = (char )(((int )*c - 97) + 65);
 8         }
 9       }
10       c ++;
11     }
12     return;
13   }
```

(a) Original Code

```
 1   void _xa(char *_k0, long _k1) {
 2     char *_k2 ;
 3     unsigned long _k3 ;
 4     int _k4 ;
 5     _k3 = 1UL;
 6     while (1) {
 7       switch (_k3) {
 8       case 4UL: ;
 9         if (97 <= (int )*_k2) {
10           _k3 = 0UL;
11         } else {
12           _k3 = 3UL;
13         }
14         break;
15         [...]
```

(b) Obfuscated Code

Fig. 1. Example Code. Figure b presents an obfuscated version of the program shown in Fig. a. This example is truncated for brevity; the full code consists of 55 lines.

```
 1   void _xa(char *_k0) {
 2     char *_k2;
 3     _k2 = _k0;
 4     while (*_k2) {
 5       if ((int )*_k2 >= 97) {
 6         if ((int )*_k2 <= 122) {
 7           *_k2 = (char )(((int )*_k2 - 97) + 65);
 8         }
 9       }
10       _k2 ++;
11     }
12     return;
13   }
```

Fig. 2. Deobfuscated code. Code recovered from the obfuscated code shown in Fig. 1b, extracted with our approach.

obfuscation as a general problem to solve. In exchange, we face two main challenges: (1) We must know which family of obfuscation has been used to leverage the corresponding deobfuscation method; (2) We must know on which scope the deobfuscation should be applied to get the best results.

Transformations. We consider five different transformation techniques. This selection aims to include transformations altering different aspects of the program code, such as complicating the control flow or increasing the number of operations in the program. Furthermore, the Tigress toolkit makes several recommendations for obfuscation chains [6]. From these, we additionally include all transformations from the first "recipe". To increase diversity, we vary the parameters for the chosen transformations using the recommendations from the Tigress documentation. In the following, we explain the five chosen transformations in more detail.

Encode Arithmetic. Mixed-Boolean-Arithmetic (MBA) translates an easy-to-understand arithmetic expression into a more obscure equivalent expression, by manipulating both arithmetic and boolean operators [13]. The following example shows a basic MBA encoding from the Tigress [7] documentation, replacing the + operator with a more complex structure:

$$x + y \longrightarrow (x \oplus y) + 2 \times (x \wedge y). \tag{3}$$

Encode Branches. This transformation disguises static jumps as return instructions to fool disassembly tools into going to the next return address instead of following the jump target [25]. Again, a manual analysis of the control flow is cumbersome and requires a great deal of effort for an analyst.

Control-Flow Flattening. The flattening protection [9, Chap 4.3.2] breaks the control-flow graph (CFG) of the code to create a loop to execute that will be dispatched over different blocks. As a result, instead of seeing an informative CFG (with branches and loops), a reverse engineer will only see a code structure that must be simplified to understand the real behavior of the code.

Opaque Predicates. The opaque predicate transformation [10] aims to break the CFG of the obfuscated code. To do so, it adds conditionals, always evaluating to true or false, to artificially increase the size of the CFG, thus obscuring which parts of the code are reachable.

Randomize Arguments. This technique randomizes the order of function arguments and adds bogus (i.e., new and semantically useless) arguments [5] requiring a reverse engineer to track them and analyze their purpose.

3   LLM-Supported Deobfuscation

We are now set to examine how LLMs can enhance the understanding and simplification of complex patterns in obfuscated code, potentially enhancing traditional deobfuscation methods. We focus on a scenario where, given an obfuscated program, we aim to retrieve a simpler and semantically equivalent version for further analysis.

Challenges. One of the main obstacles in using current deobfuscation methods is that many state-of-the-art techniques are tailored to specific transformations. To properly deobfuscate a program, it is necessary to first identify the specific obfuscation methods used in order to choose the right deobfuscation tool. Take, for instance, the obfuscated code shown in Fig. 1b: The original 13-line code from Fig. 1a has been transformed to 55 lines with a complex control flow. To deobfuscate this, we must first identify the obfuscations applied—in this case, control-flow flattening, arithmetic encoding, and argument randomization—and then apply the right tools to reverse these changes. This analysis requires expertise and can be error-prone, particularly when the code undergoes multiple chained transformations. Furthermore, deobfuscation tools often require to specify which

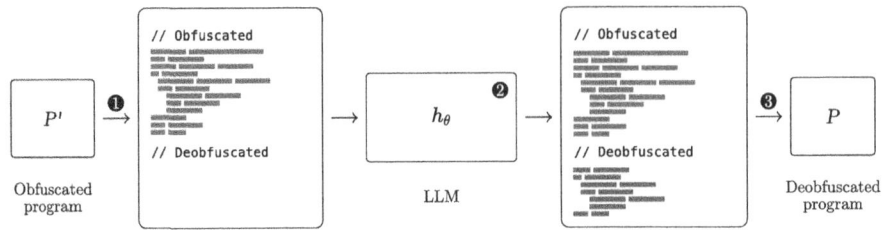

Fig. 3. LLM-based code deobfuscation. We consider the LLM (Step ❷) as a generic deobfuscator, receiving the obfuscated code embedding in its input (Step ❶). The model is trained to extend this input sequence with the deobfuscated code in its output. From the response, we extract the deobfuscated program (Step ❸). The actual format depends on the type of model. Shown here is a C comment style delimiter we use for pure code models.

part of the code to deobfuscate. For example, tools designed to simplify mixed boolean arithmetic expressions [26] demand the specific obfuscated expression as input. While some methods can process the entire code [34,41], they do not scale well with larger code bases. This limitation highlights the need for more effective deobfuscation tools that can handle code efficiently.

LLMs for Deobfuscation. We hypothesize that LLMs can help address these challenges, as they have demonstrated remarkable performance in various code tasks [14,17,19,22]. Rather than developing specialized tools for each transformation, we explore the use of LLMs that process the entire obfuscated code in their input and are trained to output the deobfuscated code in their response. The high-level idea is illustrated in Fig. 3. When, for example, the obfuscated function from Fig. 1b is fed to the LLM, the model returns a simplified version (see Fig. 2) that closely resembles the original code.

Models. The effectiveness of this approach naturally relies on the selected model. On the one hand, we consider a large instruction-tuned model like GPT-4 and design a prompt to instruct the model for deobfuscation. Although the model is not a dedicated code model but rather general-purpose, it performs surprisingly well in the code domain [1]. On the other hand, previous research indicates that models fine-tuned for code-related tasks can substantially outperform generalist models while being significantly smaller [16,33]. In light of this, we additionally explore the use of specialized coding models, namely Code Llama [33] and DeepSeek Coder [16]. We consider both the direct use of the pre-trained models and instruction-tuned versions. Instruction-tuning a code model has been shown to further enhance its coding performance across various benchmarks [16,33]. We fine-tune the code models using the dataset depicted in Fig. 4 introduced next in Sect. 3.1. Table 1 summarizes the selected models.

Prompt Format. We format samples based on the LLMs' training: We use C-style comments for pure code models as shown in Fig. 3 and conversational style for instruction-tuned models.

Table 1. LLMs considered in the study

Name	Size	Open Access	Instruction Tuned	Coding Specialist
DeepSeek Coder [16]	6.7B	✓	✓	✓
Code Llama [33]	7B	✓	✗	✓
GPT-4 [1]	n/a	✗	✓	✗

3.1 Dataset and Training

For fine-tuning and evaluation, we require a dataset that meets several criteria. First, we need a diverse and comprehensive collection of code, preferably from real-world sources. Second, we need a way to verify the semantics of the deobfuscated code to compare the functionality between the original and the deobfuscated code and to measure the models' understanding. Furthermore, to practically instantiate the dataset, we focus on a function granularity, i.e., the program P is a single source code function.

Dataset Selection. Based on the aforementioned requirements, we use the ExeBench dataset [2], a comprehensive collection of real-world C code crawled from GitHub specifically designed for machine learning purposes. The dataset is representative of real-world code based on different software metrics and provides input/output (I/O) samples for each C code, facilitating the evaluation of the semantical correctness. We use the `train-real-compilable` subset for training and the `test-real` subset for evaluation since these both are the closest to real-world code and allow for compilation, which is necessary for our checks later. These subsets comprise 885,074 and 2,134 individual functions.

Pre-processing. We pre-process the dataset and exclude `main` functions and functions that contain no arithmetic operations or branches, since such samples might lead to trivial obfuscations when applying certain transformations (e.g., encoding of arithmetic and flattening of code). In addition, we filter out functions with duplicate names to minimize the inclusion of semantically similar functions, which results in a more diverse training set. We then canonicalize the original code to reduce the variance in coding style between the original and obfuscated samples, e.g., we replace ternary operators with if-else statements. Finally, we randomize all identifier names to prevent the LLM from inferring code structures from descriptive identifier names.

Dataset Generation. The generation process is illustrated in Fig. 4. In the first step ❶, we create differently obfuscated versions of programs P_i with $i = 1, \ldots, N$ where N is defined by the number of samples to incorporate. Therefore, we construct M chains $\mathbf{T}_j = [T_1, \ldots, T_L]$ with length $L \leq L_{max}$ where L_{max} defines the maximum chain length. Transformations T_l' with $l = 1, \ldots, L$ are sampled uniformly from the set of transformations $\mathcal{T}$. For each program P_i and chain $\mathbf{T}_j$, we create an obfuscated program $P'_{i,j}$. To ensure a diverse set, we randomize

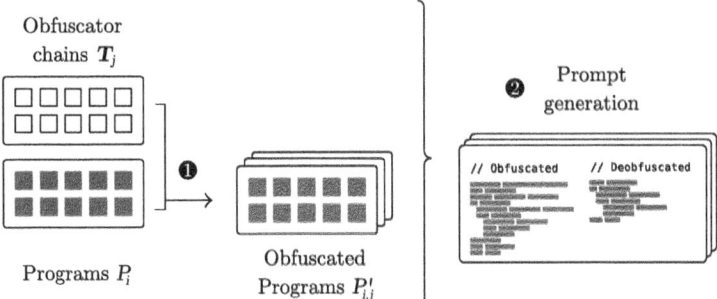

Fig. 4. Dataset generation. The dataset is created in two steps. The input for the first step ❶ is a set of programs, which are transformed into obfuscated programs using several obfuscator chains. In the second step ❷, we subsequently create data samples for the fine-tuning, consisting of pairs of the obfuscated programs and their unmodified counterparts.

the initialization seed for each chain and the parameters for the transformations which leads to $N \cdot M$ obfuscated programs $(P'_{1,1}, \ldots, P'_{1,M}, \ldots, P'_{N,1}, \ldots, P'_{N,M})$. Following this, in step ❷, we pair each obfuscated program with its corresponding original version, i.e., $(P_n, P'_{n,m})$. The resulting pairs serve as the unique samples.

3.2   Metrics

To assess the models' performance, we consider two aspects. We evaluate correctness and measure the models' effectiveness in recovering a code close to the original one from the obfuscated version.

Correctness. We distinguish between semantical and syntactical correctness.

Syntactical Correctness. We assess *syntactical correctness* using the executable wrappers from ExeBench and check for any errors during compilation. Since the wrappers are written in C++, we use the **g++** compiler for this.

Semantical Correctness. While checking for syntactical correctness is trivial using a compiler, semantical correctness is more challenging. We use the I/O samples from the Exebench dataset to approximate the program's functional correctness. We compare the behavior of the deobfuscated program to its obfuscated version. If the output is the same, we conclude that the program is semantically correct; otherwise, it is considered incorrect. We do not expect significant gains from approaches like differential testing and symbolic execution since the I/O samples were crafted specifically for correctness testing.

Deobfuscation Performance. To evaluate deobfuscation performance, we need a metric that effectively reflects the model's ability to reduce code complexity. This metric should consider the complexity of the original, obfuscated, and deobfuscated code in a single value. Therefore, we propose the following metric:

$$P_{Deobf} = 1 - \frac{C_{Deobf} - C_{Orig}}{C_{Obf} - C_{Orig}} \qquad (4)$$

Intuitively, the closer the deobfuscated code is to the original, the closer the score approaches 1. Conversely, the closer the deobfuscated code is to the obfuscated sample, the closer the score approaches 0. Scores larger than 1 imply that the model made the code less complex than the original version, indicating it found a more compact representation. Scores less than 0 imply that the code returned by the model is more complex than the obfuscated version, indicating a failure in deobfuscation.

To instantiate the complexity function C, we could use common code metrics to assess the complexity of program code, such as *cyclomatic complexity* [28] as well as the Halstead metrics [20]. Katzmarski and Koschke empirically evaluated that the Halstead metrics correlate with the programmer's perception of code complexity [23] and, therefore, are suitable for measuring performance in the deobfuscation task. For our evaluation, we focus on the Halstead program length since it can capture changes from all five transformations we chose.

4   Experimental Evaluation

We now present our empirical evaluation of the deobfuscation capabilities of large language models. All used LLMs models have been trained to manipulate C code efficiently. Hence, considering the C-level deobfuscation tasks enables focusing on the deobfuscation capabilities of LLMs *per se*. Moreover, in practice, an approximation of the C code can be retrieved from the binary through decompilation [27]. We split this investigation into three main parts.

– First, we analyze the code that has been obfuscated with a *single* transformation. This will help us infer the general capabilities of the considered LLMs in a very controlled scenario.
– Second, we move to a more complex setting and consider *chains* of transformations. We start with an experiment where the LLMs are trained on multiple obfuscation techniques simultaneously. Building on that, we also train and evaluate models on data where multiple transformations are applied on a single sample. Here, we target obfuscation chains of up to five transformations for training and up to seven for evaluation, which will allow us to learn about potential limits of deobfuscation with LLMs.
– Finally, we seek to determine if the observed performance of the language models might be due to the models detecting and recalling code memorized during training.

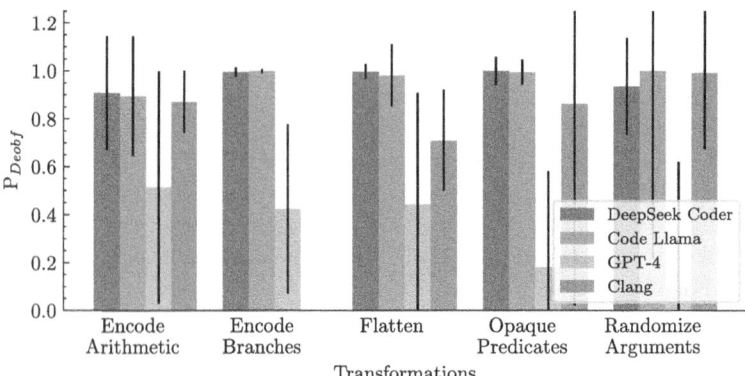

Fig. 5. Complexity reduction. We measure the average deobfuscation scores P_{Deobf} for the five transformations according to the formula introduced earlier for our fine-tuned models and GPT-4. For *encode branches*, Clang fails to produce semantically correct samples, which is why we only compare the three other models in this case.

To summarize, our investigations revolve around three main research questions:

RQ1 What are the general capabilities of LLMs in deobfuscating state-of-the-art code obfuscation transformations?

RQ2 How does the model performance evolve when chaining multiple transformations together?

RQ3 Do models use memorized code during obfuscation?

All experiments are performed on a server equipped with an A100 GPU with 40 GB of VRAM.

4.1   Single Transformations

For our first experiment, we examine how the language models can handle code obfuscated by each of the five transformation techniques individually. In this controlled scenario we can compare the LLMs' code deobfuscation capabilities grouped by different transformation types.

Setup. For the fine-tuning set, we sample 3,000 functions from the `train-real-compilable` split of the ExeBench dataset [2]. For each function, we create one obfuscated version for each transformation, resulting in a dataset of $5 \times 3,000 = 15,000$ samples in total. From these samples, we create pairs of obfuscated and original code as explained in Sect. 3 and randomize all identifier names. For the evaluation dataset, we sample 200 functions from the `test-real` split and apply each of our five transformations to each sample, resulting in a total evaluation dataset of $1,000$ samples. To evaluate the correctness of the LLMs' outputs, we use the executable wrappers from ExeBench as our test harness.

For each transformation, we fine-tune a model for both DeepSeek Coder and Code Llama. Additionally, we evaluate the performance of each transformation using GPT-4. To provide a broader context for the performance of these models, we compare their results against compiler optimization techniques. Specifically, we consider Clang [24], which was shown to be surprisingly effective for deobfuscation [34]. We therefore convert the obfuscated code into Clang's intermediate representation. We then evaluate it across all optimization levels (-O0, -O1, -O2, and -O3), selecting the best version based on the complexity metric. This step is necessary because different optimization levels balance time and memory differently, meaning the most optimized code is not always the simplest. The complexity metric is computed directly on the intermediate representation.

Results. Figure 5 presents the average deobfuscation scores P_{Deobf} together with the standard deviation, considering only samples that were successfully deobfuscated both syntactically and semantically. We observe that both code models show strong performance, with neither consistently outperforming the other across all transformations. GPT-4, on the other hand, is significantly outperformed by the code models across all transformations, with *opaque predicates* and *randomize arguments* exhibiting the highest difference and *encode arithmetic* the lowest. Clang is consistently outperformed by the fine-tuned models except for *encode arithmetic* and *randomize arguments*, where Clang is on par with the fine-tuned LLMs. This indicates that Clang has strengths at reducing complex arithmetic expressions and removing bogus arguments from *randomize arguments*, where the latter is trivial to remove for a compiler. When comparing Clang with GPT-4, we find that Clang outperforms GPT-4 across all transformations except *encode branches*, where a direct comparison is not possible as discussed next.

Table 2 shows the syntactical and semantical correctness. Similar to deobfuscation performance, there is no clear winner. Both fine-tuned models achieve high syntactical correctness and, to a lower degree, also semantical correctness (between 50 and 94.5%). Interestingly, GPT-4 outperforms the fine-tuned models in terms of semantical correctness. This suggests that larger models such as GPT-4 may possess stronger general code reasoning capabilities. As expected, Clang generally outperforms all models, with the exception of *encode branches*, where it fails to maintain semantical correctness in nearly all cases. When manually inspecting these cases, it appears that Tigress introduces subtle undefined behaviors, which Clang exploits to perform aggressive, yet *incorrect*,optimizations.

Table 2. Correctness rates for the different models by transformation type. We report semantical and syntactical correctness. For *encode branches*, the set of joint semantically correct samples is 0 when Clang is included, and thus we only compare the three other models in this case.

	Correctness	DeepSeek Coder	Code Llama	GPT-4	Clang
Encode Arithmetic	Syntactical	100.00%	100.00%	92.00%	100.00%
	Semantical	69.50%	66.00%	77.50%	98.00%
Encode Branches	Syntactical	97.50%	91.50%	97.00%	100.00%
	Semantical	62.50%	57.50%	79.00%	0.50%
Flatten	Syntactical	98.50%	96.50%	97.50%	100.00%
	Semantical	54.0%	50.00%	76.50%	94.00%
Opaque Predicates	Syntactical	100.00%	99.50%	92.00%	100.00%
	Semantical	94.50%	93.50%	89.50%	95.00%
Randomize Arguments	Syntactical	96.50%	97.00%	100.00%	100.00%
	Semantical	93.00%	96.00%	98.50%	98.50%

Finally, we compare the fine-tuned code models with their unmodified base models. DeepSeek Coder struggles with correctness, with only 21.60 % being syntactically and 16.10 % syntactically correct on average. Code Llama, on the other hand, shows better correctness, with 93.50 % being syntactically and 92.20 % semantically correct on average. However, it only achieves an average complexity reduction of 0.001.

> **Conclusion:** Models fine-tuned on specific transformations demonstrate strong deobfuscation performance. They can outperform large generalist models in reducing complexity and achieving overall correctness, but large generalist models can have some advantages in maintaining semantical correctness. LLMs outperform compiler optimizations for most obfuscation transformations. Surprisingly, considering syntactical correctness, LLMs get close to compilers — which never fail. However, their primary challenge is to ensure semantical correctness.

4.2   Multiple Transformations

So far, we have been focusing on whether the models can learn individual transformations during fine-tuning. Building upon this, we want to investigate if the models are also capable of learning *multiple* transformations simultaneously.

Setup. To do this, we fine-tune the two code models on the entire dataset consisting of multiple transformations with 15,000 samples in total. This allows us to explore if the capacity of our chosen model sizes is sufficient to exhibit enough in-depth code understanding capabilities to handle multiple transformations of different natures and various structural changes at the same time.

Results. We find that both versions of the fine-tuned models maintain strong deobfuscation performance for all the transformations, even when fine-tuned on

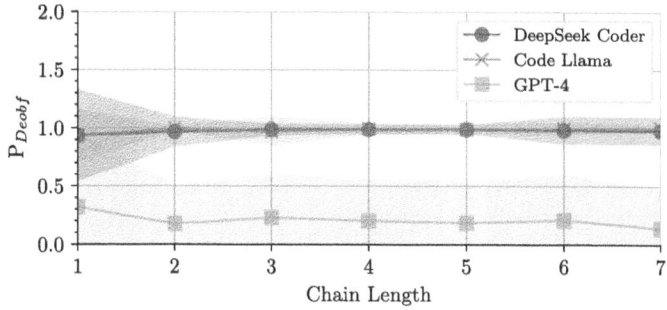

Fig. 6. Average deobfuscation scores P_{Deobf} for chained transformations of different lengths.

multiple transformations at once. However, we notice a drop in performance of around 13% for randomized arguments for both models. This indicates that the models might have a tendency to fail to recognize specific transformations when trained on multiple transformations at the same time. Also, we observe that for the models trained on all transformations, the correctness for *encode branches* increases for both models. We suspect that this is a result of the LLMs' improved general understanding of the deobfuscation task since they had seen more data during training. On the other hand, for Code Llama, specifically, the syntactical correctness for *opaque predicates* decreases when fine-tuned on multiple transformations. A possible explanation for this behavior is that the model is confusing different transformations. At the same time, the semantical correctness for *encode arithmetic* increases for Code Llama, indicating no clear trend toward improvement or degradation.

Scaling Chains of Transformations. Next, we want to scale up the experiments to better understand the potential failure points of LLMs regarding code understanding and systematically measure how much the models maintain their performance with increasing complexity.

Setup. For this purpose, we build a training data set with transformations of chain lengths from one to five, i.e., $L_{max} = 5$, consisting of 3,000 samples for each chain size, resulting in a dataset of 15,000 samples in total. We allow the same transformation to be chosen multiple times in a chain, which enables a diverse data set with $5^5 = 3125$ possible transformation chains. As before, we randomize the parameters of each transformation according to the recommendations in the Tigress documentation. For testing, we consider transformations from chain lengths one to seven, i.e., $L_{max} = 7$. With chain lengths six and seven, we evaluate the performance of out-of-training samples.

Results. Figure 6 shows the code deobfuscation performance P_{Deobf}. In three cases, specifically in the chained transformation scenario, we find that Code Llama renamed the function for deobfuscation, although not trained to do so for our models, and GPT-4 explicitly instructed not to do so. We exclude these

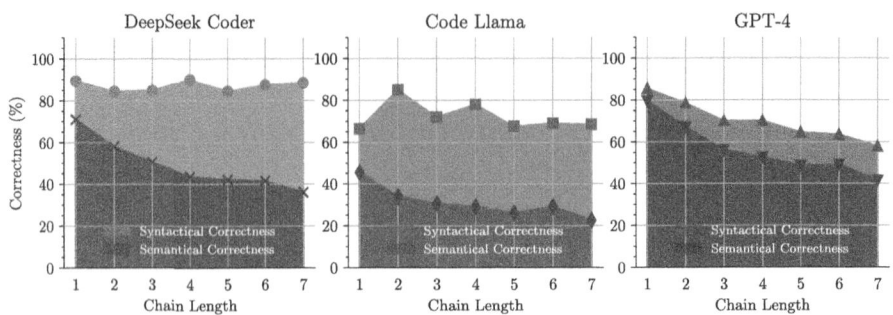

Fig. 7. Correctness rate for chained transformations.

samples from the evaluation since our pipeline relies on identical function names for obfuscated and deobfuscated samples. Renamed functions could result in erroneously computing the metrics over an auxiliary function in the code file for the deobfuscated sample, resulting in nonsensical scores.

We find that DeepSeek Coder and Code Llama maintain stable deobfuscation performance, even for chain lengths six and seven that were not part of the fine-tuning. On the other hand, the deobfuscation performance of GPT-4 starts lower and declines slightly with increasing chain lengths, and it is significantly outperformed by both code models for all possible chain lengths.

The correctness rates are reported in Fig. 7. For syntactical correctness, the rates for the code models vary with increasing chain length but still exhibit high correctness rates. On the other hand, the syntactical correctness rate for GPT-4 steadily declines. For semantical correctness, we see that GPT-4 performs the best, followed by DeepSeek Coder. GPT-4 maintains a higher semantical correctness rate for larger chain sizes. This again supports our suspicion that its larger model size can attenuate semantical correctness problems of LLMs.

> **Conclusion:** The fine-tuned models consistently maintain stable deobfuscation performance across all evaluated chain lengths, including lengths six and seven, which were not included in the training data. In contrast, the performance of GPT-4 decreases significantly as the chain length increases.

4.3 Memorization

The LLMs we consider are trained on large amounts of text and code that is publicly available online [1,16,33]. This inevitably raises the question of whether the training data contained tigress-obfuscated code, which can be found on platforms such as *Stack Overflow*. As a result, the model's deobfuscation abilities could stem from either memorization of the correct outputs or a genuine understanding of the code's structure and semantics. To investigate this, we propose the following experiment: We identify and alter constant values within a program. If the model recovers the original, unmodified constant values during deobfuscation, this would strongly suggest memorization. On the other hand, correctly recovered samples would indicate more genuine understanding capabilities.

Setup. We use samples from the previous experiment that were semantically correct and contain at least one constant, excluding constants in array declarations and references from the randomization procedure, as these likely cause the program to malfunction. For the remaining samples, we randomize all constants in the program. We exclude programs that crash or time out after ten milliseconds. The latter occurs if the randomization of constants causes a slow or infinite loop. Lastly, we update the corresponding input and output samples using the new program for reference. In total, we collect 257 programs for DeepSeek Coder, 215 samples for Code Llama, and 357 for GPT-4.

Results. We observe that the semantical correctness rate ranges between 75 and 99% across the five transformations and two models. If a deobfuscated sample is semantically correct, it is very likely to have recovered the correct constants. Therefore, we focus on the incorrect samples and manually review these.

During this analysis, we found only one sample affected by memorized constants out of the 257 for DeepSeek Coder and the 255 for Code Llama. However, we did observe that models were frequently confused by additional arithmetic complexity, as introduced by *encode arithmetic*, which was especially pronounced for DeepSeek Coder. Also, operators such as $\leq$ and $\neq$ were frequently changed. Furthermore, in several instances, the models attempted to correct nonsensical code, such as loops that are never executed or mutually exclusive logical compound conditions in if-statements, which, according to the updated I/O samples, resulted in semantically incorrect code. For GPT-4, we noticed a lower average semantical correctness over all transformations, with 83% for GPT-4 vs. 88% for DeepSeek Coder and 93.48% for Code Llama. A possible explanation is that due to the lack of training in removing specific transformations, GPT-4 might remove only part of the obfuscated code and break semantics in the process as compared to the other two. Code Llama's better score might indicate a lower tendency to try to "correct" implausible statements and thus break semantical correctness.

> **Conclusion:** We only observed minor indications of the LLMs using memorized constant code snippets for deobfuscation, which is a good indication of the inference ability of the models for deobfuscation.

5 Discussion

This paper represents an initial investigation of the capabilities of LLMs for deobfuscation. In the following, we discuss our findings and potential directions for future work in this area.

Our Semantical Correctness Check. For checking the semantical correctness, we rely on I/O sampling and, more specifically, the *rich IO* samples from the ExeBench dataset [2], which can still, in theory, miss corner cases. We did not observe failed correctness checks during our experiments. As we have full access to both the original and obfuscated samples, this check could easily be extended through standard techniques such as differential testing or symbolic execution. However, we do not expect a significant difference in results.

Semantical Incorrectness in Practice. Our study shows that, at the moment, LLMs can sometimes produce semantically incorrect results. While this is not an ideal situation and may clearly hinder some applications, these results can still be useful in some scenarios. Similarly, decompilers are known to not always be correct [27], but they are often considered useful by practitioners. We apply 8-bit quantization during fine-tuning to improve efficiency, though this can reduce output quality [12]. Furthermore, our experiments with GPT-4 suggest that larger models maintain semantical correctness better. Exploring the impact of model size and model quantization could be a promising direction for future research.

Context Size. We use a 6144-token context size to balance program length with training and inference speed, limiting the size of programs we consider. We consider compressing the programs into a more compact representation as out of scope for this work to exclude potential complications that might arise from compressing larger inputs to fit the context size.

Transformations. We focus on a subset of existing obfuscation methods. Specifically, obfuscation schemes such as virtual machine-based packing or self-modifying code are hard to deobfuscate solely with static analysis [11,35]. These types of obfuscation could be addressed using different approaches, such as dynamic techniques, to recover a dump of the code, which could then be deobfuscated using the method we presented. Additionally, our approach only includes intra-procedural obfuscation. Evaluating other obfuscation schemes, as well as inter-procedural obfuscation, is a promising direction for future work.

About Deobfuscation and LLMs. We believe that deobfuscation is a great application for LLMs for several reasons: (1) Semantical correctness of the obfuscated and the LLM-simplified code can be automatically checked (at least partially), allowing for a clear evaluation of the benefit of LLMs as well as a simple safety net against hallucinations. (2) Generating datasets is easy, as obfuscators can be naturally turned into example generators. (3) Simplifying such convoluted codes requires some form of clear understanding of the code. (4) Highly obfuscated codes combining several layers of protection are less likely to be part of the initial training dataset, reducing the risk of memorization.

General Applicability. We showed that models trained on multiple transformations still show high deobfuscation performance. While our experiment showed an increase in performance for some transformations, it showed a decrease for others. We expect that with increasing training and model size, the performance gains will outweigh potential drawbacks due to LLMs confusing different transformations. This indicates the tendency of LLMs to become universal deobfuscators, possibly rendering the necessity of dedicated deobfuscation techniques obsolete.

Open Questions. While we believe this work already gathered valuable insights regarding the potential of LLMs for code deobfuscation, several important questions remain, including generalization across different obfuscators and their different versions, other programming languages—particularly machine code or bytecode—and possible countermeasures. Another direction would be extending the memorization experiment, for example, finding different representations of the program and evaluating the representation returned by the LLM.

6 Related Work

In the following, we discuss related work on applications of code LLMs and other approaches to employing learning-based methods for code deobfuscation.

Large Language Models for Codes. LLMs have advanced various fields, including natural language processing [1, 12] and programming languages [15, 33]. Feng et al. [15] propose the CodeBERT model that utilizes an encoder-only architecture with the primary focus on code classification, code retrieval, and program repair. CodeT5 [40] and CodeT5+ [39] employ an encoder-decoder architecture with various datasets and objective functions to tackle various code generation tasks. More recently, LLMs with decoder-only architecture have shown promising performance in various code generation tasks [16, 33].

Machine Learning for Code Deobfuscation. Most deobfuscation algorithms, like symbolic deobfuscation, target specific obfuscation families. It relies on static analysis to remove, for example, opaque predicates [3] or to simplify virtual machines [34]. Closer to our work, neural networks have been used to identify obfuscated code parts, and the obfuscation used [36]. Our work is more general, as it performs obfuscation identification and simplification in a row. Hence, both aspects can benefit from prior knowledge included in our fine-tuned LLM.

7 Conclusion

In this paper, we present an exploration into the use of LLMs for the task of code deobfuscation by conducting three main experiments. First, we test the LLMs in a single transformation setting in which our models show strong performance. On the downside, we find that challenges related to maintaining semantical correctness persist, indicating areas for future improvement. Second, we consider a scenario closer to real-world conditions by employing chains of transformations. As the length of these transformation chains increases, the models' ability to produce semantically correct code decreases. However, the models' deobfuscation performance remains consistently strong across all considered chain lengths. Compared to GPT-4, our models maintain higher deobfuscation performance for longer chains. Finally, we perform a memorization experiment, which all models successfully pass. As we continue to refine these models and address their shortcomings in the future, the prospect of developing more robust, scalable, and versatile deobfuscation tools based on LLMs becomes more tangible and promises to enhance security efforts in the ever-evolving arms race in software protection.

Acknowledgments. This work was supported by the German Federal Ministry of Education and Research (BMBF) under the grant AIgenCY (16KIS2012), the Deutsche Forschungsgemeinschaft (DFG, German Research Foundation) under the project ALI-SON (492020528), the European Research Council (ERC) under the consolidator grant MALFOY (101043410), ANR Research under Plan France 2030 with reference ANR-22-PECY-0007 as well as BPI under Plan France 2030 with reference DOS0233319/00.

Disclosure of Interests. The authors have no competing interests to declare that are relevant to the content of this article.

References

1. Achiam, J., et al.: GPT-4 technical report. arXiv preprint (2023). https://doi.org/10.48550/arXiv.2303.08774
2. Armengol-Estapé, J., Woodruff, J., Brauckmann, A., Magalhães, J.W.d.S., O'Boyle, M.F.: Exebench: an ML-scale dataset of executable C functions. In: Proceedings of the 6th ACM SIGPLAN International Symposium on Machine Programming (MAPS), pp. 50–59 (2022). https://doi.org/10.1145/3520312.3534867
3. Bardin, S., David, R., Marion, J.Y.: Backward-bounded DSE: targeting infeasibility questions on obfuscated codes. In: IEEE Symposium on Security and Privacy (S&P), pp. 633–651 (2017). https://doi.org/10.1109/SP.2017.36
4. Blazytko, T., Contag, M., Aschermann, C., Holz, T.: Syntia: synthesizing the semantics of obfuscated code. In: USENIX Security Symposium, pp. 643–659 (2017). https://dl.acm.org/doi/10.5555/3241189.3241240
5. Collberg, C.: RandomizeArguments—tigress.wtf. https://tigress.wtf/randomizeArguments.html. Accessed 30 Apr 2025
6. Collberg, C.: Recipes—tigress.wtf. https://tigress.wtf/recipes.html. Accessed 30 Apr 2025
7. Collberg, C.: The Tigress C Diversifier/Obfuscator. https://tigress.wtf/index.html. Accessed 30 Apr 2025
8. Collberg, C., Thomborson, C., Low, D.: A taxonomy of obfuscating transformations. Technical report, The University of Auckland, New Zealand (1997)
9. Collberg, C.S., Nagra, J.: Surreptitious Software - Obfuscation, Watermarking, and Tamperproofing for Software Protection. Addison-Wesley (2010). https://dl.acm.org/doi/10.5555/1594894
10. Collberg, C.S., Thomborson, C.D., Low, D.: Manufacturing cheap, resilient, and stealthy opaque constructs. In: ACM Symposium on Principles of Programming Languages (POPL), pp. 184–196. https://doi.org/10.1145/268946.268962
11. Coogan, K., Lu, G., Debray, S.: Deobfuscation of virtualization-obfuscated software: a semantics-based approach. In: ACM Conference on Computer and Communications Security (CCS), pp. 275–284 (2011). https://doi.org/10.1145/2046707.2046739
12. Dettmers, T., Lewis, M., Belkada, Y., Zettlemoyer, L.: Llm.int8(): 8-bit matrix multiplication for transformers at scale. In: Advances in Neural Information Processing Systems (NeurIPS), pp. 30318–30332 (2022). https://dl.acm.org/doi/10.5555/3600270.3602468

13. Eyrolles, N., Goubin, L., Videau, M.: Defeating MBA-based obfuscation. In: Proceedings of the 2016 ACM Workshop on Software PROtection, pp. 27–38 (2016). https://doi.org/10.1145/2995306.2995308

14. Fan, Z., Gao, X., Mirchev, M., Roychoudhury, A., Tan, S.H.: Automated repair of programs from large language models. In: International Conference on Software Engineering (ICSE), pp. 1469–1481 (2023). https://doi.org/10.1109/ICSE48619.2023.00128

15. Feng, Z., et al.: CodeBERT: a pre-trained model for programming and natural languages. In: Findings of the Association for Computational Linguistics (EMNLP), pp. 1536–1547 (2020). https://doi.org/10.18653/v1/2020.findings-emnlp.139

16. Guo, D., et al.: Deepseek-coder: when the large language model meets programming–the rise of code intelligence. arXiv preprint (2024). https://doi.org/10.48550/arXiv.2401.14196

17. Hajipour, H., Hassler, K., Holz, T., Schönherr, L., Fritz, M.: Codelmsec benchmark: systematically evaluating and finding security vulnerabilities in black-box code language models. In: IEEE Conference on Secure and Trustworthy Machine Learning (SaTML), pp. 684–709 (2024). https://doi.org/10.1109/SaTML59370.2024.00040

18. Hajipour, H., Malinowski, M., Fritz, M.: IReEn: reverse-engineering of black-box functions via iterative neural program synthesis. In: Machine Learning and Principles and Practice of Knowledge Discovery in Databases (ECML PKDD), pp. 143–157 (2021). https://doi.org/10.1007/978-3-030-93733-1_10

19. Hajipour, H., Schönherr, L., Holz, T., Fritz, M.: Hexacoder: secure code generation via oracle-guided synthetic training data. arXiv preprint (2024). https://doi.org/10.48550/arXiv.2409.06446

20. Halstead, M.H.: Elements of Software Science (Operating and programming systems series). Elsevier Science Inc. (1977). https://dl.acm.org/doi/10.5555/540137

21. Hammad, M., Garcia, J., Malek, S.: A large-scale empirical study on the effects of code obfuscations on android apps and anti-malware products. In: International Conference on Software Engineering (ICSE), pp. 421–431 (2018). https://doi.org/10.1145/3180155.3180228

22. Jiang, N., Liu, K., Lutellier, T., Tan, L.: Impact of code language models on automated program repair. In: International Conference on Software Engineering (ICSE), pp. 10 pp.–54 (2023). https://doi.org/10.1109/ICSE48619.2023.00125

23. Katzmarski, B., Koschke, R.: Program complexity metrics and programmer opinions. In: 20th IEEE International Conference on Program Comprehension (ICPC), pp. 17–26 (2012). https://doi.org/10.1109/ICPC.2012.6240486

24. Lattner, C.: LLVM and clang: Next generation compiler technology. In: The BSD Conference, pp. 1–20 (2008)

25. Linn, C., Debray, S.K.: Obfuscation of executable code to improve resistance to static disassembly. In: ACM Conference on Computer and Communications Security (CCS), pp. 290–299 (2003). https://doi.org/10.1145/948109.948149

26. Liu, B., Shen, J., Ming, J., Zheng, Q., Li, J., Xu, D.: MBA-blast: unveiling and simplifying mixed Boolean-arithmetic obfuscation. In: USENIX Security Symposium, pp. 2351–2365 (2021)

27. Liu, Z., Wang, S.: How far we have come: testing decompilation correctness of C decompilers. In: Proceedings of the 29th ACM SIGSOFT International Symposium on Software Testing and Analysis (ISSTA), pp. 475–487 (2020). https://doi.org/10.1145/3395363.3397370

28. McCabe, T.J.: A complexity measure. TSE 308–320 (1976). https://doi.org/10.1109/TSE.1976.233837

29. Menguy, G., Bardin, S., Bonichon, R., Lima, C.D.S.: Search-based local black-box deobfuscation: understand, improve and mitigate. In: ACM Conference on Computer and Communications Security (CCS), pp. 2513–2525 (2021). https://doi.org/10.1145/3460120.3485250

30. Ming, J., Xu, D., Wang, L., Wu, D.: Loop: logic-oriented opaque predicate detection in obfuscated binary code. In: ACM Conference on Computer and Communications Security (CCS), pp. 757–768 (2015). https://doi.org/10.1145/2810103.2813617

31. Reichenwallner, B., Meerwald-Stadler, P.: Efficient deobfuscation of linear mixed Boolean-arithmetic expressions. In: CheckMATE Workshop, pp. 19–28 (2022). https://doi.org/10.1145/3560831.3564256

32. Reichenwallner, B., Meerwald-Stadler, P.: Simplification of general mixed Boolean-arithmetic expressions: GAMBA. In: IEEE European Symposium on Security and Privacy (EuroS&P) Workshops, pp. 427–438 (2023). https://doi.org/10.1109/EuroSPW59978.2023.00053

33. Roziere, B., et al.: Code llama: open foundation models for code. arXiv preprint (2023). https://doi.org/10.48550/arXiv.2308.12950

34. Salwan, J., Bardin, S., Potet, M.-L.: Symbolic deobfuscation: from virtualized code back to the original. In: Giuffrida, C., Bardin, S., Blanc, G. (eds.) DIMVA 2018. LNCS, vol. 10885, pp. 372–392. Springer, Cham (2018). https://doi.org/10.1007/978-3-319-93411-2_17

35. Schrittwieser, S., Katzenbeisser, S., Kinder, J., Merzdovnik, G., Weippl, E.: Protecting software through obfuscation: can it keep pace with progress in code analysis? ACM Comput. Surv. (CSUR) (2016). https://doi.org/10.1145/2886012

36. Tofighi-Shirazi, R., Asăvoae, I.M., Elbaz-Vincent, P.: Fine-grained static detection of obfuscation transforms using ensemble-learning and semantic reasoning. In: Proceedings of the 9th Workshop on Software Security, Protection, and Reverse Engineering (SSPREW), pp. 1–12 (2019). https://doi.org/10.1145/3371307.3371313

37. Udupa, S.K., Debray, S.K., Madou, M.: Deobfuscation: reverse engineering obfuscated code. In: 12th Working Conference on Reverse Engineering (WCRE'05), pp. 10 pp.–54 (2005). https://doi.org/10.1109/WCRE.2005.13

38. Wang, C., Hill, J., Knight, J., Davidson, J.: Software tamper resistance: Obstructing static analysis of programs. Technical report, University of Virginia (2000). https://dl.acm.org/doi/10.5555/900898

39. Wang, Y., Le, H., Gotmare, A., Bui, N., Li, J., Hoi, S.: Codet5+: open code large language models for code understanding and generation. In: Conference on Empirical Methods in Natural Language Processing (EMNLP), pp. 1069–1088 (2023). https://doi.org/10.18653/v1/2023.emnlp-main.68

40. Wang, Y., Wang, W., Joty, S., Hoi, S.C.: Codet5: Identifier-aware unified pre-trained encoder-decoder models for code understanding and generation. In: Conference on Empirical Methods in Natural Language Processing (EMNLP), pp. 8696–8708 (2021). https://doi.org/10.18653/v1/2021.emnlp-main.685

41. Yadegari, B., Johannesmeyer, B., Whitely, B., Debray, S.: A generic approach to automatic deobfuscation of executable code. In: IEEE Symposium on Security and Privacy (S&P), pp. 674–691 (2015). https://doi.org/10.1109/SP.2015.47

Poster: All Right Then, (Don't) Keep Your Secrets: Exposing API Hashing in Malware

Nicola Bottura[1], Giorgia Di Pietro[1], Yuya Yamada[2], Daniele Cono D'Elia[1(✉)], and Leonardo Querzoni[1]

[1] Sapienza University of Rome, Rome, Italy
{bottura,g.dipietro,delia,querzoni}@diag.uniroma1.it
[2] Nara Institute of Science and Technology, Ikoma, Japan

Abstract. Modern malware employs disparate anti-analysis techniques to complicate analysis attempts. Among them, API hashing conceals the identity of imported library functions—key indicators for understanding malware behavior—by replacing their standard names with hashed values. Currently, resolving these obfuscated calls relies heavily on manual expertise and community-maintained hash repositories, both of which are time-consuming and difficult to scale. In this work, we explore an automated approach to deobfuscate API hashing. By leveraging dynamic program analysis, we identify and map hash values back to their original function names while also extracting information about the hashing scheme. Our method can then use malware itself as a "hash oracle", enabling on-demand resolution of standard function names through the malware's hashing logic, enabling automatic updates of repositories.

1 Introduction

When analyzing untrusted software for Windows platforms, the contents of the Import Address Table (IAT) allow analysts to make educated guesses about its capabilities. In fact, the identity of the APIs the executable references can provide valuable insights into the functionality of suspicious code. Beyond aiding in capability assessment, IAT content analysis also plays a key role in clustering malware samples and correlating threat group activities. As an example, the list of imported APIs has been used to track custom backdoors employed by specific actors [5]. Therefore, IAT contents are precious for static analysis efforts.

To hinder such analysis, malware authors have developed various API obfuscation schemes [2]. With dynamic API resolution, instead of being imported, functions are resolved only at run-time, typically using standard facilities like `LoadLibrary` and `GetProcAddress` [3]. However, these techniques are well-known to experts, and many modern analysis tools can disambiguate them.

A more advanced technique that significantly complicates static analysis is *API Hashing*. Instead of using standard resolution methods, which need the sample to materialize API names as strings in memory, malware can store hashed representations of these names and use these values to look up API addresses

M. Egele et al. (Eds.): DIMVA 2025, LNCS 15747, pp. 287–293, 2025.
https://doi.org/10.1007/978-3-031-97620-9_16

covertly. During execution, a sample scans all the functions exported by the loaded libraries, computing the hash of each symbol using a custom algorithm. These hashes are then compared against one or more pre-computed values. When a match is found, the desired function address has been solved. This complicates not only automatic analysis, but also manual reverse engineering attempts.

To combat API hashing, analysts often rely on their expertise and accumulated knowledge to recognize API hashing techniques. However, comprehensive documentation remains limited, with only a handful of practitioner blog posts in recent years offering insights through case studies [4,7]. A community-sourced online repository [6] maintains a collection of hashing algorithms observed in malware, along with precomputed hash tables for Windows APIs and other common strings. While helpful, this resource lacks automation and struggles to keep pace with the evolving landscape of hashing techniques, as even minor changes to a hashing scheme can render these tables obsolete. This variability—where readable API names are replaced with seemingly arbitrary hashes—impairs traditional reverse engineering approaches and highlights the growing need for automated solutions to support deep and accurate malware analysis.

This work explores a solution that enhances the analysis of malicious code without relying on manual investigation or expert intervention. Since API hashing obstructs static analysis and complicates accurate inspection of affected samples, our approach leverages dynamic program analysis to extract valuable information in a lightweight and generalizable manner. This can enable automated, scalable deobfuscation if the dynamic analysis comes with tenable costs. Along these lines, to support the continuous updating of community-maintained hash tables—and to provide analysts with more comprehensive data—we propose leveraging the extracted information to transform the malware itself into a "hash oracle". By emulating the sample's hashing logic and injecting arbitrary strings at the point where function names are typically resolved, this method allows for computing hashes using the malware's own code. This eliminates the need for manual reverse engineering to recover the API hashing logic.

2   Background

API hashing typically involves traversing the internal structures of Dynamic-Link Libraries (DLLs) to locate their export tables [3]. Once these tables are identified, the malware iterates through the list of exported function names, applying a custom hashing algorithm to each one. When a hash matches a pre-computed target value, it indicates that the current API name corresponds to the obfuscated target function. At that point, the malware dynamically resolves the function's address by referencing its position in the export table, accessing the DLL's structure that maps function names to their corresponding addresses. A simplified version of this logic is provided in a snippet of C code in Listing 1.1.

Line 2 retrieves the name of the current API by using the DLL's base address and the name offset, while line 3 computes the hash of the API name using a custom algorithm. Line 5 checks whether the computed hash matches a pre-computed hash; if a match is found, line 6 resolves the actual address of the

```
1   for (i = 0; i < NumberOfFunctions; i++) {
2       api_name = (char *)(base + AddressOfNames[i]);
3       curr_hash = HASHING_FUNCTION(api_name);
4
5       if (curr_hash == pre_cmptd_hash) {
6           api_addr = (PDWORD)(base + AddressOfFunctions[
                AddressOfNameOrdinals[i]]);
7           return api_addr;
8       }
9   }
```

Listing 1.1. C code snippet for exemplary API hashing loop.

obfuscated API. Notably, this method is not limited to API names: malware can apply a similar approach to obfuscate DLL names, traversing the Process Environment Block (PEB) to enumerate and hash loaded module names.

A core aspect of the API hashing technique is the use of *indicative offsets* from a DLL's base address to reference the offsets—Relative Virtual Addresses (RVAs) in Windows terminology—of key fields used for resolving API calls at runtime through API hashing. When deploying API hashing, malware writers typically maintain their own IAT-like array with the API addresses that they solve dynamically upon execution startup using the method above.

3   Proposed Approach

Manually reverse-engineering a binary that employs API hashing is a highly complex task. The presence of additional obfuscation layers can further complicate the process, making a complete analysis even more challenging. Even when no extra obfuscation is present and hash tables assist in deobfuscation, the process may still fall short: malware can implement customized or entirely new hashing algorithms that evade traditional analysis methods. To address this, we are currently exploring an automated solution that operates by inspecting accesses to DLL memory that are a precondition for any hashing mechanism.

As exemplified in Listing 1.1, API hashing relies on two key operations: (1) computing the API name, where the sample retrieves each plaintext function name from DLL memory, and (2) resolving the address of the obfuscated API name once found. Our approach focuses on identifying the instructions responsible for extracting these two critical information items. The code encompassing these two operations defines a *program slice* with the following properties:

1. The hashing computation occurs within this slice;
2. At some point, a successful match between the current hash and a precomputed one occurs (e.g., a cmp instruction with identical incoming operands);
3. The slice reaches its endpoint (symbol resolution) only after a found match.

Building on these properties, we developed a prototype solution that operates in three distinct phases (Fig. 1), each needing some code (re-)execution.

Phase 1: Extracting the API Hashing Program Slice. The first step focuses on identifying the code region responsible for the API hashing mechanism. This

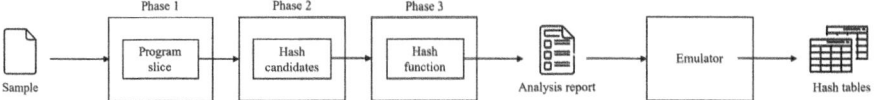

Fig. 1. Workflow of the proposed approach.

is achieved by detecting instructions that access DLL memory offsets viable to retrieve API names and addresses. By applying program slicing, we isolate a smaller, more relevant portion of the code, significantly narrowing the search space for hashing functions. This not only improves efficiency but also enables further analysis by leveraging the specific characteristics of the extracted slice.

Phase 2: Identifying Hash Candidates. With the program slice defined, the next step is to trace the operands of all comparison instructions within its boundaries. When a correct hash is eventually produced, execution reaches the slice's endpoint indicating a successful match. Although the exact hash value may still be unknown at this stage, we can postulate that one of the comparison operations in the slice must contain the correct hash as one of its operands, whereas the other operand will vary every time the samples moves to the next API name.

Phase 3: Identifying the Hash Function. Malware API hashing algorithms generally operate through a sequence of instructions that progressively modify a plaintext string, with the final iteration producing the hash value. To pinpoint the possible locations of the hashing function, we focus on arithmetic instructions commonly involved in these operations. To pursue generality, we do not assume that the hash value is returned by a function. On the contrary, by narrowing the scope to a smaller set of relevant instructions, we can map the resulting hash value back to the code sequence that generated it.

The insights gained from this analysis let us directly address a second major challenge: automating the community task of hash table creation and updating.

Currently, this process relies on manual reverse engineering to extract the hashing function, which is then submitted to the HashDB [6] service for updates. Due to the recurrent complexity of deobfuscating malware, analysts often resort to best-effort decompilation of the executable and sketch from there a Python implementation the original semantics for independent use. However, this method may incur the pitfalls of incorrect decompilation or manual follow-up translation. Additionally, it requires manual identification of hashing code.

We envision a solution that leverages the information gathered during the three phases above to orchestrate a controlled execution of the code in an emulator. This would let us use the malware itself as an on-demand hash value generator, collecting the hash value at each traversed DLL function symbol. This approach eliminates the need for costly reverse-engineering techniques to manage the hashing loop and to compute hash values.

Table 1. Outcomes from sample analysis for the different stages of API hashing.

Family	Sample	Phase 1 Slice start	Slice end	Phase 2 Compare value	Hash value	Phase 3 Hash function
BlackMatter	50c4970003a84cab1bf2634631fe39d7	✓	✓	✓	✓	✓
BlackMatter v2	d0512f2063cbd79fb0f770817cc81ab3	✓	✓	✓	✓	✓
Conti	bc92ea510a5630c770d9443be4b40fde	✓	✓	✓	✓	✓
Emotet	68c76c3403570a22cc7a60a1b68d9056	✓	✓	✓	✓	✗
Lockbit	628e4a77536859ffc2853005924db2ef	✗	✗	✗	✗	✗
Netwalker	73de5babf166f28dc81d6c2faa369379	✓	✓	✓	✓	✓
Revil	890a58f200dfff23165df9e1b088e58f	✓	✓	✓	✓	✗
Zloader	5c76c41f9d0cc939240b3101541b5475	✓	✓	✓	✓	✓

Discussion. In dynamic analysis scenarios, such as with a sandbox or antivirus, API interposition can reliably expose API call targets [3]. On the contrary, methods that rely on static information need other ways to address the challenges from API hashing. We try to bridge this gap using lightweight dynamic analysis.

The problem we investigate overlaps with what some program analysis techniques can offer. Taint analysis [8] can trace data from API names (sources) to hash comparisons (sinks), but its high overhead and susceptibility to imprecision—especially with malware code—limit its practicality. Input-to-state correspondence analysis [1] may be more tenable, but struggles when inputs do not directly map to resulting states. Hence, neither can directly untangle API hashing.

One limitation of our method lies in possibly large program slices when the hashing and comparison operations occur in separate loops. While this may impact performance, it does not reduce overall effectiveness, as the key instructions related to the hashing mechanism are still captured. Another challenging aspect may involve Phase 3, whenever multiple stack locations may be flagged as potential hash function sites (i.e., false positives). Despite this, when the correct hash value is identified in Phase 2, the method still supports effective analysis, even if the exact location of the hashing function remains uncertain.

4   Preliminary Results

To evaluate the feasibility of our approach, we conducted a preliminary assessment by developing a prototype implementation of our design, using the Dynamic Binary Instrumentation (DBI) capabilities of the DynamoRIO framework. The prototype covers the three phases of our method, whereas we are currently working on the emulation part to orchestrate executions from a snapshot.

Our initial testbed comprises eight samples drawn from popular families, such as *Emotet* and *NetWalker*, that notoriously utilize API hashing techniques. For each sample, we had access to auxiliary documentation and annotations from blog posts analyzing those specific samples, with a particular focus on their API hashing methods, enabling us to validate the accuracy of our prototype's output.

We consider a result successful if the prototype can: (1) correctly identify the relevant code slice, (2) locate the comparison instruction involving the hash values, and (3) recognize the underlying hash function.

The results are summarized in Table 1. Out of the eight samples, the prototype failed to produce any result only for *Lockbit*, due to a crash we suspect is related to a DynamoRIO bug and are further investigating. Among the remaining seven samples, the results were generally promising, with a few exceptions. In two cases, *Revil* and *Emotet*, the prototype failed to correctly identify the hash function. In both instances, the hash value was located within one of the instructions typically used by malware to transform the plaintext string during hashing that we look for in *Phase 3*. This is indicative that the hashing process is distributed across multiple functions not linked by internal calls from one another, presenting a challenge for our current analysis prototype. Since the hash value is not observed within a single isolated function, our prototype cannot reliably identify the complete hashing logic. This limitation warrants further investigation. On the bright side, for both versions of *BlackMatter*, which performs API hashing also on DLL names, our prototype operated successfully in full.

5   Conclusion

We have proposed an automated approach for analyzing API hashing mechanisms in Windows malware. Once mature, our prototype may remarkably ease ongoing community-driven efforts in this domain. As a next step, we aim to extend it by addressing its current limitations in the identification of the hash function and evaluating its performance on a larger set of malware samples.

This work has been partially supported by projects SERICS (PE00000014) and Rome Technopole (ECS00000024) under the MUR National Recovery and Resilience Plan funded by the European Union - NextGenerationEU.

References

1. Aschermann, C., Schumilo, S., Blazytko, T., Gawlik, R., Holz, T.: REDQUEEN: fuzzing with input-to-state correspondence. In: NDSS (2019)
2. Cheng, B., et al.: Obfuscation-Resilient executable payload extraction from packed malware. In: USENIX Security Symposium, pp. 3451–3468 (2021)
3. D'Elia, D.C., Nicchi, S., Mariani, M., Marini, M., Palmaro, F.: Designing robust API monitoring solutions. IEEE Trans. Dependable Secure Comput. **20**(1), 392–406 (2023)
4. Gupta, N.: API hashing - why malware loves (and you should care). https://securitymaven.medium.com/api-hashing-why-malware-loves-and-you-should-care-77c5135d9aaa
5. Mandiant: Tracking malware with import hashing. https://cloud.google.com/blog/topics/threat-intelligence/tracking-malware-import-hashing/

6. OALabs: HashDB. https://github.com/OALabs/hashdb
7. Red Team Notes: Windows API hashing in malware. https://www.ired.team/ offensive-security/defense-evasion/windows-api-hashing-in-malware
8. Schwartz, E.J., Avgerinos, T., Brumley, D.: All you ever wanted to know about dynamic taint analysis and forward symbolic execution (but might have been afraid to ask). In: IEEE Symposium on Security and Privacy, pp. 317–331 (2010)

Author Index

M. Egele et al. (Eds.): DIMVA 2025, LNCS 15747, pp. 295–297, 2025.
https://doi.org/10.1007/978-3-031-97620-9

The manufacturer's authorised representative in the EU is Springer
Nature Customer Service Centre GmbH, Europaplatz 3, 69115 Heidelberg,
Germany. If you have any concerns regarding our products, please
contact ProductSafety@springernature.com

Printed and bound by CPI Group (UK) Ltd, Croydon, CR0 4YY

28/04/2026

02098524-0004